INTERMEDIATE ALGEBRA

SECOND EDITION

INTERMEDIATE ALGEBRA

MUSTAFA A. MUNEM

WILLIAM TSCHIRHART

MACOMB COUNTY COMMUNITY COLLEGE

WORTH PUBLISHERS, INC.

INTERMEDIATE ALGEBRA

SECOND EDITION

PRINTED IN THE UNITED STATES OF AMERICA

LIBRARY OF CONGRESS CATALOG CARD NO. 76-27388

ISBN: 0-87901-064-9

FIFTH PRINTING, SEPTEMBER 1981

DESIGN BY MALCOLM GREAR DESIGNERS

WORTH PUBLISHERS, INC.

444 PARK AVENUE SOUTH

NEW YORK, NEW YORK 10016

PREFACE TO THE SECOND EDITION

PURPOSE: This textbook was written to provide every possible assistance to students who are attempting to bridge the gap between a course in beginning algebra and a more advanced course in pre-calculus mathematics. In the second edition, some topics have been added, others have been rearranged or completely rewritten. The inspiration for these improvements came not only from the authors' continued teaching of the course since publication of the first edition, but also from the many excellent recommendations received from teachers throughout the world who provided us with the insights gained from their classroom experience with the book.

PREREQUISITES: As in the first edition, it is assumed that students have completed one year of high school algebra or an equivalent course in beginning algebra.

OBJECTIVES: The objectives of the second edition remain the same: (1) to enable the average student to learn to solve routine problems and develop computational skills with little, if any, assistance from the instructor; (2) to provide the average student with an appreciation of the concepts and logical reasoning involved in this level of mathematics; and (3) to enhance the ability of the average student to apply these concepts and this orderly way of thinking to new situations and problems. All the additions, deletions, and refinements in the second edition have been made because they contribute to the realization of these basic objectives. Carried over from the first edition are certain features which have elicited a uniformly favorable reception:

1 Several examples are given to illustrate routine problem solving.
2 Geometric interpretations supplement explanations wherever possible.
3 Definitions, theorems, and properties are carefully stated.
4 A reasonable balance is maintained between theory and techniques, drill and application.
5 Review problem sets at the end of each chapter are designed to help students gain confidence in their new-found abilities, or to indicate areas requiring additional study.

NEW FEATURES: Color is used to highlight definitions, theorems, properties, important statements, and parts of graphs. Many of the examples have been reworked and reorganized and there are more of them in this edition, so that every concept is immediately illustrated by many examples. The treatment of practical applications has been expanded. More attention is given to specific techniques for solving word problems. There are more problems and they are related more closely to the examples. The problem sets have been reviewed and revised as experience with the first edition dictated. Problem sets are now arranged so that the odd-numbered problems strive for the level of understanding desired by most users, while the even-numbered problems probe for deeper understanding of the concepts. This arrangement should simplify the task of making assignments. The answers to odd-numbered problems are given in the back of the book, and the even-numbered answers are provided for the instructor in a supplemental booklet.

CONTENTS: Much of the material has been rewritten to increase clarity. In Chapter 1, we added two sections on the operations of signed numbers. Chapter 2 includes additional material on factoring. In Chapter 3, we reversed the order of presentation of fractions, introducing multiplication and division of rational expressions before addition and subtraction. The new Chapter 4 deals with first-degree equations and inequalities. The original Chapter 4 on exponents and radicals is now Chapter 5. Much of this chapter has been revised, and equations involving radicals have been added. Chapter 6 has been rewritten to increase clarity, but the basic content is the same as in the first edition. Chapter 7 has been rearranged and expanded to include material formerly treated in other chapters, so that the material on linear equations is presented in a more unified way. Chapter 8 presents logarithms and their applications. The material on exponential and logarithmic functions is now in Chapter 9, which deals with all aspects of functions. Chapter 10 covers sequences, progressions, and the binomial theorem.

PACE: The pace of a course and the selection of topics depends on the curriculum of the individual school and on the academic calendar. The following suggested pace for a three-credit, one-semester course is meant only as a general guide:

Chapters 1 and 2:   5 lectures

Chapter 3:   4 lectures

Chapter 4:   6 lectures

Chapter 5:   8 lectures

Chapter 6:   7 lectures

Chapter 7:   8 lectures (Section 7.6 is optional)

Chapter 8:   6 lectures

Chapters 9 and 10:   6 lectures

If students are well prepared, Chapters 1, 2, and 3 can be reviewed briefly in three lectures so that topics in Chapters 9 and 10 can be covered more extensively.

ADDITIONAL AIDS: A *Study Guide* is available for students who need more drill and assistance. The *Study Guide* is written in a semi-programmed format and conforms to the arrangement of topics in the textbook. It contains a list of objectives, fill-in statements and problems, true-or-false statements, and a test for each chapter. All answers are provided in the *Study Guide* to encourage self-testing at each student's own pace.

ACKNOWLEDGMENTS: By the time a book reaches its second edition, a great many people have contributed to its development. Of course, there would be no second edition if the first edition had not been so widely adopted. For this reason, we have asked our publisher to reprint below our acknowledgments for the first edition. In addition, we wish to thank those who were particularly helpful in the writing of this edition: Professors Peggy Capell of Clayton Junior College; William G. Cunningham, Walter R. Rogers, and James H. Soles of DeKalb Community College; Louise G. Fulmer of Clemson University; Janet P. Ray of Seattle Central Community College; and Raymond F. Bryant and the staff of Florissant Valley Community College.

We again would like to express our thanks to our colleagues at Macomb County Community College and to the staff of Worth Publishers for their many contributions.

ACKNOWLEDGMENTS TO THE FIRST EDITION: Discussions with colleagues at a great many two- and four-year colleges and universities about the problems they face in teaching similar courses have greatly influenced the content and pace of this text. The guidance of the Committee of Undergraduate Programs in Mathematics (CUPM) was also very helpful. A great deal is owed to the generous assistance of Professors George W. Schultz of St. Petersburg Junior College, David J. Foulis of the University of Massachusetts, James W. Snow of Lane Community College, Gerald T. Ball of Chabot College, Frank P. Prokop of Bradley University, Norman Wheeler of Schoolcraft Community College, C. R. B. Wright of the University of Oregon, and Joseph Deblassio Jr. of Allegheny County Community College.

We also wish to acknowledge Harry L. Nustad of Henry Ford Community College and Murray Peterson of the College of Marin. Joseph Schlessinger, a graduate student at the University of California at Berkeley, ferreted out many errors and suggested valuable improvements. We are also indebted to our colleagues at Macomb, especially Professor James P. Yizze who helped us in the preliminary edition and Professor John von Zellen for his valuable suggestions and criticisms. Finally, the excellent cooperation of Robert C. Andrews of Worth Publishers, Inc. has been indispensable.

Mustafa A. Munem
William Tschirhart

Warren, Michigan
January 1977

CONTENTS

CHAPTER **1   Sets and Real Numbers**   1
Section   1.1 Sets   1
Section   1.2 Sets of Numbers   9
Section   1.3 Basic Properties of Real Numbers   17
Section   1.4 Operations of Signed Numbers   25

CHAPTER **2   Algebra of Polynomials**   39
Section   2.1 Properties of Positive Integral Exponents   39
Section   2.2 Polynomials   47
Section   2.3 Sums and Differences of Polynomials   51
Section   2.4 Products of Polynomials   57
Section   2.5 Factoring Polynomials by
             Common Factors and Grouping   69
Section   2.6 Factoring Polynomials by
             Special Products   73
Section   2.7 Factoring Trinomials of the
             Form $ax^2 + bx + c, a \neq 0$ 79
Section   2.8 Division of Polynomials   85

CHAPTER **3   Algebra of Fractions**   95
Section   3.1 Rational Expressions   95
Section   3.2 Multiplication and Division of Fractions   103
Section   3.3 Addition and Subtraction of Fractions   109
Section   3.4 Complex Fractions   120

CHAPTER **4   First-Degree Equations and
             Inequalities in One Variable**   131
Section   4.1 Equations   131
Section   4.2 Literal Equations and Formulas   139
Section   4.3 Applications of First-Degree Equations   144
Section   4.4 Inequalities   153
Section   4.5 Absolute Value Equations and Inequalities   162

CHAPTER **5 Exponents, Radicals, and Complex Numbers**   175
Section   5.1 Zero and Negative Exponents   175
Section   5.2 Rational Exponents   185
Section   5.3 Radicals   191
Section   5.4 Operations of Radical Expressions   199
Section   5.5 Equations Involving Radicals   207
Section   5.6 Complex Numbers   210

CHAPTER **6 Quadratic Equations and Inequalities**   223
Section   6.1 Solving Quadratic Equations by Factoring   223
Section   6.2 Solving Quadratic Equations by
              Completing the Square   227
Section   6.3 Solving Quadratic Equations by
              the Quadratic Formula   232
Section   6.4 Equations in Quadratic Form   238
Section   6.5 Quadratic Inequalities   242
Section   6.6 Applications of Quadratic Equations   249

CHAPTER **7 First-Degree Equations in
              More Than One Variable**   259
Section   7.1 Cartesian Coordinates and the Distance Formula 259
Section   7.2 Linear Equations and Their Graphs   268
Section   7.3 Forms of Equations of a Line   280
Section   7.4 Systems of Linear Equations in Two Variables 288
Section   7.5 Systems of Linear Equations in Three Variables 296
Section   7.6 Determinants   299
Section   7.7 Cramer's Rule   305
Section   7.8 Applications of Systems of Linear Equations   311

CHAPTER **8 Logarithms**   321
Section   8.1 Logarithms   321
Section   8.2 Properties of Logarithms   325
Section   8.3 Common Logarithms   330
Section   8.4 Computation with Logarithms   337
Section   8.5 Applications of Logarithms   342

CHAPTER **9 Relations and Functions**   353
Section   9.1 Relations   353
Section   9.2 Functions   358

Section   9.3 Direct and Inverse Variations   367
Section   9.4 Polynomial Functions   371
Section   9.5 Exponential and Logarithmic Functions   379
Section   9.6 Graphs of Special Relations—The Conics   384
Section   9.7 Systems with Second-Degree Equations   389

CHAPTER **10   Sequences, Progressions, and the
                Binomial Theorem**   399
Section   10.1 Sequences and Arithmetic Progressions   399
Section   10.2 Geometric Progressions   404
Section   10.3 Finite Sums   406
Section   10.4 Binomial Theorem   409

Appendix   A   Measurement   419
Appendix   B   Formulas from Geometry   421
Section   B.1 Areas of Plane Figures   421
Section   B.2 Perimeters of Plane Figures   427
Section   B.3 Volumes and Surface Area   432
Appendix   C   Table of Common Logarithms   440

Answers to Selected Problems   443

Index   479

CHAPTER 1

Sets and Real Numbers

The study of algebra is based upon the general properties of numbers, particularly real numbers. In this chapter we shall discuss the basic properties and operations of real numbers. We begin by introducing some of the terms and symbols used to describe sets and then, with the help of this set terminology, go on to describe different sets of numbers.

1.1 Sets

A *set* is a collection of objects. Other words, such as *class, group,* or *aggregate,* can be used in nonmathematical situations, but the word "set" is usually used in modern mathematics. To better understand the use of sets in mathematics, let us recall some nonmathematical uses: we speak of a set of golf clubs or a chess set or a set of books in a library. The reader will be able to think of other examples of sets that occur in everyday life.

The objects of sets are called *elements* or *members* of the set. We use lowercase letters, a, b, c, d, etc., to represent the elements of a set and capital letters, A, B, C, D, etc., to represent the sets. For example, the set A which has as its elements b, c, d, and e, and no other elements, can be written $A = \{b,c,d,e\}$. The elements are enclosed within braces to indicate that they belong to the set. The symbol used to show that an element belongs to a set is $\in$. In our example, $b \in A$, $c \in A$, $d \in A$, and $e \in A$, read "b belongs to set A," "c belongs to set A," "d belongs to set A," and "e belongs to set A." To indicate that a particular element does not belong to a set, we use the notation $\notin$: for example, $f \notin A$ and $g \notin A$.

We call the method of representing the set A in the above example *enumeration.* For instance, we can use enumeration to write the set B of all integers from 1 to 7 as

$$B = \{1,2,3,4,5,6,7\}$$

EXAMPLES

1 Let $A = \{3,4,5,7,8,9\}$, $B = \{2,3,10,12\}$, and $C = \{2,7,9,11,15\}$. Insert in the following blanks the correct symbol, $\in$ or $\notin$.
   a)   5 _____ A b)   5 _____ B c)   9 _____ B

d) 3 _____ C	e) 8 _____ A	f) 10 _____ C
g) 15 _____ A	h) 3 _____ B	i) 11 _____ C

SOLUTION

a) $5 \in A$	b) $5 \notin B$	c) $9 \notin B$
d) $3 \notin C$	e) $8 \in A$	f) $10 \notin C$
g) $15 \notin A$	h) $3 \in B$	i) $11 \in C$

2   Use enumeration to describe the set M of the letters in the word "Mississippi."

SOLUTION. Although the word "Mississippi" contains letters that are repeated, we only show each (different) letter once, so that

$$M = \{m, i, s, p\}$$

If a set is so defined that it does not have any elements, we call it a *null set* (or *empty set*). The symbol for the null set is $\varnothing$ or $\{\ \}$. For example, the set P of all presidents of the United States who died before their thirty-fifth birthday is an empty set, that is, $P = \varnothing$, since the requirements for being president of the United States require an age of at least 35 years. *It is important to note that 0 and $\varnothing$ do not mean the same thing. That is, $\varnothing$ is not equal to 0 or $\{0\}$, since $\{0\}$ is a set with the one element 0, whereas $\varnothing$ (or $\{\ \}$) is the set with no elements; 0 represents the number of elements in $\varnothing$, although 0 is not an element of $\varnothing$.*

A set is characterized by its elements. The order in which the elements appear is of no significance. For example, the set $B = \{1, 2, 3\}$ could be enumerated as follows:

$$B = \{1, 3, 2\} = \{2, 1, 3\} = \{2, 3, 1\} = \{3, 1, 2\} = \{3, 2, 1\}$$

There are two common methods used to describe sets. One is the enumeration method, which we have discussed; the other is set description, called *set-builder notation*, which employs the identifying properties of the elements of the set. For example, the set B of all positive integers between 1 and 100 can be described using set-builder notation as follows. Let some letter, say x, represent an element of the set; then the set B is described as "the set of all x such that x is a positive integer between 1 and 100." Or, equivalently,

$$B = \{x \mid x \text{ is a positive integer between 1 and 100}\}$$

In general, the set-builder notation takes the form

$$A = \{x \mid x \text{ has property } p\}$$

which is read "*A* is the set of all elements *x* such that *x* has the property *p*."

EXAMPLE

3 Use set-builder notation to describe each of the following sets.

a) *M*, the set of all male citizens of the United States.
b) *C*, the set of odd counting numbers.

SOLUTION

a) $M = \{x \mid x$ is a male citizen of the United States$\}$
b) $C = \{x \mid x$ is an odd counting number$\}$

Finite and Infinite Sets

Consider the following three sets: *A* is the set of boys in a mathematics class, *B* is the set of positive counting numbers from 1 to 100, and *C* is the set of months in a year. Sets *A*, *B*, and *C* can be enumerated because every element of each set can be named explicitly. Such sets are examples of *finite sets*. In general, a set *M* is said to be a *finite set* if the elements in *M* can be counted, in the usual way, using counting numbers 1, 2, 3, 4 . . ., and if this counting process eventually terminates, resulting in a specific number *n* (which is then said to be the number of elements in the set *M*). A set that is neither finite nor empty is called an *infinite set*. For example, the set *N* of all positive counting numbers is an infinite set because enumeration is impossible in this case. It should be noted, however, that infinite sets whose elements form a general pattern can be described by enumeration. For instance, using enumeration, we can describe the set *N* of all positive counting numbers as

$$N = \{1, 2, 3, 4, \ldots\}$$

where the three dots mean the same as "etc."

By contrast, the set *B* of all counting numbers from 1 to 100 can be described as

$$B = \{1, 2, 3, \ldots, 100\}$$

where the three dots represent the elements of *B* between 3 and 100.

EXAMPLES

Describe each of the following sets. Indicate if the set is finite or infinite.

4   A, the set of all odd counting numbers.

SOLUTION.   The known pattern of the elements of A suggests that A can be written as $A = \{1,3,5, . . .\}$. A is an infinite set.

5   E, the set of all even counting numbers from 2 to 500.

SOLUTION.   The set E can be enumerated as follows: $\{2,4,6, . . .,500\}$; therefore, E is a finite set.

Set Relations

Suppose that S is the set of students in class and G is the set of girls in the same class. Clearly, all the members of G are also members of S. We describe the relation between G and S by saying that G is a subset of S. Symbolically, we write $G \subseteq S$. This can be read as "G is a subset of S," or "G is contained in S." In general, a set A is a *subset* of a set B, written $A \subseteq B$, if every element of A is also an element of B. That is, if $A \subseteq B$ and $x \in A$, then $x \in B$. For example, the set $A = \{1,2,3\}$ is a subset of a set $B = \{1,2,3,5\}$, since every element of the set A is also an element of the set B.

   We shall also agree that the empty set $\varnothing$ is a subset of every other set. This agreement implies that every element of $\varnothing$ must also be an element of any other set, say A. Since $\varnothing$ contains no elements, this agreement does not contradict our definition of subset, and we may write $\varnothing \subseteq A$ for any set A.

   Consider the sets $A = \{m,n,c,d\}$ and $B = \{n,m,d,c\}$. Then $A \subseteq B$ and $B \subseteq A$. (Why?) This suggests a definition of *equality of sets*, for, if $A \subseteq B$ and $B \subseteq A$, then A and B have the same elements and we write $A = B$. Note that if $A = \{1,2\}$ and $B = \{1,2,3\}$, then $A \subseteq B$, but $A \neq B$. In this case we say that A is a *proper subset* of B. In general, A is a *proper subset* of B, written $A \subset B$ (the horizontal bar is omitted), if all members of A are in B and B has at least one member not in A. That is, $A \subseteq B$, but $A \neq B$ implies that $A \subset B$.

EXAMPLES

6   List all the possible subsets of $A = \{1,2,3\}$. Indicate the proper subsets of A and the number of subsets of A.

SOLUTION. $\varnothing$, $\{1\}$, $\{2\}$, $\{3\}$, $\{1,2\}$, $\{1,3\}$, $\{2,3\}$, and $\{1,2,3\}$ are the subsets of A. Note that the subsets, with the exception of $\{1,2,3\}$, are proper subsets. Also, the set A has eight subsets.

7 List all possible subsets of $B = \{1,2,3,4\}$. Indicate the proper subsets of B and the number of subsets of B.

SOLUTION. $\emptyset$, $\{1\}$, $\{2\}$, $\{3\}$, $\{4\}$, $\{1,2\}$, $\{1,3\}$, $\{1,4\}$, $\{2,3\}$, $\{2,4\}$, $\{3,4\}$, $\{1,2,3\}$, $\{1,2,4\}$, $\{1,3,4\}$, $\{2,3,4\}$, and $\{1,2,3,4\}$ are the subsets of B. All the subsets with the exception of $\{1,2,3,4\}$ are proper subsets of B. Also, the set B has 16 subsets.

Consider a certain class in algebra; the set of girls and the set of boys in that class have no members in common. Also, the set $A = \{1,3,2\}$ has no member in common with the set $B = \{5,6,7\}$. Such sets are called *disjoint sets*. In general, if two sets have no elements in common, then they are disjoint sets.

When we form subsets, there must be some source from which the elements of these subsets are to be selected; that is, any discussion of sets must be limited to a particular set. For example, the set $A = \{a,b,c,d,e\}$ has elements that come from the alphabet. Likewise, the set $B = \{1,2,3,4,5\}$ has elements that are counting numbers. We call this larger set from which we select the elements to form subsets a *universal set* or *universe*. The universal set (designated as U) represents the complete set or largest set for the particular discussion, and all other sets in that same discussion will be made up of elements from U. The choice of U will depend on the particular problem under discussion. The universal set U will not necessarily be the same for all problems. For example, in one problem U may be the set of all counting numbers; in another, U may be the set of all fractions. We will see later in this chapter that the set of counting numbers, the set of odd integers, the set of even integers, and the set of negative integers are subsets of the universal set of integers.

EXAMPLES

Find a universal set U for each of the following.

8 $\{a,b,c\}$, $\{e,f,g,h\}$, $\{s,k,t,u\}$

SOLUTION. Since all the elements of the given sets are letters of the English alphabet, a possible universal set will be the set of the letters of the English alphabet. That is, $U = \{a,b,c,d. . .,z\}$.

9 $\{1,3,100\}$, $\{3,5,9,35,73\}$, $\{87,105,672,985\}$

SOLUTION. Since all the elements are counting numbers, a possible universal set will be the set of the counting numbers. That is, $U = \{1,2,3,4,5,. . .\}$. Note that U is an infinite set in this case.

Set Operations

Consider a universal set $U = \{1,2,3,\ldots,8\}$ and subsets of U, $A = \{1,2,3,4,5\}$, and $B = \{3,4,5,6,7\}$. How can sets A and B be combined to form other subsets of U? One way is to combine all the elements of A and B to form the set $\{1,2,3,4,5,6,7\}$. The operation suggested by this example is called *set union*. In general, the union of two sets, written $A \cup B$, is the set of all elements in set A or in set B or in both A and B. Symbolically,

$$A \cup B = \{x \mid x \in A \quad \text{or} \quad x \in B\}$$

EXAMPLES

10   If $A = \{1,3,5,7\}$ and $B = \{2,4,6,8\}$, find $A \cup B$.

SOLUTION. $A \cup B$ is the set that contains all the elements in both A and B. That is, $A \cup B = \{1,2,3,4,5,6,7,8\}$.

11   If $A = \{1,3,5\}$ and $B = \{3,5,7,9\}$, find $A \cup B$.

SOLUTION. $A \cup B$ is the set that contains all the elements in both A and B. That is, $A \cup B = \{1,3,5,7,9\}$.

The set formed by including all the elements common to both sets A and B where $A = \{1,2,3,4,5\}$ and $B = \{3,4,5,6,7\}$ is $\{3,4,5\}$. This set is called the *intersection* of A and B. In general, the intersection of two sets, written $A \cap B$, is the set of all elements common to both A and B. Symbolically,

$$A \cap B = \{x \mid x \in A \quad \text{and} \quad x \in B\}$$

EXAMPLES

12   If $A = \{3,6,9,12\}$ and $B = \{2,4,6,8,10,12\}$, find $A \cap B$.

SOLUTION. $A \cap B$ is the set whose elements are contained in both A and B. That is, $A \cap B = \{6,12\}$.

13   If $A = \{5,8,9\}$ and $B = \{7,4,2,3\}$, find $A \cap B$.

SOLUTION.   Sets A and B do not have any common elements. Therefore, $A \cap B = \varnothing$. That is, A and B are disjoint sets.

14   Determine $A \cup B$ and $A \cap B$ if $A = \{2,3,4,5,7\}$ and $B = \{5,6,3,7,8\}$.

SOLUTION

$$A \cup B = \{2,3,4,5,6,7,8\}$$
$$A \cap B = \{3,5,7\}$$

15 Determine $(A \cap B) \cap C$ and $A \cap (B \cap C)$ if $A = \{1,3\}$, $B = \{3,5,7\}$, and $C = \{3,8,7\}$.

SOLUTION

$$\begin{aligned}(A \cap B) \cap C &= (\{1,3\} \cap \{3,5,7\}) \cap \{3,8,7\} \\ &= \{3\} \cap \{3,8,7\} = \{3\}\end{aligned}$$

$$\begin{aligned}A \cap (B \cap C) &= \{1,3\} \cap (\{3,5,7\} \cap \{3,8,7\}) \\ &= \{1,3\} \cap \{3,7\} = \{3\}\end{aligned}$$

This example suggests that $(A \cap B) \cap C = A \cap (B \cap C)$.

PROBLEM SET 1.1

1 Let $A = \{4,5,6,7,8,9,10\}$, $B = \{3,6,8,9,10\}$, and $C = \{5,6,8,12\}$. Insert in the following blanks the correct symbol, $\in$ or $\notin$.

5 ____ A	5 ____ B	5 ____ C	10 ____ C
10 ____ A	10 ____ B	3 ____ C	12 ____ A
2 ____ A	8 ____ B	6 ____ C	4 ____ B

2 Let A be the set of counting numbers greater than 1 but less than 9. Describe A by enumeration.

3 Use enumeration to describe the set M of the letters in the word "Massachusetts."

4 Use enumeration to describe the set P of the letters in the word "Pennsylvania."

In problems 5–8, use set notation to describe each set.

5 The set of all months of the year whose names begin with the letter M.

6 The set of all days of the week whose names begin with the letter T.

7 The set of counting numbers greater than 2 and less than 13.

8 The set of baseball players who hit 3,000 home runs in their professional playing careers.

In problems 9–12, use set-builder notation $\{x \mid x$ has property $p\}$ to describe each set.

9 The set of all people who live in Detroit.

10 The set of the first six letters of the English alphabet.

11 The set of all people listed in the Washington, D.C. telephone directory.

12 The set of counting numbers greater than 4 and less than 17.

In problems 13–16, indicate which sets are finite and which are infinite.

13 The set of people living in the United States.

14 The set of even counting numbers.

15 The set of the letters of the English alphabet.

16 The set $A = \{1,5,9,13, . . .\}$.

In problems 17–24, indicate which statements are true and which are false.

17 $\{3\} \in \{3,4\}$ **18** $\{x,y,z\} = \{z,x,y\}$

19 $\{5\} \subseteq \{1,5,8\}$ **20** $\{7,8\} \subseteq \{5,8,9,10\}$

21 $\{1,2\} = \{2,4\}$ **22** $\{a,b,c\} \subseteq \{a,b\}$

23 $\varnothing \subseteq \{x,y,z\}$ **24** $\{a,b,c\} \subseteq \{c,b,a\}$

In problems 25–28, list all the subsets of each set. Indicate which subsets are proper subsets.

25 $\{2,3\}$ **26** $\{2\}$

27 $\{a,b,c\}$ **28** $\{5,6,7,8\}$

29 $\{a,b,c,d\}$

30 Is it true that $\varnothing \subseteq \varnothing$? Why? How many elements are in the set $\varnothing$?

31 Indicate the set relation that exists between the set of even counting numbers greater than 11 and the set $\{2,4,6,8,10\}$.

32 Given a set A with six elements, how many different subsets does the set A have? If A has n elements, how many different subsets can be formed?

In problems 33–36, make an appropriate choice for the universal set U if the sets under consideration are the following:

33 $\{1,3,5\}$, $\{7,9,11,13,15,17\}$, $\{19,21,23,25,27,29\}$

34 $\{2,4\}$, $\{6,8,10,12\}$, $\{14,16,18,20,22\}$, $\{24,26,28,30,32,34\}$

35 {football player, baseball player, basketball player}, {swimmer, surfer, scuba diver}

36 $\{x \mid x$ is a male student$\}$, $\{x \mid x$ is a female student$\}$

In problems 37–54, let $A = \{1,2,4\}$, $B = \{2,3,5,7\}$, $C = \{6,3,5,8\}$, $D = \{2,5\}$, $E = \{7,8,9\}$, and $F = \{8\}$. Form each set.

37 $A \cup B$

38 $A \cup F$

39 $A \cap B$

40 $E \cap D$

41 $B \cup C$

42 $C \cup B$

43 $D \cap E$

44 $C \cap F$

45 $F \cap C$

46 $A \cup E$

47 $B \cup E$

48 $B \cup D$

49 $C \cup F$

50 $F \cap \varnothing$

51 $C \cup \varnothing$

52 $(A \cap B) \cup F$

53 $(C \cap D) \cup E$

54 $(F \cup E) \cap D$

55 Let $A = \{6,7,8,9\}$, $B = \{7,8,10,11\}$, and $C = \{6,8,12,13\}$.
a) Find $A \cap (B \cup C)$.
b) Find $(A \cap B) \cup (A \cap C)$.
c) Is $A \cap (B \cup C) = (A \cap B) \cup (A \cap C)$?

1.2 Sets of Numbers

The language of sets can be used to describe some of the number sets of algebra. Historically, the first set of numbers developed were the *natural numbers* 1, 2, 3, etc., also called the *counting numbers*. The letter N is used to designate the set of natural numbers; thus, $N = \{1,2,3,4, \ldots\}$. This set is also called the set of positive integers. The set N is an infinite set because the process of counting its elements can never come to an end. However, there is one new "number" not obtainable by counting, which it is convenient to introduce immediately. This number is zero and is denoted symbolically by 0. The union of the set N and the set $\{0\}$ is called the set of whole numbers and is denoted by W. That is, $W = \{0,1,2,3, \ldots\}$.

The Number Line

The set of whole numbers W can be represented geometrically on a line by associating each whole number with a point on that line. Such a representation is called a *number line*. The number line can be constructed by starting with a line that extends endlessly in opposite directions. It is convenient to use arrowheads to indicate that the line extends endlessly (Figure 1.1).

Figure 1.1

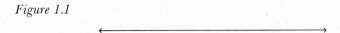

Now, choose a point on the line to associate with the number 0 (this point is called the *origin*) and another point to the right of the origin to associate with the number 1 (Figure 1.2). The distance between the points labeled 0 and 1 is called the *unit length* or the *scale unit* of the number line. By measuring one unit length to the right of the point 1, we find the point to associate with 2. Repeating this process, other points can be found to associate with 3, 4, 5, . . . (Figure 1.3).

Figure 1.2

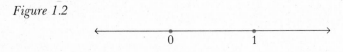

Figure 1.3

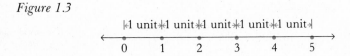

The side containing the point associated with 1 is called the *positive side* of the number line, and the direction from the origin to the point labeled 1 is called the positive direction on the number line. The point associated with a number on a number line is called the *graph* of that number, and the number is called the *coordinate* of that point. For example, the coordinate of points A, B, C, and D on the number line are 0, 2, 4, and 5, respectively (Figure 1.4).

Figure 1.4

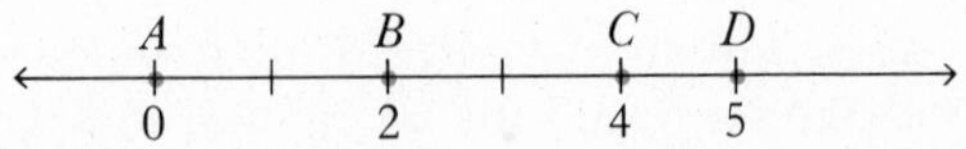

In order to assign coordinates to points to the left of the origin on the number line, we will introduce *negative numbers*. Thus, the coordinates of points *E*, *F*, *G*, and *H* on the number line (Figure 1.5) are −4, −3, −2, and −1 (which read *"negative* four," *"negative* three," *"negative* two," and *"negative* one," respectively) (Figure 1.5).

Figure 1.5

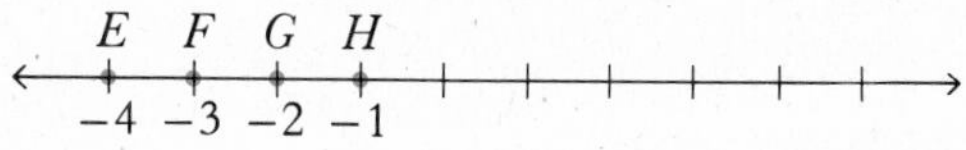

E   F   G   H
$$\longleftarrow \!\!\! \underset{-4\ -3\ -2\ -1}{+\ +\ +\ +\ \ |\ \ |\ \ |\ \ |\ \ |\ \ |} \!\!\! \longrightarrow$$

Note that a negative number is named by a numeral with a negative (minus) sign, or −, written to the left of an ordinary number symbol. The side of the line that contains the points with negative coordinates is called the *negative side* of the number line, and the direction from 0 to −1 is called the *negative* direction. The set $\{\ldots, -4, -3, -2, -1\}$ is called the set of *negative integers.* It should be pointed out that the negative integers are written with the negative sign affixed, and the positive integers are indicated by the absence of a sign. Hence, throughout the text, if a number has no sign affixed to it, it is positive. Thus, the coordinate of a point that corresponds to the number +5 is written as 5. (*Note:* Some authors write the number names preceded by the *raised symbol* + for the points to the right of zero and the *raised symbol* − for the points to the left of zero. For example, a point corresponding to the number "positive five" is denoted by ⁺5, and a point corresponding to the number "negative five" is denoted by ⁻5. However, in this text, we denote these numbers by 5 and −5.)

Real Numbers

The set *W*, together with the set of negative integers, is called the set of *integers* and is denoted by *I*. Thus, $I = \{\ldots, -4, -3, -2, -1, 0, 1, 2, 3, 4, \ldots\}$. Investigating the graph of the set *I* (Figure 1.5), we note that there are other points not associated with an element of *I*. For example, the point halfway between 0 and 1 is not associated with an integer. Instead, the fraction $\frac{1}{2}$ can be associated with this point (Figure 1.6). We could go on marking more points because there is at least one more point between any two that we marked. For instance, to find a point between *A* and *C*, we could take 0 and $\frac{1}{2}$ and find a number between the two, say $\frac{1}{4}$.

Figure 1.6

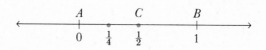

A   C   B
$$\longleftarrow \!\!\! \underset{0\qquad \frac{1}{4}\qquad \frac{1}{2}\qquad\quad 1}{|\qquad \bullet \qquad \bullet \qquad\quad |} \!\!\! \longrightarrow$$

Note that there are infinitely many other points on the number line that require fractions to indicate their location with respect to the origin. Numbers such as those described above are called *rational numbers*. A rational number, then, is a number that can be expressed as a quotient of two integers. That is, a rational number can be expressed in the form a/b, where a and b are integers and $b \neq 0$. Examples of rational numbers are $\frac{3}{4}, \frac{21}{11}, -\frac{2}{3}$, and $-\frac{11}{7}$. The set of rational numbers is denoted by Q. Figure 1.7 shows the location of a few points which correspond to some rational numbers on the number line.

Figure 1.7

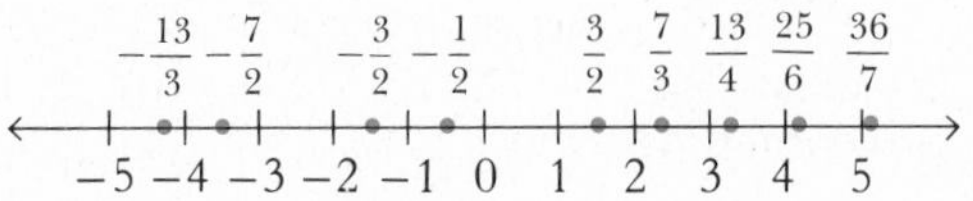

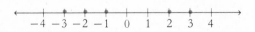

Since any integer can be expressed as the quotient of itself and 1 (that is, $3 = \frac{3}{1}$ and $-5 = -\frac{5}{1}$), the set I is a subset of the set Q. Symbolically, it is written $I \subseteq Q$.

EXAMPLES

Represent the following sets on the number line.

1   $\{-3, -2, -1, 2, 3\}$

SOLUTION.   The members of the set are located on a number line (Figure 1.8).

Figure 1.8

2   $\{\frac{1}{3}, \frac{2}{3}, \frac{5}{3}, \frac{6}{3}, \frac{7}{3}\}$

SOLUTION.   The members of the set are located on a number line (Figure 1.9).

Figure 1.9

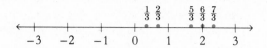

The rational numbers can also be described by investigating their decimal representations. Thus, rational numbers such as 3, $\frac{1}{2}$, and $\frac{13}{4}$, which can be written with an integral number of decimal places as

$$3 = \tfrac{3}{1} = 3.0 \qquad \tfrac{1}{2} = 0.5 \qquad \text{and} \qquad \tfrac{13}{4} = 3.25$$

are sometimes called *terminating decimals.* In general, any terminating decimal can be written in the form a/b, where a and b are integers and $b \neq 0$. For example, $2.37 = 2 + \frac{3}{10} + \frac{7}{100} = \frac{237}{100}$.

EXAMPLES

3 Express each of the following rational numbers in decimal notation.
   a) $\frac{3}{4}$
   b) $-\frac{2}{5}$
   c) $\frac{1}{8}$

SOLUTION.   The rational numbers $\frac{3}{4}$, $-\frac{2}{5}$, and $\frac{1}{8}$ can be written with an integral number of decimal places, so that
   a) $\frac{3}{4} = 0.75$
   b) $-\frac{2}{5} = -0.4$
   c) $\frac{1}{8} = 0.125$

4 Express each of the following terminating decimals in the form $\frac{a}{b}$, where $b \neq 0$.
   a) 0.3
   b) 2.7
   c) 3.68

SOLUTION

   a) $0.3 = \frac{3}{10}$
   b) $2.7 = 2 + \frac{7}{10} = \frac{27}{10}$
   c) $3.68 = 3 + \frac{6}{10} + \frac{8}{100} = \frac{368}{100}$

The rational numbers can also be expressed as repeating decimals. For example, the rational number $\frac{2}{3}$ is written 0.6666666. . . . The shortest complete repeating pattern is indicated by drawing a horizontal bar over it, so that $\frac{2}{3} = 0.\overline{6}$.

The rational number $\frac{3}{7}$, which can be interpreted as $3 \div 7$, is expressed as a decimal representation:

$$
\begin{array}{r}
0.4\,28571428571 \ldots \\
\hline
\end{array}
$$

```
            0.4 28571428571 . . .
         ┌─────────────────────────
  7  │  |3.0|000000
         |2 8|
         |_ 2|0
            1 4
            ──
            60
            56
            ──
            40
            35
            ──
            50
            49
            ──
            10
             7
            ──
           | 30 |
           | 28 |
           |_ 2 |
```

Note that in each step within the division, the remainder must be either 0, 1, 2, 3, 4, 5, or 6. Therefore, if enough zeros are annexed after the decimal of the dividend (this does not affect the value of the dividend) and the division by 7 is performed more than seven times, one of the remainders must reoccur; but as soon as a remainder appears again (in this example it is 2), the digits in the quotient repeat. In this example

$$\frac{3}{7} = 0.428571\overline{428571}$$

This concept can be generalized; for if a/b is a rational number, where $a \in I$, $b \in I$, and $b \neq 0$,

$$b\overline{)a.0000 \ldots}$$

can be performed until a remainder repeats [there are only b or $-b$ remainders possible, depending on whether b is positive or negative (why?)]. When a remainder repeats, the digits in the quotient repeat. Hence, it is true that every rational number can be represented by an eventually repeating decimal. The converse of this statement also holds. That is, every eventually repeating decimal represents a rational number.

EXAMPLES

5 Express each of the following rational numbers as a decimal.
   a) $\frac{10}{3}$
   b) $\frac{5}{6}$
   c) $\frac{7}{9}$

SOLUTION

a) $\frac{10}{3} = 3.\overline{3}$ has a repeating block (the digit 3).
b) $\frac{5}{6} = 0.8\overline{3}$ has a repeating block (the digit 3).
c) $\frac{7}{9} = 0.\overline{7}$ has a repeating block (the digit 7).

6 Express each of the following rational numbers in the form a/b, where a and b are integers and $b \neq 0$.
   a) $0.\overline{7}$
   b) $0.\overline{31}$
   c) $1.35\overline{28}$

SOLUTION

a) Let $x = 0.777\overline{7}$; then

$$10x = 7.777\overline{7}$$
$$\underline{-x = -0.777\overline{7}}$$
$$9x = 7$$

so that $x = \frac{7}{9}$.

b) Let $x = 0.3131\overline{31}$; then

$$100x = 31.31\overline{31}$$
$$\underline{-x = -0.31\overline{31}}$$
$$99x = 31$$

so that $x = \frac{31}{99}$.

c) Let $x = 1.352828\overline{28}$; then

$$10{,}000x = 13{,}528.28\overline{28}$$
$$\underline{-100x = -135.28\overline{28}}$$
$$9{,}900x = 13{,}393$$

so that $x = \frac{13{,}393}{9{,}900}$.

Although the set of rational numbers is the set of numbers represented by repeating decimals, there are decimals that do not repeat. For example, the decimal 1.010010001000010001 . . . , where there is one more 0 after each 1 than there is before the 1, is a nonrepeating decimal. Another example of a nonrepeating decimal is a number used in plane geometry called π, where $\pi = 3.14159265358$ . . . . Also, all numbers, such as $\sqrt{2}$, $\sqrt{3}$, and $\sqrt{5}$, which can be approximated by decimals: $\sqrt{2} = 1.14142136$ . . . , $\sqrt{3} = 1.7320508$ . . . , and $\sqrt{5} = 2.23606797$ . . . , have nonrepeating decimal representations. Such numbers are called *irrational numbers.*

The set of all numbers used to represent all the points on a number line is called the set of *real numbers.* That is, the set of real numbers consists of the union of the set of rational numbers and the set of irrational numbers. The number line that results from associating each of its points with a real number is called a *real line* or *real axis.*

In summary, we say that the set of real numbers can be characterized by their decimal representations. That is, the set of real numbers consists of two subsets, one subset that contains real numbers with repeating decimals, called rational numbers, and the other subset containing nonrepeating decimals, called irrational numbers. If L represents the set of irrational numbers, Q the set of rational numbers, and R the set of real numbers, we have

$$R = Q \cup L \quad \text{and} \quad Q \cap L = \varnothing$$

PROBLEM SET 1.2

In problems 1–6, represent each set on a number line.

1 $\{-4,-3,-2,-1,2\}$ **2** $\{-5,-3,0,3\}$

3 $\{-\frac{2}{3},-\frac{1}{3},0,\frac{1}{3},\frac{2}{3}\}$ **4** $\{\frac{1}{5},\frac{2}{5},\frac{3}{5},\frac{4}{5}\}$

5 $\{-2,-1,0,\frac{1}{2},\frac{3}{2},\frac{5}{2}\}$ **6** $\{-\frac{5}{9},-\frac{4}{9},-\frac{3}{9},-\frac{2}{9},-\frac{1}{9}\}$

In problems 7–10, find each set.

7 $\{0,1,2,3,. . .\} \cup \{1,2,3,4,. . .\}$

8 $\{0,1,2,3,4,. . .\} \cap \{2,3,4,5,. . .\}$

9 $\{. . .,-3,-2,-1,0\} \cap \{x | x \text{ is a whole number}\}$

10 $\{-1,-2,-3,. . .\} \cup \{x | x \text{ is a positive integer}\}$

In problems 11–20, express each rational number in decimal notation.

11 $\frac{3}{5}$ **12** $-\frac{7}{4}$

13 $\frac{3}{2}$ 14 $-\frac{7}{2}$

15 $\frac{4}{5}$ 16 $\frac{9}{100}$

17 $-\frac{5}{4}$ 18 $\frac{5}{9}$

19 $-\frac{7}{3}$ 20 $\frac{6}{7}$

In problems 21–38, express each decimal in the form a/b, where a and b are integers, $b \neq 0$.

21 0.27 22 1.72

23 2.64 24 7.15

25 -0.125 26 -0.008

27 -0.0527 28 -0.0098

29 0.000329 30 -0.000052

31 $0.\overline{5}$ 32 $1.\overline{3}$

33 $0.\overline{46}$ 34 $0.05\overline{3}$

35 $0.4\overline{9}$ 36 $0.\overline{128}$

37 $-7.\overline{362}$ 38 $-8.5\overline{821}$

1.3 Basic Properties of Real Numbers

In Section 1.2 we considered the procedure of locating real numbers on the number line. Here we shall list a set of properties of real numbers. These properties serve as a foundation for justifying the algebraic steps in later chapters. Before we encounter these properties, we shall assume that the reader knows from arithmetic how to add, subtract, multiply, and divide positive numbers in general. (The rules of signed numbers are given in Section 1.4.) With this idea in mind, we can characterize the sum and the product (throughout the text, the product of a and b is denoted $a \cdot b$ or ab, although it is customary to write the product of the numbers 5 and 3 as 5×3) of real numbers as follows:

1 **CLOSURE PROPERTY**

a) *Closure property for addition:* The sum of two real numbers is always a real number. That is, if a and b are real numbers, then $a + b$ is a real number. For example, the sum of the real numbers 2.3 and 7.9 is a real number 10.2. That is, $2.3 + 7.9 = 10.2$. Also, since $\sqrt{3}$ and $\sqrt{5}$ are real numbers, then

$$\sqrt{3} + \sqrt{5} = 1.732050808\ldots + 2.236067977\ldots$$
$$= 3.968118785\ldots$$

is a real number.

b) *Closure property for multiplication:* The product of two real numbers is a real number. That is, if a and b are real numbers, then $a \cdot b$ is a real number. For example, the product of the real numbers 8.7 and 5.3 is the real number 46.11. That is, $8.7 \times 5.3 = 46.11$. Also, since $\sqrt{5}$ and $\sqrt{7}$ are real numbers, then

$$5 \times \sqrt{7} = 5 \times 2.645751311\ldots = 13.22875656\ldots$$

is a real number.

EXAMPLE

1   Consider the set $A = \{1,2\}$. Does the set A possess the closure properties for addition and for multiplication?

SOLUTION.   Since $1 + 2 = 3$ and 3 does not belong to A, the set A does not possess the closure property for addition. Also, since $2 \times 2 = 4$, and 4 does not belong to set A, then A does not possess the closure property for multiplication.

2   **COMMUTATIVE PROPERTY**

a) *Commutative property for addition:* For all real numbers a and b,

$$a + b = b + a$$

For example, $13 + 20 = 20 + 13$, since

$$13 + 20 = 33 \qquad \text{and} \qquad 20 + 13 = 33$$

Also,

$$85.76 + 19.33 = 19.33 + 85.76 = 105.09$$

b) *Commutative property for multiplication:* For all real numbers a and b,

$$a \cdot b = b \cdot a$$

For example, $3 \times 2 = 2 \times 3$, since

$$3 \times 2 = 6 \qquad \text{and} \qquad 2 \times 3 = 6$$

Also,

$$97.46 \times 51.93 = 51.93 \times 97.46 = 5{,}061.0978$$

EXAMPLES

Verify the commutative property by performing the actual computations.

2    $73.84 + 39.67$ and $39.67 + 73.84$

SOLUTION

$$73.84 + 39.67 = 113.51$$
$$39.67 + 73.84 = 113.51$$

Therefore,

$$73.84 + 39.67 = 39.67 + 73.84$$

3    $\frac{9}{4} \times \frac{8}{6}$ and $\frac{8}{6} \times \frac{9}{4}$

SOLUTION

$$\frac{9}{4} \times \frac{8}{6} = 3$$
$$\frac{8}{6} \times \frac{9}{4} = 3$$

Therefore,

$$\frac{9}{4} \times \frac{8}{6} = \frac{8}{6} \times \frac{9}{4}$$

3   ASSOCIATIVE PROPERTY

a)   *Associative property for addition:* For all real numbers a, b, and c,

$$a + (b + c) = (a + b) + c$$

For example, $7 + (3 + 9) = (7 + 3) + 9$, since

$$7 + (3 + 9) = 7 + 12 = 19 \qquad \text{and}$$
$$(7 + 3) + 9 = 10 + 9 = 19$$

b) *Associative property for multiplication:* For all real numbers, a, b, and c,

$$a \cdot (b \cdot c) = (a \cdot b) \cdot c$$

For example, $0.3 \times (5.2 \times 1.74) = (0.3 \times 5.2) \times 1.74$, since

$$0.3 \times (5.2 \times 1.74) = 0.3 \times 9.048 = 2.7144 \qquad \text{and}$$
$$(0.3 \times 5.2) \times 1.74 = 1.56 \times 1.74 = 2.7144$$

EXAMPLES

Verify the associative property by performing the actual computations.

4    $77 + (19 + 2)$ and $(77 + 19) + 2$

SOLUTION.   Since $77 + (19 + 2) = 77 + 21 = 98$ and $(77 + 19) + 2 = 96 + 2 = 98$, then

$$77 + (19 + 2) = (77 + 19) + 2$$

5    $14 \times (13 \times 7)$ and $(14 \times 13) \times 7$

SOLUTION.   Since $14 \times (13 \times 7) = 14 \times 91 = 1{,}274$ and $(14 \times 13) \times 7 = 182 \times 7 = 1{,}274$, then

$$14 \times (13 \times 7) = (14 \times 13) \times 7$$

4   DISTRIBUTIVE PROPERTY

If a and b are real numbers, then

a) $a \cdot (b + c) = a \cdot b + a \cdot c$
b) $(a + b) \cdot c = a \cdot c + b \cdot c$

For example, $6 \times (5 + 7) = (6 \times 5) + (6 \times 7)$, since

$$6 \times (5 + 7) = 6 \times 12 = 72 \qquad \text{and}$$
$$(6 \times 5) + (6 \times 7) = 30 + 42 = 72$$

Also, $(5 + 7) \times 2 = (5 \times 2) + (7 \times 2)$, since

$$(5 + 7) \times 2 = 12 \times 2 = 24 \qquad \text{and}$$
$$(5 \times 2) + (7 \times 2) = 10 + 14 = 24$$

EXAMPLES

Verify the distributive property by performing the actual computations.

6 $0.64 \times (0.31 + 0.8)$ and $(0.64 \times 0.31) + (0.64 \times 0.8)$

SOLUTION

$$0.64 \times (0.31 + 0.8) = 0.64 \times (1.11) = 0.7104$$
$$(0.64 \times 0.31) + (0.64 \times 0.8) = 0.1984 + 0.512 = 0.7104$$

Therefore,

$$0.64 \times (0.31 + 0.8) = (0.64 \times 0.31) + (0.64 \times 0.8)$$

7 $\left(\dfrac{3}{14} + \dfrac{5}{28}\right) \times 7$ and $\left(\dfrac{3}{14} \times 7\right) + \left(\dfrac{5}{28} \times 7\right)$

SOLUTION

$$\left(\frac{3}{14} + \frac{5}{28}\right) \times 7 = \left(\frac{6+5}{28}\right) \times 7 = \frac{11}{28} \times 7 = \frac{11}{4}$$

$$\left(\frac{3}{14} \times 7\right) + \left(\frac{5}{28} \times 7\right) = \frac{3}{2} + \frac{5}{4} = \frac{6+5}{4} = \frac{11}{4}$$

Therefore,

$$\left(\frac{3}{14} + \frac{5}{28}\right) \times 7 = \left(\frac{3}{14} \times 7\right) + \left(\frac{5}{28} \times 7\right)$$

5 IDENTITY PROPERTY

a) *Identity property for addition:* If a is a real number, then

$$a + 0 = 0 + a = a$$

For example,

$$5 + 0 = 0 + 5 = 5$$

b) *Identity property for multiplication:* If a is a real number, then

$$a \cdot 1 = 1 \cdot a = a$$

For example,

$$6 \times 1 = 1 \times 6 = 6$$

6 INVERSE PROPERTY

a) *Additive inverse:* For each real number a, there is a real number, called the *additive inverse* and denoted by $-a$, such that

$$a + (-a) = (-a) + a = 0$$

For example,

$$3 + (-3) = (-3) + 3 = 0$$

b) *Multiplicative inverse:* For each real number a, where $a \neq 0$, there is a real number a, called the *multiplicative inverse* or *reciprocal*, denoted by $1/a$, such that

$$a \cdot 1/a = 1/a \cdot a = 1$$

For example,

$$7 \times \tfrac{1}{7} = \tfrac{1}{7} \times 7 = 1$$

EXAMPLES

8   Find the additive inverse of:

a) 7
b) $(-\tfrac{2}{3})$

SOLUTION

a) The additive inverse of 7 is -7, since

$$7 + (-7) = (-7) + 7 = 0$$

b) The additive inverse of $(-\tfrac{2}{3})$ is $\tfrac{2}{3}$, since

$$(-\tfrac{2}{3}) + \tfrac{2}{3} = \tfrac{2}{3} + (-\tfrac{2}{3}) = 0$$

9   Find the multiplicative inverse of:
a) 8
b) $\tfrac{3}{7}$

SOLUTION

a) The multiplicative inverse of 8 is $\tfrac{1}{8}$, since

$$8 \times \tfrac{1}{8} = \tfrac{1}{8} \times 8 = 1$$

b)   The multiplicative inverse of $\frac{3}{7}$ is $\frac{7}{3}$, since

$$\tfrac{3}{7} \times \tfrac{7}{3} = \tfrac{7}{3} \times \tfrac{3}{7} = 1$$

7   EQUALITY

The properties and definitions introduced so far have all involved the use of "equality," which is denoted by the symbol $=$. The statement $a = b$ means that a and b are two names for the same number. Here we list additional properties of real numbers.

Let a, b, and c be elements of the set of real numbers R. Then

a)   *Reflexive property:* $a = a$.
b)   *Symmetric property:* If $a = b$, then $b = a$.
c)   *Transitive property:* If $a = b$ and $b = c$, then $a = c$.

Another important and frequently used property is the substitution property.

8   THE SUBSTITUTION PROPERTY

If $a = b$, then a can be substituted for b in any statement involving b without affecting the truthfulness of the statement. For example, if $a = b$ and $b + 3 = 10$, then $a + 3 = 10$.

The properties of the real numbers and the substitution property can be combined to form other true statements about the real numbers. These statements have applications to topics introduced later, such as solving equations.

Consider a, b, and c real numbers. Then:

a)   *Addition property of equality:* If $a = b$ and $c = d$, then $a + c = b + d$.
b)   *Multiplication property of equality:* If $a = b$ and $c = d$, then $ac = bd$.

9   THE CANCELLATION PROPERTIES

a)   *Cancellation property for addition:* If a and b are real numbers and if $a + c = b + c$, then $a = b$.
b)   *Cancellation property for multiplication:* If a and b are real numbers and if $ac = bc$, with $c \neq 0$, then $a = b$.

Now we list additional properties that can be useful throughout the text.

If a and b are real numbers, then:

10   PROPERTY: $a \cdot 0 = 0 \cdot a = 0$

11 PROPERTY

    a) $-(-a) = a$

    b) $(-a)(b) = a(-b) = -(ab)$

    c) $(-a)(-b) = ab$

12 PROPERTY: If $a \cdot b = 0$, then either $a = 0$ or $b = 0$ or both $a = 0$ and $b = 0$.

    Other properties of real numbers will be discussed as we encounter them.

EXAMPLE

10 State the properties that justify each of the following equalities.

    a) $\frac{5}{3} \times (-\frac{7}{2}) = (-\frac{7}{2}) \times \frac{5}{3}$ b) $(\frac{2}{3} \times \frac{5}{4}) \times \frac{1}{7} = \frac{2}{3} \times (\frac{5}{4} \times \frac{1}{7})$

    c) $14 \times (2 + \sqrt{3}) = 28 + 14 \cdot \sqrt{3}$ d) $11 \times \frac{1}{11} = 1$

    e) $\frac{2}{3} \times (5 + \frac{3}{8}) = \frac{2}{3} \times (5 + \frac{6}{16})$ f) $7 + (-7) = 0$

SOLUTION

    a) $\frac{5}{3} \times (-\frac{7}{2}) = (-\frac{7}{2}) \times \frac{5}{3}$ (commutative property for multiplication)

    b) $(\frac{2}{3} \times \frac{5}{4}) \times \frac{1}{7} = \frac{2}{3} \times (\frac{5}{4} \times \frac{1}{7})$ (associative property for multiplication)

    c) $14 \times (2 + \sqrt{3}) = 28 + 14 \cdot \sqrt{3}$ (distributive property)

    d) $11 \times \frac{1}{11} = 1$ (multiplicative inverse property)

    e) $\frac{2}{3} \times (5 + \frac{3}{8}) = \frac{2}{3} \times (5 + \frac{6}{16})$ (substitution property)

    f) $7 + (-7) = 0$ (additive inverse property)

PROBLEM SET 1.3

1 a) Is the set $A = \{1,3,5,7,9\}$ closed under the operation of addition? Is it closed under the operation of multiplication?

    b) Is the fact $(3 + 7) + 9 = 3 + (7 + 9)$ an example of the commutative property or the associative property or both?

    c) Is the fact $(7 + 2) + 13 = 7 + (13 + 2)$ an example of the commutative property or the associative property or both?

2 Indicate which of the following sets is closed under the operations of addition and multiplication.

    a) $\{0,1\}$

    b) $\{0,2,4,6,8 \ldots\}$

    c) $\{1,\frac{1}{2},\frac{1}{4},\frac{1}{8}, \ldots\}$

In problems 3–22, state which property (properties) justifies each equality.

3 $5 + \sqrt{7}$ is a real number.   **4** $8\sqrt{7}$ is a real number.

5 $\frac{7}{8} \times (-\frac{2}{3}) = (-\frac{2}{3}) \times \frac{7}{8}$ **6** $\frac{9}{16} + (-\frac{8}{11}) = (-\frac{8}{11}) + \frac{9}{16}$

7 $\frac{3}{11} + (\frac{7}{12} + \frac{16}{25}) = (\frac{3}{11} + \frac{7}{12}) + \frac{16}{25}$ **8** $(\frac{2}{7} \times \frac{5}{3}) \times (-\frac{3}{4}) = \frac{2}{7} \times [\frac{5}{3} \times (-\frac{3}{4})]$

9 $8 \times (\sqrt{7} + 5) = 8\sqrt{7} + 40$ **10** $(1 \times \sqrt{11}) = \sqrt{11}$

11 $(5 + \sqrt{3}) \times 19 = 95 + 19\sqrt{3}$ **12** $\sqrt{7} + (-\sqrt{7}) = 0$

13 $(1/\sqrt{3}) \times \sqrt{3} = 1$ **14** $1 \times \frac{7}{9} = \frac{7}{9}$

15 $(-2) \times (-3) = 2 \times 3 = 6$ **16** $-2 \times 5 = -10$

17 If $5a = 0$, then $a = 0$.   **18** If $ax = ay$ and $a \neq 0$, then $x = y$.

19 If $7x = 7y$, then $x = y$.   **20** $\sqrt{13} + (\sqrt{13} \times 0) = \sqrt{13}$

21 $5 \times 0 = 0$ **22** If $x + z = y + z$, then $x = y$.

In problems 23–34, simplify each expression. (Use the properties discussed.)

23 $(16 + 3) + 7$ **24** $(5 + 3) + 2$

25 $(8 \times 3) + (3 \times 7)$ **26** $3 \times (4 + 7)$

27 $14 \times (\frac{5}{7} + \frac{14}{21})$ **28** $35 \times (\frac{3}{7} + \frac{14}{21})$

29 $(8 \times 15) + (8 \times 11)$ **30** $(74 \times 3) + (74 \times 5)$

31 $(\frac{1}{4} \times \frac{2}{5}) + (\frac{1}{4} \times \frac{3}{5})$ **32** $\frac{1}{6} \times (\frac{1}{3} + \frac{1}{4})$

33 $(\frac{2}{3} + \frac{7}{4}) + \frac{7}{12}$ **34** $21.6 + (23.7 + 5.2)$

In problems 35–37, use the properties of real numbers to prove each statement. Assume that all variables represent real numbers.

35 If $a = b$, then $ac = bc$.

36 If $ac = bc$ and $c \neq 0$, then $a = b$.

37 $(a + b)(c + d) = ac + ad + bc + bd$.

1.4 Operations of Signed Numbers

The rules for adding, subtracting, multiplying, and dividing positive integers were discussed in arithmetic. However, in order to extend these operations to include negative numbers, we shall discuss the relationship of any particular number and its negative relative to the origin on the number line. Hence, we begin with the notion of absolute value.

Absolute Value

Associated with each number, there is exactly one nonnegative number called the *absolute value* of the number. On the number line, the absolute value of a number is the distance from the origin to that number. For example, we notice that both 5 and −5 have the same displacements (distances) from the origin. Also, 6 and −6 lie 6 units from the origin, and $\frac{1}{2}$ and $-\frac{1}{2}$ lie $\frac{1}{2}$ unit from the origin (Figure 1.10). To express this idea, we say that 5 and −5 have the same absolute value, which is expressed symbolically as $|-5| = 5$ and $|5| = 5$, where the two vertical bars, $|\ \ |$, denote the absolute value of a number. Then $|-6| = 6$, $|6| = 6$, $|-\frac{1}{2}| = \frac{1}{2}$, and $|\frac{1}{2}| = \frac{1}{2}$. The absolute value of 0 is defined to be 0, that is, $|0| = 0$. In general, if x is a real number, we denote the absolute value of x by $|x|$.

Figure 1.10

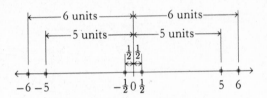

EXAMPLE

1   Find the value of the following expressions.

a) $|8|$ b) $|-\frac{2}{3}|$ c) $|\frac{7}{5}|$ d) $-|-17|$

SOLUTION

a) $|8| = 8$, since 8 is a positive number that lies 8 units from the origin.

b) $|-\frac{2}{3}| = \frac{2}{3}$, since $-\frac{2}{3}$ is a negative number that lies $\frac{2}{3}$ unit from the origin.

c) $|\frac{7}{5}| = \frac{7}{5}$, since $\frac{7}{5}$ is a positive number that lies $\frac{7}{5}$ units from the origin.

d) $-|-17| = -(17) = -17$

Addition and Subtraction of Signed Numbers

Addition of real numbers can be illustrated on the number line. With the aid of the properties of the real numbers we can represent these numbers by directed moves (changes of position) on the line. For example, to add 2 and 3 on the number line (Figure 1.11), we start at the origin and move 2 units to the right; thus, the number 2 is represented by an arrow from 0 to 2. We then start at the number 2 and

Figure 1.11

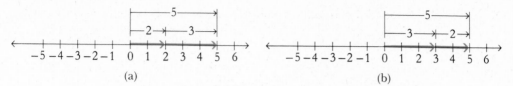

(a)   (b)

move 3 units to the right, so that the number 3 is represented by an arrow between 2 and 5. Together, the sum of the two directed moves is $2 + 3 = 5$ (Figure 1.11a). It can also be seen that $3 + 2 = 5$ (Figure 1.11b). Therefore, $3 + 2 = 2 + 3$, which also illustrates the commutative property of addition.

Now consider the sum $(-2) + (-3)$. By interpreting the addition of a negative number as a movement to the left on the number line, the sum can be found; that is, $(-2) + (-3) = -5$ (Figure 1.12).

Figure 1.12

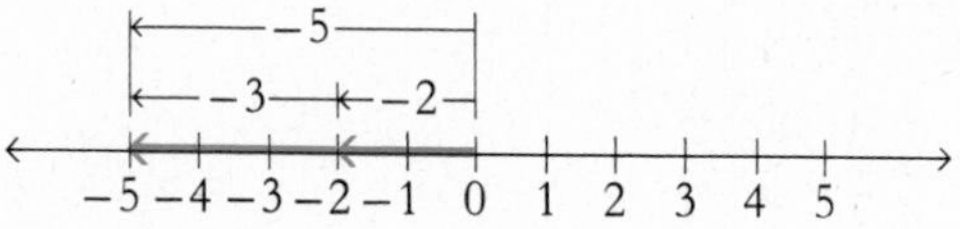

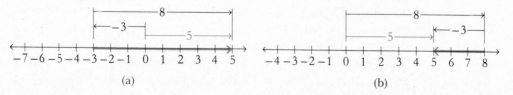

Note that in both cases illustrated above, the sums can be found by adding the absolute values of the numbers and retaining the common sign of the numbers. Thus, we have the following rule: To add like signed numbers, add their absolute values and keep their common sign. For example,

$$5 + 7 = |5| + |7| = 12$$

$$(-5) + (-8) = -(|-5| + |-8|) = -(5 + 8) = -13$$

The addition of unlike signed numbers can also be conveniently illustrated on a number line. For example, the sum $(-3) + 8$ can be interpreted as a movement of 8 units to the right of the number -3 (Figure 1.13a). The sum $(-3) + 8$ can also be interpreted as a movement of 3 units to the left of 8 (Figure 1.13b). In both approaches, the same sum is obtained; that is, $(-3) + 8 = 5$.

Figure 1.13

(a)   (b)

Another way to find the sum $(-3) + 8$ is to rewrite 8 as $3 + 5$ and proceed as follows:

$$
\begin{aligned}
(-3) + 8 &= (-3) + 3 + 5 &&\text{(substitution property)} \\
&= [(-3) + 3] + 5 &&\text{(associative property)} \\
&= 0 + 5 &&\text{(additive inverse)} \\
&= 5 &&\text{(identity element of addition)}
\end{aligned}
$$

Now consider the sum $(-7) + 5$. By use of the number line (Figure 1.14) we see that $(-7) + 5 = -2$. This sum can also be found as follows:

$$
\begin{aligned}
(-7) + 5 &= (-2) + (-5) + 5 \\
&= -2 + [(-5) + 5] \\
&= -2 + 0 = -2
\end{aligned}
$$

Figure 1.14

Note that in both examples, that is, $-3 + 8 = 5$ and $-7 + 5 = -2$, the sum can be found by subtracting the absolute values of the numbers and retaining the sign of the number with the largest absolute value. Thus, we have the following rule: To add two unlike signed numbers, subtract their absolute values and retain the sign of the number with the greatest absolute value. For example,

$$2 + (-2) = 0$$

$$(-2) + 5 = +(|5| - |-2|) = +(5 - 2) = 3$$

$$(-5) + 2 = -(|-5| - |2|) = -(5 - 2) = -3$$

To subtract two signed numbers, we appeal to the following definition. The differences of two real numbers a and b, denoted by $a - b$, is defined by $a + (-b)$. That is, to subtract b from a, add the additive inverse of b to a. This definition, together with the rules for adding signed numbers, provides a method for subtracting signed numbers. For example, the difference $(-7) - 3$ can be found as follows:

$$(-7) - 3 = (-7) + (-3) = -10$$

We can illustrate this subtraction on a number line by interpreting the subtractions of 3 from -7 as a movement of 3 units to the left of -7; that is, $(-7) - 3 = -10$ (Figure 1.15).

Figure 1.15

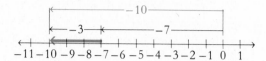

The rule for subtracting signed numbers can be stated as follows: To subtract signed numbers, change the sign of the number to be subtracted, and then add by following the rules for adding signed numbers. For example,

$$7 - 4 = 7 + (-4) = 3$$

$$4 - 7 = 4 + (-7) = -3$$

$$(-9) - (6) = (-9) + (-6) = -15$$

EXAMPLES

2　Find the following sums.

a)　$9 + 7$　　　　b)　$(-5) + (-7)$　　　c)　$13 + (-13)$

d)　$(-5) + 19$　　　e)　$7 + (-23)$

SOLUTION

a)　$9 + 7 = 16$

b)　$(-5) + (-7) = -(5 + 7) = -12$

c)　$13 + (-13) = 0$

d)　$(-5) + 19 = +(19 - 5) = 14$

e)　$7 + (-23) = -(23 - 7) = -16$

3　Find the following differences.

a)　$5 - 2$　　　　　　　　b)　$(-11) - 6$

c)　$7 - (-4)$　　　　　　　d)　$(-15) - (-8)$

SOLUTION

a)　$5 - 2 = 5 + (-2) = 3$

b)　$(-11) - 6 = (-11) + (-6) = -17$

c)　$7 - (-4) = 7 + 4 = 11$

d)　$(-15) - (-8) = (-15) + 8 = -7$

Multiplication and Division of Signed Numbers

In arithmetic, multiplication of two positive integers is often described as repeated addition. For example, 3×2 can be interpreted as $3 + 3 = 6$, or $2 + 2 + 2 = 6$. The same approach can be used to describe the multiplication of any real number by a positive number. Thus, $(-6) \times 3$ can be interpreted as $(-6) + (-6) + (-6) = -18$. Also, $(-5) \times (2) = (-5) + (-5) = -10$.

Note that in both cases the product of a negative number and a positive number is the negative of the product of their absolute values. That is,

$$(-6) \times 3 = -(|-6| \times |3|) = -(6 \times 3) = -18$$

$$(-5) \times 2 = -(|-5| \times |2|) = -(5 \times 2) = -10$$

In general, the product of two unlike signed numbers is the negative of the product of their absolute values.

EXAMPLE

4   Find the indicated products.
   a) $(-7) \times 6$ b) $(-8) \times \frac{1}{2}$ c) $6 \times (-1.5)$

SOLUTION

   a) $(-7) \times 6 = -(7 \times 6) = -42$
   b) $(-8) \times \frac{1}{2} = -(8 \times \frac{1}{2}) = -4$
   c) $6 \times (-1.5) = -(6 \times 1.5) = -9$

Now consider the multiplication of any real number by a negative number; for instance, let us consider the product $3 \times (-2)$. If we interpret this multiplication as the repeated addition of the negative of 3, we obtain $3 \times (-2) = (-3) + (-3) = -6$, which is consistent with the results obtained above for finding the product of two unlike signed numbers. This approach can also be applied to the product of two negative numbers. For example, $(-4) \times (-3)$ is obtained as follows:

$$(-4) \times (-3) = [-(-4)] + [-(-4)] + [-(-4)]$$
$$= 4 + 4 + 4 = 12$$

Hence, we see that the product of the two negative numbers is the product of their absolute values. That is, $(-4) \times (-3) = |-4| \times |-3| = 4 \times 3 = 12$. In general, the product of two like signed numbers is the product of their absolute values.

EXAMPLE

5   Find the indicated products.
   a) 5×3 b) $(-6) \times (-2)$ c) $(-\frac{2}{3}) \times (-\frac{1}{2})$

SOLUTION

a) $5 \times 3 = 15$

b) $(-6) \times (-2) = 6 \times 2 = 12$

c) $(-\frac{2}{3}) \times (-\frac{1}{2}) = \frac{2}{3} \times \frac{1}{2} = \frac{1}{3}$

Rules for multiplying signed numbers are as follows:

1. If two numbers have like signs, their product is positive.

2. If two numbers have unlike signs, their product is negative.

When multiplying more than two signed numbers, the factors can be paired to determine the sign of the product. For example, the product $(-3) \times 2 \times (-5) \times 6$ can be written

$$[(-3) \times (-5)] \times [2 \times 6] = 15 \times 12 = 180$$

Here, there were two negative factors, an even number, and so the product is positive. In the case of the product $(-4) \times (-7) \times (-2) \times 3$, we have

$$[(-4) \times (-7)] \times [(-2) \times 3] = 28 \times (-6) = -168$$

In this example, there are three negative factors, an odd number, and so the product is negative. Thus, we can state a more general rule for finding the product of signed numbers: The product of signed numbers is positive if there is an even number of negative factors, and the product is negative if there is an odd number of negative factors.

EXAMPLE

6   Determine the indicated products.

a) $3 \times (-2) \times 6 \times (-7) \times (-8) \times 2 \times (-4)$

b) $(-\frac{1}{2}) \times 7 \times (-\frac{3}{4}) \times (-8) \times (-4) \times (-7)$

SOLUTION

a) By counting, we have an even number of negative factors, so the product is positive. Thus,

$$3 \times (-2) \times 6 \times (-7) \times (-8) \times 2 \times (-4) = 16{,}128$$

b) Since there are an odd number of negative factors, the product is negative. Thus,

$$(-\frac{1}{2}) \times 7 \times (-\frac{3}{4}) \times (-8) \times (-4) \times (-7) = -588$$

To find the quotient of two signed numbers, we appeal to the following definition: The quotient of real numbers a and b, for $b \neq 0$, denoted by $a \div b$, or a/b, is defined by: $a/b = a(1/b)$. That is, to divide a by b, where $b \neq 0$, multiply a by the multiplicative inverse of b. This definition, together with the rules for multiplying signed numbers, provides a method for dividing signed numbers. For example, to divide -8 by 4, we proceed as follows:

$$(-8) \div 4 = (-8) \times \tfrac{1}{4} = -(8 \times \tfrac{1}{4}) = -2$$

Thus, the rules for dividing signed numbers can be stated as follows:

1.   If two numbers have like signs, their quotient is positive.

2.   If two numbers have unlike signs, their quotient is negative.

EXAMPLE

7   Find the indicated quotients.
   a) $12 \div (-3)$ b) $(-15) \div (-5)$ c) $(-7) \div 21$

SOLUTION

   a) $12 \div (-3) = -(12 \div 3) = -(12 \times \tfrac{1}{3}) = -4$
   b) $(-15) \div (-5) = 15 \div 5 = 15 \times \tfrac{1}{5} = 3$
   c) $(-7) \div 21 = -(7 \div 21) = -(7 \times \tfrac{1}{21}) = -\tfrac{1}{3}$

PROBLEM SET 1.4

In problems 1–12, find the value of each expression.

1   $|-4| + |4|$ 2 $|5| + |-5|$

3   $|-(-7)|$ 4 $|\tfrac{2}{3}| + |-\tfrac{1}{3}|$

5   $|-4| \times |-3|$ 6 $|-5| \times |7 - 3|$

7   $-|-\tfrac{4}{5} + \tfrac{2}{5}|$ 8 $-|\tfrac{6}{7}| - |-\tfrac{1}{7}|$

9   $|10 - 1| + |10 + 1|$ 10 $|-\tfrac{13}{4}| - |\tfrac{13}{4}|$

11   $|14 - 5| - |8 - 5|$ 12 $|-3 - 2| \times |3 + 2|$

In problems 13–68, find the value of each expression.

13   $7 + (-3)$ 4   14 $(-23) + 14$ 9

15   $(-8) + (-6)$ -14 16 $(-11) + (-7)$ -18

17   $27 + (-39)$ -12 18 $(-100) + 43$ -57

19   $(-8) + 8$ 0   20 $(-\tfrac{3}{2}) + \tfrac{3}{2}$ 0

21 $10 + (-101)$ -91 **22** $(-119) + 227$ 108

23 $(-21) + 7$ -14 **24** $(-120) + (-117)$ -237

25 $(-21) + (-18)$ -39 **26** $(-750) + (-625)$ -1375

27 $18 - (-19)$ 37 **28** $(-25) - 8$ -33

29 $(-25) - (-45)$ 20 **30** $(-97) - (-39)$ -136

31 $(-22) - (-7)$ -15 **32** $16 - (-30)$ 46

33 $(-8) - 8$ -16 **34** $(-8\frac{1}{3}) - (-5\frac{1}{6})$ $-3\frac{1}{3}$

35 $11.1 - (-0.9)$ 12 **36** $(-16\frac{1}{2}) - (3\frac{1}{2})$ -20

37 $0.052 - (-0.007)$ $.059$ **38** $(-5.03) - (-4.83)$ $-.2$

39 $\frac{2}{3} - (-\frac{5}{6})$ $1\frac{1}{2}$ **40** $(-\frac{3}{7}) - (-\frac{5}{7})$ $\frac{2}{7}$

41 3×19 57 **42** $(-7) \times 2$ -14

43 $(-2) \times (-3)$ 6 **44** $\frac{2}{3} \times (-\frac{3}{4})$ $-\frac{1}{2}$

45 $(-3) \times (-\frac{1}{9})$ $\frac{1}{3}$ **46** $-(-2) \times (-2)$ -4

47 $(-3) \times (-3) \times (-3)$ -27 **48** $(-4) \times (-5) \times (-2)$ -40

49 $(-3) \times (-\frac{1}{3}) \times (-7)$ -7 **50** $(-11) \times (-\frac{1}{11}) \times (-7) \times (-\frac{2}{7})$ 2

51 $(-1) \times (-2) \times (-3) \times (-5)$ 30 **52** $(-2) \times (-3) \times (-4) \times (-5)$ 120

53 $5 \times (-7) \times (-2) \times 3$ 210 **54** $3 \times (-2) \times (-10) \times 4$ 240

55 $10 \div 5$ 2 **56** $27 \div 9$ 3

57 $(-18) \div 9$ -2 **58** $(-10) \div (-2)$ 5

59 $(-57) \div (-19)$ 3 **60** $(-32) \div 8$ -4

61 $(-3.9) \div 0.3$ -13 **62** $\frac{5}{4} \div 2$ $\frac{5}{8}$

63 $(-75) \div (-5)$ 15 **64** $12.5 \div (-1.25)$ -10

65 $(-22.5) \div (-0.015)$ 1500 **66** $(-49) \div (-7)$ 7

67 $0 \div \frac{4}{3}$ 0 **68** $(-\frac{15}{17}) \div 5$ $-\frac{25}{17} = -4\frac{7}{17}$

REVIEW PROBLEM SET

In problems 1–10, use the following sets for each case.

N, the set of natural numbers
I, the set of integers
F, the set of quotient numbers
    (rational numbers which are not integers)
Q, the set of rational numbers
L, the set of irrational numbers

R, the set of real numbers

$\emptyset$, the empty set

Each description corresponds to one or more than one of the sets above. Identify these sets:

1 A set that contains 4×9 but not -4×9.

2 A set that contains both $\frac{2}{3}$ and π.

3 A set that contains $-\frac{3}{4}$ but not π.

4 The union of the natural numbers and the integers.

5 The intersection of the quotient numbers and the rational numbers.

6 The union of the irrational numbers and the real numbers.

7 A subset of the quotient numbers which contains $\frac{3}{2}$ but not 3.

8 A subset of the real numbers which does not contain 1 but which is not the empty set.

9 A subset of the irrational numbers which does not contain π.

10 A set which is disjoint from the natural numbers and which is not a subset of the integers.

In problems 11–20, if A is a set that contains five elements and B is a set that contains three elements, which of the following are true?

11 $A \cap B$ contains exactly five elements.

12 $A \cup B$ contains at least five elements.

13 $A \cup B$ contains four elements.

14 $A \cap B$ is a subset of A.

15 A is a subset of B.

16 $A \cup B$ can contain no more than eight elements.

17 If $A \cap B = \emptyset$, then $A \cup B = \emptyset$.

18 If $x \in A$ and $x \in B$, then $A \cap B$ is not an empty set.

19 If $A \cap B$ contains three elements, then B is a subset of A.

20 If $A \cup B$ contains six elements, then $A \cap B$ contains two elements.

In problems 21–24, what set would you select as a universal set U from which each pair of sets has been formed?

21 {all single women}; {all married women}

22 {all even integers}; {all odd integers}

23 {all Southern states}; {all Western states}

24 {all boys}; {all girls}

In problems 25–34, let $A = \{a,b,c\}$, $B = \{c,d,e,f\}$, and $C = \{a,c,d,g\}$. Find each of the following.

25 $A \cap B$ **26** $B \cup A$

27 $B \cup C$ **28** $B \cap C$

29 $A \cap C$ **30** $C \cup B$

31 $C \cap A$ **32** $A \cap \varnothing$

33 $C \cup A$ **34** $B \cup \varnothing$

In problems 35 and 36, represent the set on a number line.

35 $A = \{-4,-2,0,2,4\}$ **36** $A = \{-\frac{4}{7},-\frac{2}{7},0,\frac{2}{7},\frac{4}{7}\}$

In problems 37–46, express each rational number in decimal representation.

37 $\frac{7}{40}$ **38** $\frac{3}{11}$

39 $\frac{17}{25}$ **40** $\frac{7}{45}$

41 $\frac{7}{9}$ **42** $\frac{2}{7}$

43 $\frac{5}{14}$ **44** $\frac{6}{13}$

45 $\frac{11}{22}$ **46** $\frac{7}{6}$

In problems 47–58, express each decimal in the form a/b, where a and b are integers, $b \neq 0$.

47 0.16   **48** 0.48

49 0.125   **50** 0.0035

51 0.87   **52** 0.0019

53 $0.03\overline{1}$ **54** $0.000\overline{571}$

55 $3.0\overline{49}$ **56** $8.00\overline{7}$

57 $0.\overline{629}$ **58** $0.00\overline{1576}$

In problems 59–61, write, at most, four members of the set described, where m and $n \in I$.

59 $\left\{ x \mid x = \dfrac{m}{n} \text{ and } m + n = 3 \right\}$

60 $\left\{ x \mid x = \dfrac{m}{n} \text{ and } m \cdot n = m + n \right\}$

61 $\left\{ x \mid x = \dfrac{m}{n} \text{ and } m = n \right\}$

62 Let $x = 0.12345610112134\ldots$, and find a number y such that $x + y$ is a rational number.

63 Notice that $\frac{1}{9} = 0.11\overline{1}$ repeats in one-digit blocks, and $\frac{1}{3} = 0.333\overline{3}$ repeats in one-digit blocks. Express the decimal representations of $\frac{1}{9} + \frac{1}{3}$.

64 Which of the subsets of real numbers N, I, Q, and L are closed with respect to addition? With respect to multiplication?

In problems 65–78, justify each statement by giving the appropriate property. Assume that all letters represent real numbers.

65 $x + 5$ is a real number.

66 $a + (b + 3) = (a + b) + 3$

67 $3x = x3$

68 $a(b + c)$ is a real number.

69 $2(xy) = (2x)y$

70 $1 \times 4 = 4$

71 $5 + 0 = 5$

72 If $x = y$, then $2x = 2y$.

73 $0 \cdot x = 0$

74 $(-2) \times (-3) = 2 \times 3$

75 If $x + 9 = y + 9$, then $x = y$.

76 If $6a = 0$, then $a = 0$.

77 $(-2)x = -(2x)$

78 $x \cdot \dfrac{1}{x} = 1$, for $x \neq 0$.

In problems 79–107, evaluate each expression.

79 $|-(-13)|$

80 $|(-7)| + |13|$

81 $|(-5)| - |3|$

82 $0 + |(-9)|$

83 $-|(-5)|$

84 $|-(-7)|$

85 $(-5) + 7$

86 $(-13) - (-21)$

87 $(-6) + (-8)$

88 $(-5) + (9) + (-2)$

89 $5 - (-9)$

90 $(-5) + (-7)$

91 $17 - (-13)$

92 $0 - (-3)$

93 $(-7) + (-3) + (-2)$

94 $(277 + 531) - 89$

95 $[531 - (-32)] + 71$

96 $(-11) \times (-10)$

97 $8 \times (-5)$

98 $(-6) \times (-8)$

99 $(-16) \times (-1) \times (-49)$

100 $(-7) \times 2 \times (-15) \times 6$

101 $(-3) \times (-7) \times 5$

102 $25 \div (-5)$

103 $(-63) \div 7$

104 $(-36) \div (-6)$

105 $(-40) \div (-5)$

106 $(-56) \div (-7)$

107 $21 \div (-7)$

CHAPTER 2

Algebra of Polynomials

The properties of real numbers considered in Chapter 1 also apply to expressions that represent real numbers. In this chapter we shall apply these properties in performing addition, subtraction, multiplication, and division of expressions such as $5x + 7$, $3x - 9$, $(2x + 3)(x - 2)$, and $(2x + 1)/(5x - 4)$, where x represents a real number. Such expressions are called *algebraic expressions,* and the letter x involved is called a *variable.* The set of numbers that the variable represents is called the *replacement set* of the expression. Throughout the text we shall assume that the replacement set of our algebraic expressions is the set of real numbers, unless otherwise stated.

Algebraic expressions such as $2x + 3$, $x^2 + 3x + 5$, and $7x^3 - 2x^2 + 5x - 7$ are called *polynomials.* To discuss polynomials, we shall begin with the properties of "exponents," a notational convenience helpful in finding the product of polynomials.

2.1 Properties of Positive Integral Exponents

If two or more identical numbers (or variables) are used in a given product, an abbreviation is used to indicate this multiplicity, and each number is called a *factor* of the product. For instance, to show that 5 is used as a factor three times, we write 5^3, which reads "the cube of five" or "five cubed," rather than write $5 \cdot 5 \cdot 5$; similarly, $5^4 = 5 \cdot 5 \cdot 5 \cdot 5$ and $x^5 = x \cdot x \cdot x \cdot x \cdot x$. In general, we have the following definition:

DEFINITION 1 EXPONENTS

The shorthand exponential notation x^n is used to show x as a factor n times, where x is a real number. That is,

$$x^n = \overbrace{x \cdot x \cdot x \cdot \cdots \cdot x}^{n \text{ factors}} \qquad \text{where } n \text{ is a positive integer}$$

The product of n factors is called the nth *power of x,* the variable x is called the *base,* and n is called the *exponent* of the expression x^n.

For example, $x \cdot x$ is written as x^2, $y \cdot y \cdot y$ is written as y^3, and $3x \cdot x \cdot x \cdot x \cdot x \cdot x$ is written as $3x^6$. An exponent applies only to the base to which it is attached. For example,

$$x^4y^2 = x \cdot x \cdot x \cdot x \cdot y \cdot y \qquad \text{and} \qquad x^2y^3 = x \cdot x \cdot y \cdot y \cdot y$$

If $n = 1$ in the definition, we have $x^1 = x$. Thus, when no exponent is shown, we assume it to be the number 1. For example,

$$3 = 3^1 \qquad a = a^1 \qquad \text{and} \qquad x^2y = x^2y^1$$

We should note the difference in meaning between the two expressions 3^4 and $4(3)$. By definition, 3^4 means $3 \times 3 \times 3 \times 3$, whereas $4(3)$ means $3 + 3 + 3 + 3$. In general, we have

$$x^n = \underbrace{x \cdot x \cdot x \cdots x}_{n \text{ factors}} \qquad \text{whereas} \qquad nx = \underbrace{x + x + \cdots + x}_{n \text{ addends}}$$

EXAMPLES

1 Find the value of each of the following expressions. In parts (a) and (b), indicate the base and exponent.
   a) 10^4 b) 7^3 c) $2^4 + 4(2)$

   SOLUTION
   a) $10^4 = 10 \times 10 \times 10 \times 10 = 10{,}000$
       Notice here that the base is 10 and the exponent is 4.
   b) $7^3 = 7 \times 7 \times 7 = 343$
       Note here that the base is 7 and the exponent is 3.
   c) $2^4 + 4(2) = (2 \times 2 \times 2 \times 2) + (2 + 2 + 2 + 2)$
       $$= 16 + 8$$
       $$= 24$$

2 Rewrite the following expressions, using exponents.
   a) $a \cdot a \cdot a \cdot a \cdot a \cdot b \cdot b$
   b) $3y \cdot y \cdot y \cdot z \cdot z$
   c) $(3x) \cdot (3x) \cdot (3x) \cdot (3x)$

   SOLUTION
   a) $a \cdot a \cdot a \cdot a \cdot a \cdot b \cdot b = a^5b^2$
   b) $3y \cdot y \cdot y \cdot z \cdot z = 3y^3z^2$
   c) $(3x) \cdot (3x) \cdot (3x) \cdot (3x) = (3 \times 3 \times 3 \times 3)(x \cdot x \cdot x \cdot x) \qquad$ (why?)
       $$= 3^4x^4 = 81x^4$$

Suppose that we wish to find the product of x^2 and x^3. We know that $x^2 = x \cdot x$ and $x^3 = x \cdot x \cdot x$. Therefore,

$$x^2 \cdot x^3 = (x \cdot x)(x \cdot x \cdot x)$$

and since this expression contains five factors of x, we have

$$x^2 \cdot x^3 = x^5$$

Notice that the exponent of the product obtained by applying the definition of positive integral exponents could also have been obtained by adding the exponents of the factors x^2 and x^3. That is,

$$x^2 \cdot x^3 = x^{2+3} = x^5$$

Similarly, the expression $(x^2)^3$ can be written as a single power of x by applying the definition of positive exponents as follows:

$$
\begin{aligned}
(x^2)^3 &= x^2 \cdot x^2 \cdot x^2 && \text{(3 factors of } x^2) \\
&= (x \cdot x)(x \cdot x)(x \cdot x) && \text{(replacing } x^2 \text{ by } x \cdot x) \\
&= x^6
\end{aligned}
$$

The same result could have been obtained by simply multiplying the exponents 2 and 3, that is,

$$(x^2)^3 = x^{2 \times 3} = x^6$$

We can generalize the preceding results in the first two properties below; and by using the same methods as above, we can show that Properties 3, 4, and 5 hold true.

PROPERTIES OF EXPONENTS

If a and b are real numbers and m and n are positive integers, then:

1 *Multiplication property:* $a^m \cdot a^n = a^{m+n}$

2 *Power-of-a-power property:* $(a^m)^n = a^{mn}$

3 *Power-of-a-product property:* $(ab)^n = a^n b^n$

4 *Power-of-a-quotient property:* $\left(\dfrac{a}{b}\right)^n = \dfrac{a^n}{b^n}; \ b \neq 0$

5 *Division property:* $\dfrac{a^m}{a^n} = a^{m-n}; \ a \neq 0$ and m is greater than n

In Property 5 we have assumed that m is greater than n, so that $m - n$ represents a positive exponent. However, if $m = n$, we have

$$\frac{a^m}{a^n} = \frac{a^m}{a^m} = 1$$

For example, $\dfrac{x^3}{x^3} = 1$, for $x \neq 0$. Also, if m is less than n, we have

$$\frac{a^m}{a^n} = \frac{1}{a^{n-m}} \qquad \text{for } a \neq 0$$

where $n - m$ represents a positive exponent. For instance,

$$\frac{x^2}{x^5} = \frac{1}{x^{5-2}} = \frac{1}{x^3} \qquad \text{for } x \neq 0$$

These properties can be proved by applying the definition of positive integral exponents. We shall verify Properties 1, 3, and 5 here and leave the verifications of 2 and 4 to the reader as an exercise. (See problems 72 and 74 of Problem Set 2.1.)

PROOF OF 1

$$a^m a^n = a^{m+n}$$

$$a^m a^n = \underbrace{(a \cdot a \cdots a)}_{m \text{ factors}} \underbrace{(a \cdot a \cdot a \cdots a)}_{n \text{ factors}} \qquad \text{(definition)}$$

$$= \underbrace{a \cdot a \cdot a \cdot a \cdot a \cdots a \cdot a}_{m + n \text{ factors}}$$

$$= a^{m+n} \qquad \text{(definition)}$$

PROOF OF 3

$$(ab)^n = a^n b^n$$

$$(ab)^n = \underbrace{(ab)(ab) \cdots (ab)}_{n \text{ factors}} \qquad \text{(definition)}$$

$$= \underbrace{(a \cdot a \cdots a)}_{n \text{ factors}} \underbrace{(b \cdot b \cdots b)}_{n \text{ factors}} \qquad \text{(why?)}$$

$$= a^n b^n \qquad \text{(definition)}$$

PROOF OF 5

$$\frac{a^m}{a^n} = a^{m-n} \qquad \text{if } m \text{ is greater than } n \text{ and } a \neq 0$$

$$\frac{a^m}{a^n} = \frac{\overbrace{a \cdot a \cdot a \cdot a \cdot a \cdots a}^{m \text{ factors}}}{\underbrace{a \cdot a \cdots a}_{n \text{ factors}}} = \frac{\overbrace{(a \cdot a \cdots a)}^{n \text{ factors}} \overbrace{(a \cdot a \cdot a \cdots a)}^{m-n \text{ factors}}}{\underbrace{a \cdot a \cdots a}_{n \text{ factors}}}$$

$$= \underbrace{a \cdot a \cdots a}_{m-n \text{ factors}}$$

$$= a^{m-n}$$

EXAMPLES

3   Use Property 1 to simplify the following expressions.

a) $2^3 \cdot 2^4$ b) $x^4 \cdot x^6$ c) $(-a)^2 \cdot (-a)^4$

SOLUTION.   To apply Property 1 we write the common base and add exponents. Thus,

a) $2^3 \cdot 2^4 = 2^{3+4} = 2^7$
b) $x^4 \cdot x^6 = x^{4+6} = x^{10}$
c) $(-a)^2 \cdot (-a)^4 = (-a)^{2+4}$
$$= (-a)^6 = a^6 \quad \text{(why?)}$$

4   Use Property 2 to simplify the following expressions.

a) $(3^2)^3$ b) $(x^4)^3$ c) $(y^7)^n$, n any positive integer

SOLUTION.   To apply Property 2 write the base and multiply exponents, so that

a) $(3^2)^3 = 3^{2 \times 3} = 3^6 = 729$
b) $(x^4)^3 = x^{4 \times 3} = x^{12}$
c) $(y^7)^n = y^{7 \cdot n} = y^{7n}$

5   Use Property 3 to simplify the following expresssions.

a) $(3x)^3$ b) $(-x)^5$ c) $(xyz)^4$

SOLUTION.   To apply Property 3 we raise each factor to the indicated power, so that

a) $(3x)^3 = 3^3 \cdot x^3 = 27x^3$

b) $(-x)^5 = [(-1)x]^5 = (-1)^5 x^5 = (-1)x^5 = -x^5$

c) $(xyz)^4 = [x(yz)]^4$ (Associative property)

$\qquad = x^4(yz)^4 \quad = x^4 y^4 z^4$

6   Use Property 4 to simplify the following expressions.

a) $\left(\dfrac{2}{3}\right)^3$
b) $\left(\dfrac{x}{y}\right)^5$
c) $\left(\dfrac{-a}{3}\right)^2$

SOLUTION.   To apply Property 4 we raise both terms of the quotient to the indicated power. Thus,

a) $\left(\dfrac{2}{3}\right)^3 = \dfrac{2^3}{3^3} = \dfrac{8}{27}$

b) $\left(\dfrac{x}{y}\right)^5 = \dfrac{x^5}{y^5}$

c) $\left(\dfrac{-a}{3}\right)^2 = \dfrac{(-a)^2}{3^2} = \dfrac{a^2}{9}$

7   Use Property 5 to simplify the following expressions.

a) $\dfrac{2^5}{2^2}$
b) $\dfrac{x^7}{x^2}$
c) $\dfrac{(-a)^5}{(-a)^2}$

SOLUTION.   To apply Property 5, that is, $\dfrac{a^m}{a^n} = a^{m-n}$, where m is larger than n, we write the common base and subtract exponents, so that

a) $\dfrac{2^5}{2^2} = 2^{5-2} = 2^3 = 8$

b) $\dfrac{x^7}{x^2} = x^{7-2} = x^5$

c) $\dfrac{(-a)^5}{(-a)^2} = (-a)^{5-2} = (-a)^3 = -a^3$

8   Use Properties 1–5 to simplify the given expressions.

a) $(3x^2)^3$
b) $\left(\dfrac{x^3}{y^2}\right)^4$
c) $\dfrac{x^7}{x^3 \cdot x^2}$
d) $\left(\dfrac{x^3 y^5}{2xy^2}\right)^7$

SOLUTION

a) $(3x^2)^3 = 3^3(x^2)^3$ (Property 3)

$\qquad = 27x^{2 \times 3}$ (Property 2)

$\qquad = 27x^6$

b) $\left(\dfrac{x^3}{y^2}\right)^4 = \dfrac{(x^3)^4}{(y^2)^4}$ (Property 4)

$= \dfrac{x^{3\times4}}{y^{2\times4}}$ (Property 2)

$= \dfrac{x^{12}}{y^8}$

c) $\dfrac{x^7}{x^3\cdot x^2} = \dfrac{x^7}{x^{3+2}} = \dfrac{x^7}{x^5}$ (Property 1)

$= x^{7-5}$ (Property 5)

$= x^2$

d) $\left(\dfrac{x^3y^5}{2xy^2}\right)^7 = \left(\dfrac{x^{3-1}y^{5-2}}{2}\right)^7$ (Property 5)

$= \left(\dfrac{z^2y^3}{2}\right)^7$

$= \dfrac{(x^2)^7(y^3)^7}{2^7}$ (Properties 3 and 4)

$= \dfrac{x^{2\times7}y^{3\times7}}{128}$ (Property 2)

$= \dfrac{x^{14}y^{21}}{128}$

PROBLEM SET 2.1

In problems 1–6, find the value of each expression.

1 $2^4\cdot3^2$ = 144 ~~~~~

2 4^2 16

3 $(-3)^3(-2)^2$ −108

4 $(-2)^4\cdot(-3)^2$ 144

5 $2^5+5(2)$ 42

6 $(-3)^3+3(-3)$ −36

In problems 7–10, write each expression in equivalent exponential form.

7 $x\cdot x\cdot x\cdot x$ x^4

8 $y\cdot y\cdot y\cdot y\cdot y\cdot y$ y^6

9 $2\cdot x\cdot x\cdot y-3\cdot x\cdot x\cdot x$ $2x^2y-3x^3$

10 $5\cdot x\cdot x\cdot x\cdot y\cdot y+7\cdot y\cdot z\cdot z$ $5x^3y^2+7yz^2$

In problems 11–20, use Property 1 to simplify each expression.

11 $3^2\cdot3^3$ 3^5

12 $2^4\cdot2^2$ 2^6

13 $(-2)^3 \cdot (-2)^2$ $(-2)^5$ **14** $(-2) \cdot (-2)^2 \cdot (-2)^3$ $(-2)^6$

15 $x^5 \cdot x^7$ x^{12} **16** $y^6 \cdot y^7$ y^{13}

17 $x^3 \cdot x^4 \cdot x^5$ x^{12} **18** $(-x)^9 \cdot (-x)^3 \cdot (-x)$ $(-x)^{13}$

19 $(-x)^n \cdot (-x)^n$; n a positive integer $(-x)^n$ **20** $(-y)^n \cdot (-y)^{2n}$; n an odd positive integer $(-y)^{3n}$

In problems 21–30, use Property 2 to simplify each expression.

21 $(2^2)^3$ 2^6 **22** $(3^3)^2$ 3^6

23 $[(-2)^3]^2$ $(-2)^6$ **24** $[(-3)^2]^2$ $(-3)^4$

25 $(x^7)^5$ x^{35} **26** $(x^3)^{12}$ x^{36}

27 $(y^2)^{11}$ y^{22} **28** $(y^4)^5$ y^{20}

29 $[(-x)^3]^4$ $(-x)^{12}$ **30** $[(x^2)^3]^5$ x^{30}

In problems 31–40, use Property 3 to simplify each expression.

31 $(2x)^4$ $2^4 x^4$ **32** $(5y)^2$ $5^2 y^2$

33 $(xy)^5$ $x^5 y^5$ **34** $(ab)^4$ $a^4 b^4$

35 $(xyz)^7$ $x^7 y^7 z^7$ **36** $(3xy)^5$ $3^5 x^5 y^5$

37 $(-2x)^3$ $-2^3 x^3$ **38** $(-2xyz)^4$ $-2^4 x^4 y^4 z^4$

39 $[-3(-x)y]^n$; n an odd positive integer $-3^n x^n y^n$ **40** $(-3xy)^n$; n an even positive integer $-3^n x^n y^n$

In problems 41–50, use Property 4 to simplify each expression.

41 $\left(\dfrac{3}{4}\right)^2$ $\dfrac{3^2}{4^2}$ **42** $\left(\dfrac{1}{2}\right)^5$ $\dfrac{1^5}{2^5}$

43 $\left(\dfrac{-2}{3}\right)^3$ $\dfrac{-2^3}{3^3}$ **44** $\left(-\dfrac{3}{2}\right)^3$ $\dfrac{-3^3}{2^3}$

45 $\left(\dfrac{x}{y}\right)^4$ $\dfrac{x^4}{y^4}$ **46** $\left(\dfrac{y}{z}\right)^7$ $\dfrac{y^7}{z^7}$

47 $\left(\dfrac{a}{-b}\right)^5$ $\dfrac{a^5}{-b^5}$ **48** $\left(-\dfrac{x}{y}\right)^8$ $\dfrac{-x^8}{y^8}$

49 $\left(\dfrac{-x}{y}\right)^6$ $\dfrac{-x^6}{y^6}$ **50** $\left(-\dfrac{x}{y}\right)^n$; n an odd positive integer $\dfrac{-x^n}{y^n}$

In problems 51–60, use Property 5 to simplify each expression.

51 $\dfrac{3^5}{3^2}$ 3^{5-2}

52 $\dfrac{(-2)^6}{(-2)^3}$ -2^{6-3}

53 $\dfrac{4^9}{4^6}$ 4^{9-6}

54 $\dfrac{(-3)^4}{(-3)^6}$ -3^{6-4}

55 $\dfrac{x^8}{x^3}$ x^{8-3}

56 $\dfrac{x^{25}}{x^{11}}$ x^{25-11}

57 $\dfrac{y^{25}}{y^{20}}$ y^{25-20}

58 $\dfrac{(-x)^{25}}{(-x)^{25}}$ $-x^{25-25}$ 1

59 $\dfrac{x^n}{x^{10}}$; n a positive integer greater than 10 x^{n-10}

60 $\dfrac{x^n}{x^2}$; n a positive integer greater than 2 x^{n-2}

In problems 61–70, use Properties 1–5 to simplify each expression.

61 $(3x^2y^3)^4$ $3x^8y^{12}$

62 $(-2x^4)^3$

63 $\dfrac{x^4y^7}{xy^3}$ x^3y^4

64 $\dfrac{x^{16}y^5}{x^5y^2}$

65 $\left(\dfrac{x^2}{y}\right)^3$ $\dfrac{x^6}{y^3}$

66 $\left(\dfrac{3x^2}{4y}\right)^4$

67 $\left(\dfrac{4a^5b^3}{2a^2b}\right)^4$ b^4

68 $\left(\dfrac{4x^2y^3}{8x^3y}\right)^5$

69 $\dfrac{(-4xy^4z^5)^3}{(2x^2yz^3)^2}$

70 $\dfrac{(x^4y^7)^3}{(4x^2y^3)^2}$

71 Find the value of $(-1)^n$ if n is an even positive integer.

72 Verify Property 2, page 41.

73 Find the value of $(-1)^n$ if n is an odd positive integer.

74 Verify Property 4, page 41.

2.2 Polynomials

As previously stated in the introduction, a polynomial is a particular type of algebraic expression. $4x$, $3x - 5$, $y^2 + 4y + 7$, and $-z^5 - 3z^3 + z$ are examples of polynomials in one variable, whereas algebraic expressions such as $3/x^2$ and $-1/x$ are not polynomials. In general, a *polynomial in one variable*, say x, is any number of the set of real numbers and x, or any expression formed from this set by using only the operations of addition, subtraction, and multiplication. Those parts of the polynomials separated by either a $+$ or $-$ sign are called *terms* of the polynomial. If a polynomial contains only one term, it is called a *monomial.* A polynomial of two terms is called a *binomial,* while a polynomial of three terms is called a *trinomial.* Thus, in our examples, $4x$ is a monomial, $3x - 5$ is a binomial, and $y^2 + 4y + 7$ is a trinomial. Polynomials in one variable are also described by the highest power of the variable that it contains. We call this number the *degree* of the polynomial. For example, the degree of $4x$ is 1, the degree of $y^2 + 4y + 7$ is 2, and the degree of $-z^5 - 3z^3 + z$ is 5. A nonzero real number, such as 8, is called a *polynomial of degree zero,* while the real number zero is not assigned a degree. Instead, it is described as the *zero polynomial.* The numerical factor of any term of a polynomial is called its *numerical coefficient.* For example, the numerical coefficients of $y^2 + 4y + 7$ are 1, 4, and 7, while the numerical coefficients of $-z^5 - 3z^3 + z$ are $-1, -3,$ and 1.

EXAMPLE

1   Find the degree and the numerical coefficients of each of the following polynomials.

a) $3x - 2$
b) $-x^2 + 3x - 4$
c) $7x^3 + 13x^2 + 5x + 6$
d) $5x^4 - 17x^3 + 10x^2 + 7x - 13$

SOLUTION

a)   The degree is 1 and the numerical coefficients are 3 and -2.
b)   The degree is 2 and the numerical coefficients are $-1, 3,$ and -4.
c)   The degree is 3 and the numerical coefficients are 7, 13, 5, and 6.
d)   The degree is 4 and the numerical coefficients are 5, -17, 10, 7, and -13.

Polynomials such as $2x + 3$ and $y^2 + 2y - 4$ can be evaluated for specific values of x and y. For example, when x is replaced by 4, we have

$$2x + 3 = 2 \times 4 + 3$$

and when y is replaced by 1, we have

$$y^2 + 2y - 4 = 1^2 + 2 \times 1 - 4$$

Note that the order in which we complete the above operations will determine the final results. For example, the expression $2 \times 4 + 3$ can be evaluated as $(2 \times 4) + 3 = 8 + 3 = 11$, if we multiply first and then add, or $2 \times 4 + 3$ can be evaluated as $2 \times (4 + 3) = 2 \times 7 = 14$, if we add first and then multiply. To avoid this confusion, we agree that the correct order of operations is to multiply first and then perform the additions or subtractions. Thus, for $x = 4$ we have

$$
\begin{aligned}
2x + 3 &= 2 \times 4 + 3 \\
&= 8 + 3 \\
&= 11
\end{aligned}
$$

and for $y = 1$, we obtain

$$
\begin{aligned}
y^2 + 2y - 4 &= 1^2 + 2 \times 1 - 4 \\
&= 1 + 2 - 4 \\
&= -1
\end{aligned}
$$

In general, when evaluating polynomials, we perform the operations in the following order:

1 Find indicated powers.
2 Multiply powers by their numerical coefficients.
3 Perform the indicated additions or subtractions.

EXAMPLE

2 Evaluate each of the following polynomials for $x = -3$.
   a) $x^2 - 3$
   b) $3x + 4$
   c) $3x^3 - x$

SOLUTION. Replacing x by -3 in parts (a), (b), and (c), we have
a) $x^2 - 3 = (-3)^2 - 3 = 9 - 3 = 6$
b) $3x + 4 = 3(-3) + 4 = -9 + 4 = -5$
c) $3x^3 - x = 3(-3)^3 - (-3) = -81 + 3 = -78$

We can extend the notation of polynomials to include algebraic expressions of more than one variable. For example, $3xy + 5$ is a polynomial of two variables, $x^2y^3z + x^3y^2z^4 - 4$ is a polynomial of three variables, and $xy + yzw - 3$ is a polynomial of four variables. The *degree of a polynomial of more than one variable* is the largest sum of exponents in any one term of the polynomial. For example, the degree of $3xy + 5$ is 2, the degree of $x^2y^3z + x^3y^2z^4 - 4$ is 9, and the degree of $xy + yzw - 3$ is 3.

In evaluating a polynomial of more than one variable, we simply replace each variable by the number assigned to it. For example, the value of the expression $3xy + 5$ at $x = 2$ and $y = 7$ is given by $3xy + 5 = 3(2)(7) + 5 = 42 + 5 = 47$.

EXAMPLES

3 Determine the degree of $3x^3y - 2x^2y + 7xy^2 - y^3$.

SOLUTION.   The sum of the exponents of each term of the polynomial $3x^3y - 2x^2y + 7xy^2 - y^3$ is:

First term, $3 + 1 = 4$ Third term, $1 + 2 = 3$
Second term, $2 + 1 = 3$ Fourth term, 3

Since the highest sum is 4, the degree of the polynomial is 4.

4 Evaluate the polynomial $3x^2 - xy + y - 3$ if
   a) $x = 1$ and $y = -1$
   b) $x = 2$ and $y = 2$
   c) $x = -1$ and $y = 3$

SOLUTION

a) $3x^2 - xy + y - 3 = 3(1)^2 - (1)(-1) + (-1) - 3$
$$= 3 + 1 - 1 - 3 = 0$$
b) $3(2)^2 - 2(2) + 2 - 3 = 12 - 4 + 2 - 3 = 7$
c) $3(-1)^2 - (-1)3 + 3 - 3 = 3 + 3 + 3 - 3 = 6$

PROBLEM SET 2.2

In problems 1–10, identify the polynomials as monomial, binomial, or trinomial. Also, find the degree of the polynomial and list its numerical coefficients.

1   $3x - 2$ 2 -3

3   $4x^2$ 4 $5x^3 + 5$

5   $x^2 - 5x + 6$ 6 $x^3 - x - 1$

7   $2x^7 - 13$ 8 $3 - 2x$

9   $-x^4 - x^2 + 13$ 10 $x^4 + 4x^2 + 8$

In problems 11–20, evaluate the polynomial for the given value of the variable.

11   $4x - 1$ for $x = 2$ 12 $3z + 2$ for $z = -2$

13 $2x^2 - x + 4$ for $x = 3$ **14** $3y^3 - 5y + 3$ for $y = 4$

15 $x^3 - 2x^2 - x + 1$ for $x = -1$ **16** $-2x^4 - x$ for $x = -3$

17 $4 - y - y^2$ for $y = -2$ **18** $3x^4 + 3x^2 + 15$ for $x = -1$

19 $x^5 - x^4 + x^3 - x^2 + x - 1$ for $x = 2$ **20** $x^4 - 3x^2 + 17x + 23$ for $x = 3$

In problems 21–26, find the degree of each polynomial of more than one variable.

21 $5x^2y + 16x$ **22** $3x^2 + 5xyz - 1$

23 $x^2 - 5xy + 2z^2$ **24** $3x^4 - 3xy + 5xz^4$

25 $7x^2 + 13x^2y^3 + 9x^4$ **26** $4x^2 - 3xy^2 + 5xyz^3$

In problems 27–32, evaluate the given polynomial of more than one variable for $x = 1$, $y = -1$, and $z = 2$.

27 $2xy^2 - yz$ **28** $3x^2 + xy - z^2y$

29 $x^3y^2z + 2xy^2z^3$ **30** $4yz^2 - 2x^3y$

31 $3xy^2 - 2yz^2 + xyz$ **32** $xyz + 2x^2z - yz^3$

In problems 33–36, indicate whether or not the algebraic expressions are polynomials.

33 $\dfrac{13}{x} + 75$ **34** $7x^5 - 3x + 11$

35 $-\dfrac{1}{17}x^8 + \dfrac{1}{35}x^5 + 1$ **36** $\dfrac{5}{x^3} + 9x + 2$

2.3 Sums and Differences of Polynomials

We have studied the addition properties of real numbers in Chapter 1. Since polynomials are expressions that represent real numbers, these properties also apply to polynomials. Assume that P, Q, and R represent polynomial expressions. Then the following properties are true:

1   *Commutative property of addition:* $P + Q = Q + P$

2   *Associative property of addition:* $P + (Q + R) = (P + Q) + R$

3   *Distributive properties*
    a) $P(Q + R) = (PQ) + (PR)$
    b) $(P + Q)R = (PR) + (QR)$

Now, we can use the above properties to add polynomials. For example, to find the sum of the monomials $3x^2$ and $4x^2$, we use the distributive property, so that $3x^2 + 4x^2 = (3 + 4)x^2 = 7x^2$. On the other hand, the sum of the monomials $2x$ and $3y$ can only be expressed as $2x + 3y$ or $3y + 2x$, since the distributive property does not apply in this case.

The terms $3x^2$ and $4x^2$ are called *similar terms*. Also, xy^2 and $3xy^2$ are similar terms, whereas $2x^2$ and $3x$ are not similar terms. In general, if terms have the same variables and each variable in one term has the same exponent as the identical variable in the other term, they are called similar terms.

EXAMPLE

1 Find the sum of the following monomial expressions.

a) $3x, 5x$
b) $4x^2, -2x^2$
c) $2x^3, 4x^3, -3x^3$
d) $3xy^2, 7xy^2, 2x^2y$

SOLUTION

a) $3x + 5x = (3 + 5)x = 8x$
b) $4x^2 + (-2x^2) = [4 + (-2)]x^2 = 2x^2$
c) $2x^3 + 4x^3 + (-3x^3) = [2 + 4 + (-3)]x^3 = 3x^3$
d) $3xy^2 + 7xy^2 + 2x^2y = (3 + 7)xy^2 + 2x^2y = 10xy^2 + 2x^2y$

We shall see that the addition of polynomials can always be reduced to the simple addition of monomials. For example, consider the sum

$$(3x^3 + 2x^2 + 4) + (5x^2 + 7x^3 + 8)$$

By using the commutative and associative properties of addition, we can rearrange the terms into groups of similar terms:

$$(3x^3 + 2x^2 + 4) + (5x^2 + 7x^3 + 8)$$

$$= (3x^3 + 7x^3) + (2x^2 + 5x^2) + (4 + 8)$$

$$= (3 + 7)x^3 + (2 + 5)x^2 + (4 + 8) \qquad \text{(Distributive property)}$$

$$= 10x^3 + 7x^2 + 12$$

Often the rearrangement and reordering of the terms is done by use of the "vertical scheme."

Associative property

	$3x^3$	$+$	$2x^2$	$+$	4
$(+)$	$7x^3$	$+$	$5x^2$	$+$	8
	$10x^3$	$+$	$7x^2$	$+$	12

Although the vertical arrangement makes the *mechanics* of polynomial addition easier, it is important to note that the vertical arrangement implicitly applies the same properties we used when we added the polynomials "horizontally."

To perform the polynomial addition

$$(3x^3 - 7x^2 + 4) + (5x + 8x^2 - 7)$$

we use the property of additive inverse for real numbers that we introduced in Chapter 1 [that is, if b is a real number, then $b+(-b)=0$], together with the above properties, so that

$$(3x^3 - 7x^2 + 4) + (5x + 8x^2 - 7)$$
$$= [3x^3 + (-7x^2) + 4] + [5x + 8x^2 + (-7)]$$
$$= 3x^3 + [(-7x^2) + 8x^2] + 5x + [4 + (-7)]$$
$$= 3x^3 + (-7 + 8)x^2 + 5x + [4 + (-7)]$$
$$= 3x^3 + 1 \cdot x^2 + 5x + (-3)$$
$$= 3x^3 + x^2 + 5x - 3$$

Using the vertical scheme and the rules for adding signed numbers, we have

$$\begin{array}{l} 3x^3 - 7x^2 + 4 \\ (+) + 8x^2 + 5x - 7 \\ \hline 3x^3 + x^2 + 5x - 3 \end{array}$$

EXAMPLE

2 Perform the following additions.
a) $(4x^3 + 7x - 3) + (2 + x^3 - x)$
b) $(2x - 7x^2 + 8) + (-x^3 - x^2 - 1)$
c) $(7 - 3x^2 + x) + (2x^2 + 8 + 13x) + (3x - 7 + 2x^2)$

SOLUTION. Arranging the polynomials in a vertical scheme, we have
a) $ 4x^3 + 7x - 3$
$ (+) x^3 - x + 2$
$ \overline{5x^3 + 6x - 1}$

b)
$$\begin{array}{r} -7x^2 + 2x + 8 \\ (+)\ \underline{-x^3 -\ \ x^2 \qquad - 1} \\ -x^3 - 8x^2 + 2x + 7 \end{array}$$

c)
$$\begin{array}{r} -3x^2 +\ \ \ x + 7 \\ 2x^2 + 13x + 8 \\ (+)\ \underline{\ 2x^2 +\ \ 3x - 7} \\ x^2 + 17x + 8 \end{array}$$

To subtract polynomials, recall the definition of subtraction from Chapter 1. That is, for any real numbers a and b, $a - b = a + (-b)$. For example, to subtract the monomial $2x^2$ from $5x^2$, we have

$$5x^2 - 2x^2 = 5x^2 + (-2x^2) = 5x^2 + (-2)x^2$$
$$= [5 + (-2)]x^2 = 3x^2$$

The same result could be obtained by simply subtracting the numerical coefficients. Thus,

$$5x^2 - 2x^2 = (5 - 2)x^2 = 3x^2$$

EXAMPLE

3   Perform the following subtractions.

a) $6y^3 - 3y^3$ b) $-2xy - 3xy$ c) $4x^4 - (-5x^4)$

SOLUTION

a) $6y^3 - 3y^3 = (6 - 3)y^3 = 3y^3$
b) $-2xy - 3xy = (-2 - 3)xy = -5xy$
c) $4x^4 - (-5x^4) = [4 - (-5)]x^4 = 9x^4$

Subtractions of any polynomials can be accomplished by first grouping similar terms and then performing the subtraction of the similar monomials as illustrated above. For example, to subtract $2x^2 + 3x - 4$ from $3x^2 + 5x + 7$, we write

$$(3x^2 + 5x + 7) - (2x^2 + 3x - 4)$$
$$= (3x^2 - 2x^2) + (5x - 3x) + [7 - (-4)]$$
$$= (3 - 2)x^2 + (5 - 3)x + 11$$
$$= x^2 + 2x + 11$$

Using the vertical scheme and the rules for subtracting signed numbers, we have

$$\begin{array}{r} 3x^2 + 5x +\ \ 7 \\ (-)\ \underline{2x^2 + 3x -\ \ 4} \\ x^2 + 2x + 11 \end{array}$$

EXAMPLES

4   Subtract $2x^2 - 3x + 4$ from $4x^2 - x + 2$.

SOLUTION.   Using the vertical scheme, we have

$$
\begin{array}{r}
4x^2 - x + 2 \\
(-)\ \underline{2x^2 - 3x + 4} \\
2x^2 + 2x - 2
\end{array}
$$

5   Subtract $-4x^3 + 5x^2 - 7x + 3$ from $2x^3 - x^2 - 8x$.

SOLUTION

$$
\begin{array}{r}
2x^3 - x^2 - 8x + 0 \\
(-)\ \underline{-4x^3 + 5x^2 - 7x + 3} \\
6x^3 - 6x^2 - x - 3
\end{array}
$$

6   Subtract $x^2 - 5x + 8$ from the sum of $x^2 - 3x + 4$ and $2x^2 - x + 2$.

SOLUTION.   First, we perform the addition of $x^2 - 3x + 4$ and $2x^2 - x + 2$, so that

$$
\begin{array}{r}
x^2 - 3x + 4 \\
(+)\ \underline{2x^2 - x + 2} \\
3x^2 - 4x + 6
\end{array}
$$

Then, we perform the subtraction, so that

$$
\begin{array}{r}
3x^2 - 4x + 6 \\
(-)\ \underline{x^2 - 5x + 8} \\
2x^2 + x - 2
\end{array}
$$

PROBLEM SET 2.3

In problems 1–20, perform the additions of the polynomial expressions.

1     $5x^2 + 7x^2$

2     $3y^3 + 8y^3$

3     $2x^3 + 9x^3$

4     $4x^2 + 5x^2$

5     $xy + 3xy + 8xy$

6     $5xyz + xyz + 3xyz$

7     $-3x^2 + 7x^2$

8     $(-8y) + (-5y)$

9     $5xy^2 + (-3xy^2) + 2xy^2$

10     $-7y^4 + (-3y^4) + (-4y^4)$

11     $(3x + 4) + (5x + 3)$

12     $(-5x - 3) + (7x + 1)$

13 $(2x^2 + 3x + 1) + (5x^2 + 2x + 4)$

14 $(8x^3 + 3x + 5) + (2x^2 + 7x + 5)$

15 $(7x^2 - 4x + 3) + (-3x^2 + 2x - 5)$

16 $(-8x^2y + 6xy - 7xy^2) + (3x^2y - 4xy + 3xy^2)$

17 $(5x^2 - 3x^3 - x + 2x^4) + (4x^3 + 3x^4 - x^2 + 2x)$

18 $(1 + 2x - 3x^2 + 4x^3) + (5x^3 - x^2 + 3x - 7)$

19 $(10x^3 + 4x^2y - 5xy^2 - 8y^3) + (5y^3 + 6x^3 - 3x^2y + 4xy^2)$

20 $(11xy - 9x^2y^2 + 13x^3y^3) + (3x^3y^3 + 5xy + 7x^2y^2 + 13)$

In problems 21–40, perform the subtractions of the polynomial expressions.

21 $7x - 3x$ **22** $10y - 3y$

23 $3x^2 - x^2$ **24** $4x^3 - x^3$

25 $10xy - 7xy$ **26** $12x^2y - 9x^2y$

27 $3x^3 - (-2x^3)$ **28** $5y^4 - (-2y^4)$

29 $-5x^2y - (-3x^2y)$ **30** $-11x^3y^2 - (-12x^3y^2)$

31 $(12x + 4y + 7z) - (3x + 7y + 5z)$

32 $(3x^2 + 5x + 8) - (x^2 + 3x + 4)$

33 $(4x^3 - 7x - 8) - (-2x^3 + 4x - 2)$

34 $(-5x^2 + 3x - 4) - (-6x^2 - 2x + 1)$

35 $(3x^2y + 4xy - 7xy^2) - (-2x^2y + xy + 3xy^2)$

36 $(-5x^3y^3 - 8xy + 7) - (-2x^3y^3 + 2xy + 4)$

37 $(3x^4 - 4x^3 + 6x^2 + x - 1) - (4 - x + 2x^2 - 3x^3 - x^4)$

38 $(5y^3 - 3y^2 + 2y - 8) - (y - y^3 + 3y^2 - 1)$

39 $(4x^4y^4 - 3x^3y^3 + 2x^2y^2 + xy) - (4x^3y^3 - 3x^2 y^2 - 2xy - 3)$

40 $(x^6 - 2x^4 - 3x^2) - (x^5 - 2x^3 - 3x - 4)$

In problems 41–50, perform the indicated operations.

41 $(8x^2 + 3x - 7) + (-5x^2 + 2x + 1) + (2x^2 - 3x + 4)$

42 $(-2x^2 + 5x + 2) + (3x^2 - 2x + 3) + (x^2 - 3x - 7)$

43 $(x^3 - 2x^2 + 3x + 1) + (2x^3 + x^2 - 2x + 2) + (-x^3 + 3x^2 - 2x - 1)$

44 $(3y^4 - 4y^3 + y^2 - 2y + 3) + (7y^4 + 5y^3 + 2y^2 - y - 7) + (6y^3 - 6y + 5)$

45 $(2x^2 - 3x + 4) + (x^2 + 5x - 1) - (2x^2 + x - 6)$

46 $(-x^2 + 8x - 11) + (3x^2 - 10x + 3) - (x^2 - 3x - 2)$

47 $(2xy - 3xz + 4yz) - (5xz - 3yz + xy) - (-2yz + 3xy - xz)$

— 48 $(7x^3y^2 - 3x^2y + 2x) - (4x + x^2y - x^3y^2) - (-2x^2y + 5x^3y^2 + 3x)$

49 $(4x^3 + 2x^2 - x - 3) - (x^3 - x^2 + 3x + 4) + (-2x^3 + 3x^2 + 5x + 6)$

50 $(x^2 + 5x - 16) - (13x^2 + 7) + (3x^2 + 7x + 6)$

2.4 Products of Polynomials

As with the properties for addition, the multiplication properties of real numbers introduced in Chapter 1 are applicable to polynomial expressions. Suppose that P, Q, and R are polynomial expressions. Then:

1 *Commutative property of multiplication:* $PQ = QP$

2 *Associative property of multiplication:* $P(QR) = (PQ)R$

Let us consider the example of finding the product of the monomials $8x^3$ and $2x^2$ by applying these properties. We have

$$(8x^3)(2x^2) = (8 \times 2)(x^3 \cdot x^2)$$
$$= 16x^{3+2}$$
$$= 16x^5$$

Another example is finding the product of $-5x^2y$ and $3x^3y^4$:

$$(-5x^2y)(3x^3y^4) = (-5 \times 3)(x^2 \cdot x^3)(y \cdot y^4)$$
$$= -15x^5y^5$$

Note that in both examples the products were essentially found by multiplying the numerical coefficients and applying the rule for multiplying like bases by adding exponents.

EXAMPLE

1 Find the following products of monomials.

a) $(4x^2)(7x)$ b) $(-3x)(2x^3)$ c) $(-5x^2y^3)(-2x^2y)$

SOLUTION

a) $(4x^2)(7x) = (4 \times 7)x^{2+1} = 28x^3$

b) $(-3x)(2x^3) = (-3 \times 2)x^{1+3} = -6x^4$

c) $(-5x^2y^3)(-2x^2y) = (-5) \times (-2)x^{2+2}y^{3+1} = 10x^4y^4$

In order to multiply a monomial by a polynomial, we apply the distributive property to reduce the process to a multiplication of monomials. For example, to find the product of $3x^2$ and $2x^3 + 4x$ we have

$$3x^2(2x^3 + 4x) = (3x^2)(2x^3) + (3x^2)(4x)$$
$$= 6x^5 + 12x^3$$

EXAMPLE

2   Find the following products.

a) $4x(3x - 2)$ b) $2x^3(x^2 - 3x + 4)$ c) $-2xy(-3x^2y - 4xy^3)$

SOLUTION

a) $4x(3x - 2) = (4x)(3x) + (4x)(-2)$
$$= 12x^2 - 8x$$

b) $2x^3(x^2 - 3x + 4) = (2x^3)(x^2) + (2x^3)(-3x) + (2x^3)(4)$
$$= 2x^5 - 6x^4 + 8x^3$$

c) $-2xy(-3x^2y - 4xy^3) = (-2xy)(-3x^2y) + (-2xy)(-4xy^3)$
$$= 6x^3y^2 + 8x^2y^4$$

The distributive property can also be used to reduce the multiplication of a polynomial by a polynomial to a multiplication of monomials. For example, to multiply $x + 2$ by $x + 4$, we have

$$(x + 2)(x + 4) = x(x + 4) + 2(x + 4) \qquad \text{(distributive property)}$$
$$= x^2 + 4x + 2x + 8 \qquad \text{(distributive property)}$$
$$= x^2 + 6x + 8$$

EXAMPLES

Determine the following products.

3   $(x + y)(2x - 3y)$

SOLUTION

$$(x + y)(2x - 3y) = (x + y)2x + (x + y)(-3y)$$
$$= 2x^2 + 2xy - 3xy - 3y^2$$
$$= 2x^2 - xy - 3y^2$$

4 $(x^2 + 2xy - y^2)(x + xy + y)$

SOLUTION

$$(x^2 + 2xy - y^2)(x + xy + y)$$
$$= (x^2 + 2xy - y^2)x + (x^2 + 2xy - y^2)xy + (x^2 + 2xy - y^2)y$$
$$= x^3 + 2x^2y - xy^2 + x^3y + 2x^2y^2 - xy^3 + x^2y + 2xy^2 - y^3$$
$$= x^3 + 3x^2y + xy^2 + x^3y + 2x^2y^2 - xy^3 - y^3$$

Since in many multiplications involving polynomials the actual computations can become quite tedious, as seen in the examples above, any device to help perform the computations and simplify the work is desirable. One such device, the vertical scheme, is illustrated by the following example. To find the product of $2x - 1$ and $x^3 + x^2 - 2x - 1$, we arrange the polynomial in a vertical scheme, so that

$$
\begin{array}{l}
x^3 + x^2 - 2x - 1 \\
2x - 1 \\
\hline
2x^4 + 2x^3 - 4x^2 - 2x \qquad [(2x)(x^3 + x^2 - 2x - 1)] \\
 - x^3 - x^2 + 2x + 1 \qquad [(-1)(x^3 + x^2 - 2x - 1)] \\
\hline
2x^4 + x^3 - 5x^2 + 1 \qquad \text{(product)}
\end{array}
$$

This shortcut involves arranging the partial products so that similar terms are in the same column, ready for the final step of addition. As with the vertical scheme for addition, it is important to note that the "mechanics" of this method are based on the properties already learned.

EXAMPLES

Use the vertical scheme to multiply each of the following.

5 $(x + 4)(2x - 3)$

SOLUTION

$$
\begin{array}{l}
x + 4 \\
2x - 3 \\
\hline
2x^2 + 8x \\
 - 3x - 12 \\
\hline
2x^2 + 5x - 12
\end{array}
$$

6 $(3x^3 - 5xy + 7y^2)(x - y)$

SOLUTION

$$
\begin{array}{l}
3x^3 - 5xy + 7y^2 \\
\underline{x - y} \\
3x^4 - 5x^2y + 7xy^2 \\
5xy^2 - 3x^3y - 7y^3 \\
\overline{3x^4 - 5x^2y + 12xy^2 - 3x^3y - 7y^3}
\end{array}
$$

7    $(x^2 - 2x + 1)(x^2 + x + 2)$

SOLUTION

$$
\begin{array}{l}
x^2 - 2x + 1 \\
\underline{x^2 + x + 2} \\
x^4 - 2x^3 + x^2 \\
 x^3 - 2x^2 + x \\
 2x^2 - 4x + 2 \\
\overline{x^4 - x^3 + x^2 - 3x + 2}
\end{array}
$$

We can also apply a device for simplifying the computations required when multiplying two binomials, where the first terms and the second terms of the two binomials are similar. For example, consider the product $(x + 2)(x + 3)$. By applying the distributive properties, we have

$$
\begin{aligned}
(x + 2)(x + 3) &= x(x + 3) + 2(x + 3) \\
&= x^2 + 3x + 2x + 6 \\
&= x^2 + 5x + 6
\end{aligned}
$$

The result of the above multiplication is a trinomial whose terms are determined as follows:

First term: $(x + 2)(x + 3)$ $=$ $\boxed{x^2} + 5x + 6$

$(x)(x)$

Middle term: $(x + 2)(x + 3)$ $=$ $x^2 + \boxed{5x} + 6$

$\oplus$ $3x + 2x$

Last term: $(x + 2)(x + 3)$ $=$ $x^2 + 5x + \boxed{6}$

$(2)(3)$

This method enables us to multiply two binomials and write the product directly, without having to show any intermediate steps.

EXAMPLES

Find the following products by using the method just illustrated.

8   $(x + 4)(2x + 1)$

SOLUTION

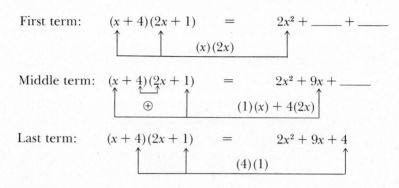

First term: $(x + 4)(2x + 1)$ $=$ $2x^2 +$ ____ $+$ ____

$(x)(2x)$

Middle term: $(x + 4)(2x + 1)$ $=$ $2x^2 + 9x +$ ____

$\oplus$ $(1)(x) + 4(2x)$

Last term: $(x + 4)(2x + 1)$ $=$ $2x^2 + 9x + 4$

$(4)(1)$

Therefore, $(x + 4)(2x + 1) = 2x^2 + 9x + 4$.

9   $(x + 7)(x - 3)$

SOLUTION

$(7)(-3)$

$(x)(x)$

$(x + 7)(x - 3)$ $=$ $x^2 + 4x - 21$

$\oplus$ $-3x + 7x$

10   $(3x - 4)(2x + 3)$

SOLUTION

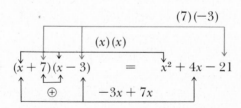

$(-4)(3)$

$(3x)(2x)$

$(3x - 4)(2x + 3)$ $=$ $6x^2 + x - 12$

$\oplus$ $9x - 8x$

11   $(4x - 3y)(x - y)$

SOLUTION. We can mentally obtain this product, using the following scheme:

$$(4x - 3y)(x - y) \quad = \quad 4x^2 - 7xy + 3y^2$$

First term: $(4x)(x) = 4x^2$

Middle term: $(4x)(-y) + (-3y)(x) = -7xy$

Last term: $(-3y)(-y) = 3y^2$

Special Products

Certain products of polynomials occur often enough in algebra to be worthy of special consideration. Learning these special products has the same advantage as learning the multiplication tables in arithmetic: It enables us to find many products in less time than would otherwise be required. Here we shall develop these special products. Assume that a and b represent the terms of the polynomials.

SPECIAL PRODUCT 1

The square of the binomial $a + b$: $(a + b)^2 = a^2 + 2ab + b^2$. To verify Special Product 1, we have

$$(a + b)^2 = (a + b)(a + b)$$
$$= a(a + b) + b(a + b)$$
$$= a^2 + ab + ab + b^2$$
$$= a^2 + 2ab + b^2$$

EXAMPLES

Use Special Product 1 to find the following products.

12   $(x + 3)^2$

SOLUTION. By substituting x for a and 3 for b in Special Product 1, we have

$$(x + 3)^2 = (x)^2 + 2(x)(3) + (3)^2$$
$$= x^2 + 6x + 9$$

13   $(x + 2y)^2$

SOLUTION. By substituting x for a and $2y$ for b, we have

$$(x + 2y)^2 = (x)^2 + 2(x)(2y) + (2y)^2$$
$$= x^2 + 4xy + 4y^2$$

SPECIAL PRODUCT 2

The square of the binomial $a - b$: $(a - b)^2 = a^2 - 2ab + b^2$. To verify Special Product 2, replace b by $-b$ in Special Product 1, and use $a - b = a + (-b)$. Thus,

$$(a - b)^2 = [a + (-b)]^2$$
$$= a^2 + 2a(-b) + (-b)^2$$
$$= a^2 - 2ab + b^2$$

EXAMPLES

Use Special Product 2 to find the following products.

14 $(x - 2)^2$

SOLUTION. Substituting x for a and 2 for b in Special Product 2, we have

$$(x - 2)^2 = (x)^2 - 2(x)(2) + (2)^2$$
$$= x^2 - 4x + 4$$

15 $(2x - 5y)^2$

SOLUTION. Substituting $2x$ for a and $5y$ for b, we have

$$(2x - 5y)^2 = (2x)^2 - 2(2x)(5y) + (5y)^2$$
$$= 4x^2 - 20xy + 25y^2$$

SPECIAL PRODUCT 3

The product of the sum and difference of the same two numbers: $(a + b)(a - b) = a^2 - b^2$. To verify Special Product 3, we have

$$(a + b)(a - b) = a(a - b) + b(a - b)$$
$$= a^2 - ab + ab - b^2$$
$$= a^2 - b^2$$

EXAMPLES

Use Special Product 3 to find the following products.

16 $(x + 5)(x - 5)$

SOLUTION. Letting $a = x$ and $b = 5$ in Special Product 3, we have

$$(x - 5)(x + 5) = (x)^2 - (5)^2$$
$$= x^2 - 25$$

17    $(2x + 3y)(2x - 3y)$

SOLUTION

$$(2x + 3y)(2x - 3y) = (2x)^2 - (3y)^2$$
$$= 4x^2 - 9y^2$$

SPECIAL PRODUCT 4

The cube of the binomial $a + b$: $(a + b)^3 = a^3 + 3a^2b + 3ab^2 + b^3$. To verify Special Product 4, we have

$$(a + b)^3 = (a + b)(a + b)^2$$
$$= (a + b)(a^2 + 2ab + b^2)$$
$$= a(a^2 + 2ab + b^2) + b(a^2 + 2ab + b^2)$$
$$= a^3 + 2a^2b + ab^2 + a^2b + 2ab^2 + b^3$$
$$= a^3 + 3a^2b + 3ab^2 + b^3$$

EXAMPLES

Use Special Product 4 to find the following products.

18    $(x + 2)^3$

SOLUTION.   Letting $a = x$ and $b = 2$ in Special Product 4, we have

$$(x + 2)^3 = (x)^3 + 3(x)^2(2) + 3(x)(2)^2 + (2)^3$$
$$= x^3 + 6x^2 + 12x + 8$$

19    $(4x + 3y)^3$

SOLUTION.   Letting $a = 4x$ and $b = 3y$, we have

$$(4x + 3y)^3 = (4x)^3 + 3(4x)^2(3y) + 3(4x)(3y)^2 + (3y)^3$$
$$= 64x^3 + 3(16x^2)(3y) + 3(4x)(9y^2) + 27y^3$$
$$= 64x^3 + 144x^2y + 108xy^2 + 27y^3$$

SPECIAL PRODUCT 5

The cube of the binomial $a - b$: $(a - b)^3 = a^3 - 3a^2b + 3ab^2 - b^3$. We can verify Special Product 5 by applying Special Product 4, and $a - b = a + (-b)$. Thus,

$$(a - b)^3 = [a + (-b)]^3 = a^3 + 3a^2(-b) + 3a(-b)^2 + (-b)^3$$
$$= a^3 - 3a^2b + 3ab^2 - b^3$$

EXAMPLES

Use Special Product 5 to find the following products.

20   $(x-4)^3$

SOLUTION.   Letting $a = x$ and $b = 4$ in Special Product 5, we have

$$(x-4)^3 = (x)^3 - 3(x)^2(4) + 3(x)(4)^2 - (4)^3$$
$$= x^3 - 12x^2 + 48x - 64$$

21   $(2x-y)^3$

SOLUTION.   Let $a = 2x$ and $b = y$, so that

$$(2x-y)^3 = (2x)^3 - 3(2x)^2(y) + 3(2x)(y)^2 - (y)^3$$
$$= 8x^3 - 12x^2y + 6xy^2 - y^3$$

SPECIAL PRODUCT 6

Product of the form: $(a+b)(a^2-ab+b^2) = a^3+b^3$. To verify Special Product 6, we have

$$(a+b)(a^2-ab+b^2) = (a+b)a^2 + (a+b)(-ab) + (a+b)b^2$$
$$= a^3 + a^2b - a^2b - ab^2 + ab^2 + b^3$$
$$= a^3 + b^3$$

EXAMPLES

22   Find the product of $2x+y$ and $4x^2-2xy+y^2$.

SOLUTION.   Let $a = 2x$ and $b = y$, so that $a+b = 2x+y$ and

$$a^2 - ab + b^2 = (2x)^2 - (2x)y + y^2 = 4x^2 - 2xy + y^2$$

Applying Special Product 6, we have

$$(2x+y)(4x^2-2xy+y^2) = (2x)^3 + (y)^3 = 8x^3 + y^3$$

23   Determine if Special Product 6 can be used to find the product $(x+y)(x^2-2xy+y^2)$.

SOLUTION.   Letting $a = x$ and $b = y$, we have $a+b = x+y$ and

$$a^2 - ab + b^2 = (x)^2 - (x)(y) + (y)^2 = x^2 - xy + y^2$$

Since $x^2 - 2xy + y^2$ is not of the form $a^2 - ab + b^2$, Special Product 6 does not apply in this case.

SPECIAL PRODUCT 7

Product of the form: $(a - b)(a^2 + ab + b^2) = a^3 - b^3$. We verify Special Product 7 by replacing b by $-b$ in Special Product 6, so that

$$[a + (-b)][a^2 - a(-b) + (-b)^2] = a^3 + (-b)^3$$
$$= a^3 - b^3$$

EXAMPLES

24   Find the product $(x - 2)(x^2 + 2x + 4)$.

SOLUTION.   Letting $a = x$ and $b = 2$, we have $a - b = x - 2$ and $a^2 + ab + b^2 = x^2 + 2x + 4$, so that Special Product 7 applies. Hence,

$$(x - 2)(x^2 + 2x + 4) = (x)^3 - (2)^3$$
$$= x^3 - 8$$

25   Find the product of $(3x - 2y)$ and $(9x^2 + 6xy + 4y^2)$.

SOLUTION.   Letting $a = 3x$ and $b = 2y$, we have $a - b = 3x - 2y$ and $a^2 + ab + b^2 = 9x^2 + 6xy + 4y^2$, and so, by Special Product 7, we have

$$(3x - 2y)(9x^2 + 6xy + 4y^2) = (3x)^3 - (2y)^3$$
$$= 27x^3 - 8y^3$$

PROBLEM SET 2.4

In problems 1–10, find the products of the given monomials.

1   $(2x^2)(3x^4)$ 2 $(3y^3)(5y^2)$

3   $(-5x^3)(6x^4)$ 4 $(7x^2)(-8x^5)$

5   $(7x^2y^3)(-4x^3y^4)$ 6 $(-3xy^7)(6x^5y)$

7   $(-3x^2yz^3)(-4xy^2z)$ 8 $(-10x^3y^4z^5)(-2xy^2z^3)$

9   $(2xy)(3x^2z)(-4y^2z^3)$ 10 $(3x^2yz^3)(-4xy^2z^3)(-2xz^2)$

In problems 11–20, use the distributive property to find each product.

11   $x(x + 1)$ $-$12 $-y(2y - 3)$

13   $x^2(x + 2)$ 14 $x^3(2x - 4)$

15 $3x(2x^2 - 4)$

16 $5x(3x^4 + 7)$

17 $-xy^2(2x^2 + 3y^2 + 2)$

18 $y^2z^3(3y^2z - 4yz + 5yz^2)$

19 $2x^2y(4x^3y - 3x^2y^2 + xy^3 - 1)$

20 $-3x^4z^3(-2xz^2 - 4x^2y + 5x^3 - 3)$

In problems 21–30, use the vertical scheme to find each product.

21 $(x + y)(x + y - 1)$

22 $(x - y)(x - y + 1)$

23 $(2x + y)(x^2 + 2xy + y^2)$

24 $(x - 2y)(x^2 - 3xy - y^2)$

25 $(x^2 + 3)(x^3 + 2x^2 - 3x + 4)$

26 $(4 - x^2)(x^4 - 2x^2 + 3)$

27 $(x^2 - 5x + 6)(x^2 + 4x + 4)$

28 $(y^2 - 2y + 1)(y^2 + 2y + 1)$

29 $(x^2 + 2xy + y^2)(x^3 - 3x^2y + 3xy^2 - y^3)$

—30 $(2x^2 - xy + y^2)(x^3 - x^2y + xy^2 - y^3)$

In problems 31–50, use the method illustrated in Examples 8–11 to find each product.

31 $(x + 1)(x + 2)$

32 $(x + 3)(x + 4)$

33 $(x + 4)(x + 5)$

34 $(x + 6)(x + 8)$

35 $(2x + y)(x + 3y)$

36 $(x + 3y)(3x + y)$

37 $(3x - 1)(2x + 1)$

38 $(4x - 7)(x + 4)$

39 $(5x + 4)(x - 1)$

40 $(9x + 2)(x - 3)$

41 $(6x - 5y)(4x + 3y)$

42 $(11x - 3y)(2x + y)$

43 $(7x + 3y)(4x - 3y)$

44 $(3x - 2y)(2x + 3y)$

45 $(2x - 3)(x - 4)$

46 $(2y - 1)(y - 3)$

47 $(5x - y)(2x - 5y)$

~ 48 $(12y - 5x)(3y - 2x)$

49 $(10x - 7)(5x - 8)$

50 $(9x - 7y)(5x - 4y)$

In problems 51–60, use Special Products 1 and 2 to find each product.

51 $(x + 1)^2$

52 $(x + 5)^2$

53 $(2x + y)^2$

54 $(x + 3y)^2$

55 $(x - y)^2$

56 $(y - 2x)^2$

57 $(3x - 5)^2$

58 $(7z - 3)^2$

59 $(4x + 5y)^2$

60 $(9y + 8x)^2$

In problems 61–70, use Special Product 3 to find each product.

61 $(x + y)(x - y)$ **62** $(y - x)(y + x)$

63 $(x - 7)(x + 7)$ **64** $(x + 11)(x - 11)$

65 $(2x + 9)(2x - 9)$ **66** $(10 - 3y)(10 + 3y)$

67 $(8x - y)(8x + y)$ **68** $(11x - 10y)(11x + 10y)$

69 $(5x + 6y)(5x - 6y)$ **70** $(7z - 2y)(7z + 2y)$

In problems 71–80, use Special Products 4 and 5 to find each product.

71 $(x + 1)^3$ **72** $(x + y)^3$

73 $(2x + y)^3$ **74** $(4x + y)^3$

75 $(x - y)^3$ **76** $(x - 3y)^3$

77 $(3x - 2y)^3$ **78** $(2x - 5y)^3$

79 $(5x + 2y)^3$ **80** $(3x - 4)^3$

In problems 81–90, use Special Products 6 and 7 to find each product.

81 $(x + 1)(x^2 - x + 1)$ **82** $(x + y)(x^2 - xy + y^2)$

83 $(3 + y)(9 - 3y + y^2)$ **84** $(x + 7)(x^2 - 7x + 49)$

85 $(x - y)(x^2 + xy + y^2)$ **86** $(5 - x)(25 + 5x + x^2)$

87 $(3x + 2y)(9x^2 - 6xy + 4y^2)$ **88** $(3x - 4y)(9x^2 + 12xy + 16y^2)$

89 $(4x - 5y)(16x^2 + 20xy + 25y^2)$ **90** $(6x - y)(36x^2 + 6xy + y^2)$

In problems 91–100, find each product. Use special products when they apply.

91 $(x^2 + y^3)(x^2 - y^3)$ **92** $(x^2 - y^2)(x^2 + y^2)$

93 $(x^2 + 3)(x^2 + 3)$ **94** $(2x + y + z)(2x - y - z)$

95 $(x + y + 1)(x + y - 1)$ **96** $(x^2 - y^2)(x^4 + x^2y^2 + y^4)$

97 $(x + 2y)(x^2 - 4xy + 4y^2)$ **98** $(x^2 + y^2)^3$

99 $(x^3 - 2)^3$ **100** $(x + y + z)^2$

2.5 Factoring Polynomials by Common Factors and Grouping

In Section 2.4 we discussed methods for finding the product of polynomials using the distributive property. In the following three sections we shall explore ways to reverse this process. That is, we shall start with a given polynomial and see if we can express it as the product of two or more polynomials. This process is called *factoring* a polynomial. If this process can be accomplished, we say that the polynomial is *factorable*. If a polynomial has no factors other than itself and 1, or its negative and −1, the polynomial is called *prime*. To factor a polynomial, we express it as the product of prime polynomials, or as the product of a monomial and prime polynomials. Once we accomplish this objective, we refer to the final result as a *complete factorization* of the given polynomial. We should note that throughout the text, unless otherwise specified, we shall agree to accept as factors of any polynomial with integral coefficients only those factors which also contain integral coefficients. We begin by using the distributive property $PQ + PR = P(Q + R)$ to factor polynomials whose terms contain a common factor.

Common Factors

Consider the polynomial $x^2 + x$. This expression can be written as $x \cdot x + x \cdot 1$, so that, by applying the distributive property, we have the following factorization:

$$x^2 + x = x(x + 1)$$

Now consider the polynomial $3x^3y - 15xy^2 + 21xy$. Using the distributive property, we may "factor out" or "remove" the common factor $3xy$ and write

$$3x^3y - 15xy^2 + 21xy = 3xy(x^2 - 5y + 7)$$

The monomial factor of highest degree and largest integral coefficient that is common to every term of a polynomial is called the *highest common factor* of the polynomial. Thus, the highest common factors of the polynomials $x^2 + x$ and $3x^3y - 15xy^2 + 21xy$ are x and $3xy$, respectively.

Every factorization may be checked by finding the product of the factors. Thus, one can show that x and $x + 1$ are factors of $x^2 + x$ by finding the product $x(x + 1)$, so that

$$x(x + 1) = x \cdot x + x \cdot 1 = x^2 + x$$

Therefore, $x(x + 1)$ is the correct factorization of $x^2 + x$.

EXAMPLES

Express each of the following polynomials as the product of the highest common factor and a polynomial.

1   $2x^3 + 4x^2$

SOLUTION.   Factoring out the highest common factor, $2x^2$, in each term, we have

$$2x^3 + 4x^2 = 2x^2(x) + 2x^2(2)$$
$$= 2x^2(x + 2) \qquad \text{(distributive property)}$$

2   $3x^2y - 5xy^3$

SOLUTION.   The highest common factor is xy, so that

$$3x^2y - 5xy^3 = xy(3x) + xy(-5y^2)$$
$$= xy(3x - 5y^2)$$

3   $3x^3y^4 - 9x^2y^3 - 6xy^2$

SOLUTION.   The highest common factor is $3xy^2$, so that

$$3x^3y^4 - 9x^2y^3 - 6xy^2 = 3xy^2(x^2y^2 - 3xy - 2)$$

Some algebraic expressions are written so that all the terms have a common binomial factor which can be factored out by the methods above. The following example will illustrate this case.

EXAMPLE

4   Factor $5x(y + z) + 2(y + z)$.

SOLUTION.   Since the binomial $y + z$ is a common factor of each term, we have

$$5x(y + z) + 2(y + z) = (5x + 2)(y + z)$$

Factoring by Grouping

Polynomials of a particular form that do not contain factors common to every term may still be factorable by the method of common factors. In this case we first group the terms of the polynomials, then look for common polynomial factors in each group. For example, consider the polynomial $3xm + 3ym - 2x - 2y$. We note that two terms of the expression contain a factor of x and the other two terms contain a factor of y. Grouping these terms accordingly, we obtain

$$3xm + 3ym - 2x - 2y = (3xm - 2x) + (3ym - 2y)$$

Factoring out the common monomial x from the first group and the common monomial y from the second group, we have

$$3xm + 3ym - 2x - 2y = (3m - 2)x + (3m - 2)y$$

Removing the common binomial $(3m - 2)$, we have

$$3xm + 3ym - 2x - 2y = (3m - 2)(x + y)$$

EXAMPLES

Factor the given polynomials by grouping.

5   $3x + xy + 3x^2 + y$

SOLUTION

$$\begin{aligned} 3x + xy + 3x^2 + y &= (3x + 3x^2) + (y + xy) \\ &= 3x(1 + x) + y(1 + x) \\ &= (3x + y)(1 + x) \end{aligned}$$

6   $2ax^2 + 2ay^2 - bx^2 - by^2$

SOLUTION

$$\begin{aligned} 2ax^2 + 2ay^2 - bx^2 - by^2 &= (2ax^2 + 2ay^2) + (-bx^2 - by^2) \\ &= 2a(x^2 + y^2) - b(x^2 + y^2) \\ &= (2a - b)(x^2 + y^2) \end{aligned}$$

7   $3ax + 3ay + 3az - 2bx - 2by - 2bz$

SOLUTION

$$3ax + 3ay + 3az - 2bx - 2by - 2bz$$
$$= (3ax + 3ay + 3az) + (-2bx - 2by - 2bz)$$
$$= 3a(x + y + z) - 2b(x + y + z)$$
$$= (3a - 2b)(x + y + z)$$

PROBLEM SET 2.5

In problems 1–20, factor each expression by finding the highest common factor.

1	$x^2 - x$	**2**	$4x^2 + 2x$
3	$9x^2 + 3x$	**4**	$10x^2 - 5x$
5	$4x^2 + 7xy$	**6**	$9x^3 + 3x^4$
7	$a^2b - ab^2$	**8**	$17x^3y^2 - 34x^2y$
9	$6p^2q + 24pq^2$	**10**	$12a^3b^2 + 36a^2b^3$
11	$6ab^2 + 30a^2b$	**12**	$5abc + 20abc^2$
13	$12x^3y - 48x^2y^2$	**14**	$4x^3 - 2x^2 + x$
15	$2a^3b - 8a^2b^2 - 6ab^3$	**16**	$4xy^2z + x^2y^2z^2 - x^3y^3$
17	$x^3y^2 + x^2y^3 + 2xy^4$	**18**	$3x^2y^2 + 6x^2z^2 - 9x^2$
19	$9m^2n + 18mn^2 - 27mn$	**20**	$8xy^2 + 24x^2y^3 + 4xy^3$

In problems 21–30, factor out the common binomial factor in each expression.

21	$3x(2a + b) + 5y(2a + b)$	**22**	$(2m + 3)x - (2m + 3)y$
23	$5x(a + b) + 9y(a + b)^2$	**24**	$x(y - z) - (z - y)$
→**25**	$m(x - y) + (y - x)$	**26**	$(x - y) + 5(x - y)^2$
27	$7x(2a + 7b) + 14(2a + 7b)^2 + (2a + 7b)^3$		
28	$(x + y)^3 + x(x + y)^2 + 5y(x + y)$		
29	$y(xy + 2)^3 - 5x(xy + 2)^2 + 7(xy + 2)$		
30	$x^2(a + b)^3 - 4xy(a + b)^2 + 3(a + b)$		

In problems 31–45, factor each expression by grouping.

31	$ax + ay + bx + by$	**32**	$x^2a + x^2b - a - b$

33 $x^5 + 3x^4 - x - 3$

34 $ax^5 + b - bx^5 - a$

35 $yz + 2y - z - 2$

36 $a^2x + 1 - a^2 - x$

37 $ab^2 - b^2c - ad + cd$

38 $2x^2 - yz^2 - x^2y + 2z^2$

39 $2ax + by - 2ay - bx$

40 $x^3 + x^2 - 4x - 4$

41 $x^2 - ax + bx - ab$

42 $x^2 + ax + bx + ab$

43 $ax + bx + ay + by + a + b$

44 $2ax - b + 2bx - c + 2cx - a$

45 $2x^3 + x^2y - x^2 + 2xy + y^2 - y$

2.6 Factoring Polynomials by Special Products

In Section 2.4 we introduced special products of polynomials which occur often enough in algebra to be worth memorizing. These special products can also be used to factor those polynomials that fit the forms of the special products. For example, to factor $x^2 - y^2$, we apply Special Product 3, $(a - b)(a + b) = a^2 - b^2$, to obtain the factorization $x^2 - y^2 = (x - y)(x + y)$.

The complete list of these special products is as follows:

SQUARES OF A BINOMIAL

1 $a^2 + 2ab + b^2 = (a + b)^2$

2 $a^2 - 2ab + b^2 = (a - b)^2$

DIFFERENCE BETWEEN TWO SQUARES

3 $a^2 - b^2 = (a - b)(a + b)$

CUBES OF A BINOMIAL

4 $a^3 + 3a^2b + 3ab^2 + b^3 = (a + b)^3$

5 $a^3 - 3a^2b + 3ab^2 - b^3 = (a - b)^3$

SUM OR DIFFERENCE OF TWO CUBES

6 $a^3 + b^3 = (a + b)(a^2 - ab + b^2)$

7 $a^3 - b^3 = (a - b)(a^2 + ab + b^2)$

The following examples will illustrate how these special products are used to factor certain polynomials.

EXAMPLES

In Examples 1 and 2, use Special Product 1 or 2 to factor the given polynomials.

1   $x^2 + 4x + 4$

SOLUTION. $x^2 + 4x + 4 = x^2 + 2(2)x + 2^2$, which is in the form of Special Product 1, $a^2 + 2ab + b^2 = (a + b)^2$, where $a = x$ and $b = 2$, so that

$$x^2 + 2(2)x + 2^2 = (x + 2)^2 \quad\text{or}\quad x^2 + 4x + 4 = (x + 2)^2$$

2   $25x^2 - 30xy + 9y^2$

SOLUTION. $25x^2 - 30xy + 9y^2 = (5x)^2 - 2(5x)(3y) + (3y)^2$. Using Special Product 2, $a^2 - 2ab + b^2 = (a - b)^2$, with $a = 5x$ and $b = 3y$, we have

$$25x^2 - 30xy + 9y^2 = (5x - 3y)^2$$

In Examples 3 and 4, use Special Product 3 to factor the given polynomials.

3   $x^2 - 25$

SOLUTION. $x^2 - 25 = x^2 - 5^2$, so that, using Special Product 3, $a^2 - b^2 = (a - b)(a + b)$, with $a = x$ and $b = 5$, we have

$$x^2 - 25 = (x - 5)(x + 5)$$

4   $16x^4 - y^4$

SOLUTION. $16x^4 - y^4 = (4x^2)^2 - (y^2)^2$ so that, by Special Product 3, we have

$$16x^4 - y^4 = (4x^2 - y^2)(4x^2 + y^2)$$

However, $4x^2 - y^2$ can also be factored using Special Product 3, so that

$$4x^2 - y^2 = (2x - y)(2x + y)$$

Hence, the complete factorization is

$$16x^4 - y^4 = (2x - y)(2x + y)(4x^2 + y^2)$$

In Examples 5 and 6, use Special Product 4 or 5 to factor the given polynomials.

5   $x^3 + 3x^2y + 3xy^2 + y^3$

SOLUTION. $x^3 + 3x^2y + 3xy^2 + y^3$ is, in the form of Special Product 4,

$$a^3 + 3a^2b + 3ab^2 + b^3 = (a + b)^3$$

where $a = x$ and $b = y$, so that

$$x^2 + 3x^2y + 3xy^2 + y^3 = (x + y)^3$$

6   $x^3 - 6x^2 + 12x - 8$

SOLUTION. $x^3 - 6x^2 + 12x - 8 = x^3 - 3x^2(2) + 3x(2)^2 - (2)^3$, which is, in the form of Special Product 5,

$$a^3 - 3a^2b + 3ab^2 - b^3 = (a - b)^3$$

where $a = x$ and $b = 2$, so that

$$x^3 - 6x^2 + 12x - 8 = (x - 2)^3$$

In Examples 7 and 8, use Special Product 6 or 7 to factor the given polynomials.

7   $x^3 + 27$

SOLUTION. $x^3 + 27 = x^3 + 3^3$, which is, in the form of Special Product 6,

$$a^3 + b^3 = (a + b)(a^2 - ab + b^2)$$

where $a = x$ and $b = 3$, so that

$$x^3 + 27 = (x + 3)(x^2 - 3x + 9)$$

8   $8x^3 - y^3$

SOLUTION. $8x^3 - y^3 = (2x)^3 - y^3$, which is, in the form of Special Product 7,

$$a^3 - b^3 = (a - b)(a^2 + ab + b^2)$$

where $a = 2x$ and $b = y$, so that

$$8x^3 - y^3 = (2x - y)[(2x)^2 + (2x)(y) + (y)^2]$$
$$= (2x - y)(4x^2 + 2xy + y^2)$$

Factoring by Combined Methods

Up to now we have discussed factoring by common factors, factoring by grouping, and factoring by special products. Factoring some polynomials may require the application of more than one of these methods. To accomplish this, the following steps are suggested:

1   Factor out common factors.

2   Examine remaining polynomial to see if it is prime or if a special product can be applied directly.

3   If step 2 fails, determine if factoring by grouping can be applied.

The following examples illustrate this procedure.

EXAMPLES

Factor the given polynomials by following the steps above.

9   $x^3y - xy^3$

SOLUTION

$$x^3y - xy^3 = xy(x^2 - y^2) \qquad \text{(step 1)}$$
$$= xy(x - y)(x + y) \qquad \text{(step 2)}$$

10   $x^2 - 2xy + y^2 - a^2 - 2ab - b^2$

SOLUTION.   Steps 1 and 2 fail. Using step 3, we have

$$x^2 - 2xy + y^2 - a^2 - 2ab - b^2 = (x^2 - 2xy + y^2) + (-a^2 - 2ab - b^2)$$
$$= (x^2 - 2xy + y^2) - (a^2 + 2ab + b^2)$$

Applying Special Products 1 and 2 to these groupings, we obtain

$$x^2 - 2xy + y^2 - a^2 - 2ab - b^2 = (x - y)^2 - (a + b)^2$$

Since this latter form is the difference between two squares, we can use Special Product 3 to obtain

$$x^2 - 2xy + y^2 - a^2 - 2ab - b^2$$
$$= [(x - y) - (a + b)][(x - y) + (a + b)]$$
$$= (x - y - a - b)(x - y + a + b)$$

11 $x^4 + 2x^2y^2 + 9y^4$

SOLUTION. Steps 1 and 2 do not apply in this case. To apply step 3, we first note that the given trinomial will be a perfect square if the middle term is $6x^2y^2$. Thus, we have

$$x^4 + 2x^2y^2 + 9y^4$$
$$= x^4 + 6x^2y^2 + 9y^4 - 4x^2y^2$$
$$= (x^4 + 6x^2y^2 + 9y^4) - (4x^2y^2) \qquad \text{(step 3)}$$
$$= (x^2 + 3y^2)^2 - (2xy)^2$$
$$= [(x^2 + 3y^2) - 2xy][(x^2 + 3y^2) + 2xy] \qquad \text{(Special Product 3)}$$
$$= (x^2 - 2xy + 3y^2)(x^2 + 2xy + 3y^2)$$

PROBLEM SET 2.6

In problems 1–20, use the squares of a binomial (Special Products 1 and 2) to factor each expression.

1 $x^2 + 2xy + y^2$ 2 $x^2 + 10x + 25$

3 $x^2 - 8x + 16$ 4 $y^2 - 12y + 36$

5 $4x^2 + 4x + 1$ 6 $36x^2 - 12x + 1$

7 $9y^2 - 6y + 1$ 8 $25x^2 + 30x + 9$

9 $4x^2 + 12xy + 9y^2$ 10 $16y^2 - 56y + 49$

11 $9x^2 - 42xy + 49y^2$ 12 $25x^2 - 40xy + 16y^2$

13 $16y^2 + 24y + 9$ 14 $4(x - a)^2 + 4(x - a) + 1$

15 $x^2y^2 - 4xyz + 4z^2$ 16 $4x^4 + 20x^2y^2 + 25y^4$

17 $x^4 + 14x^2 + 49$ 18 $x^6 - 10x^3 + 25$

19 $a^2x^6 + 9 - 6ax^3$ 20 $b^4x^2 - 2ab^2x + a^2$

In problems 21–40, use the difference between two squares (Special Product 3) to factor each expression.

21 $x^2 - 4$ 22 $100 - x^2$

23 $1 - 9y^2$ 24 $25 - 4a^2$

25 $36 - 25a^2$

26 $16x^2y^2 - 9$

27 $16x^2 - 25y^2$

28 $25x^2 - 49y^2$

29 $a^2b^2 - c^2$

30 $49a^2 - 81b^2$

31 $1 - 100a^2b^2c^2$

32 $144 - x^2z^2$

33 $4x^4 - 1$

34 $256x^4 - y^4$

35 $x^4 - 81y^4$

36 $64x^4 - 49y^4$

37 $x^8 - y^8$

38 $625x^4 - 81y^4$

39 $(x + y)^2 - (a - b)^2$

40 $(3x + 2y)^2 - 25b^2$

In problems 41–48, use the cubes of a binomial (Special Products 4 and 5) to factor each expression.

$A^3 - 3A^2B + 3AB^2 + B^3$

41 $x^3 + 3x^2 + 3x + 1$

42 $x^3 + 12x^2 + 48x + 64$

43 $x^3 - 3x^2y + 3xy^2 - y^3$

44 $8x^3 + 12x^2 + 6x + 1$

45 $x^3 + 6x^2 + 12x + 8$

46 $27x^3 - 27x^2 + 9x - 1$

47 $x^3 - 9x^2 + 27x - 27$

48 $8x^3 - 36x^2 + 54x - 27$

In problems 49–66, use the sum or difference of two cubes (Special Products 6 and 7) to factor each expression.

49 $x^3 + 1$

50 $y^3 + 125$

51 $64 - y^3$

52 $27x^3 - y^3$

53 $64x^3 + y^3$

54 $x^3y^3 - 125$

55 $8x^3 - 27y^3$

56 $125t^3 - 216$

57 $x^3 - 8y^3z^3$

58 $64a^3 + 27b^3$

59 $x^6 + 8$

60 $x^6 - 27$

61 $x^9 - 1$

62 $x^9 + 512$

63 $(x + 2)^3 - 1$

64 $(a + 3)^3 + 27$

65 $(y + 1)^3 + 8$

66 $64 - (x - 1)^3$

In problems 67–76, use common factors, grouping, or special products to factor each expression.

67 $8x^3 - 2xy^2$

68 $2ax^2 + 4ax + 2a$

69 $x^3y - 4x^2y + 4xy$

70 $25x^2y^2 + 20xy^2 + 4y^2$

71 $8x^2 - x^5$ **72** $3a^5 + 3a^2$

73 $x^2 + y^2 - z^2 - 9 + 2xy - 6z$ **74** $x^2 - y^2 + 2by - 2ax + a^2 - b^2$

75 $x^4 + x^2y^2 + y^4$ **76** $9x^4 + 2x^2y^2 + y^4$

2.7 Factoring Trinomials of the Form $ax^2 + bx + c, a \neq 0$

In this section we shall discuss methods of factoring trinomials of the form $ax^2 + bx + c$, where a, b, and c are integral coefficients and $a \neq 0$. Such an expression is called a *second-degree trinomial* or a *quadratic trinomial*. To factor a quadratic trinomial of the form $ax^2 + bx + c$, we first consider the case where $a = 1$, so that

$$ax^2 + bx + c \quad \text{will become} \quad x^2 + bx + c$$

Examples of such trinomials are $x^2 + 9x + 10$, $x^2 - 5x + 6$, and $x^2 - x - 2$.
One way of factoring expressions such as $x^2 + 7x + 10$ is to write

$$x^2 + 7x + 10 = x^2 + 2x + 5x + 10$$

Then, using the method of factoring by grouping terms, we have

$$x^2 + 7x + 10 = (x^2 + 2x) + (5x + 10)$$
$$= x(x + 2) + 5(x + 2)$$
$$= (x + 5)(x + 2)$$

so that

$$x^2 + 7x + 10 = (x + 5)(x + 2)$$

To factor trinomials by grouping can be tedious, even for such relatively simple expressions as in the example above. Therefore, let us consider an alternative factoring procedure. If the binomials $x + p$ and $x + q$ are factors of the quadratic trinomial $x^2 + bx + c$, we can write

$$x^2 + bx + c = (x + p)(x + q) \quad (p \text{ and } q \text{ are integers})$$
$$= x^2 + px + qx + pq$$
$$= x^2 + (p + q)x + pq$$

so that

$$x^2 + bx + c = x^2 + (p + q)x + pq$$

The corresponding coefficients are indicated in the diagram.

Therefore, to factor a quadratic trinomial of the form $x^2 + bx + c$, we must be able to find integers p and q such that

$$b = p + q \qquad \text{and} \qquad c = pq$$

For example, in factoring the above expression $x^2 + 7x + 10$, we try to find integers p and q so that

$$x^2 + 7x + 10 = (x + p)(x + q) = x^2 + (p + q)x + pq$$

We start with the fact that $pq = 10$. There are four possibilities for p and q. They are

$$10 = 1 \times 10 \qquad 10 = (-1) \times (-10) \qquad 10 = 2 \times 5 \qquad \text{and}$$
$$10 = (-2) \times (-5)$$

Since $5 + 2 = 7$, we try the factorization $(x + 2)(x + 5)$, which yields

$$(x + 2)(x + 5) = x^2 + 7x + 10$$

Therefore, the correct factorization is

$$x^2 + 7x + 10 = (x + 2)(x + 5)$$

Now, consider the factoring of the general quadratic trinomial $ax^2 + bx + c$, where $a \neq 1$. If the binomials $(sx + p)$ and $(rx + q)$ are factors of $ax^2 + bx + c$, we can write

$$
\begin{aligned}
ax^2 + bx + c &= (sx + p)(rx + q) \\
&= srx^2 + sxq + prx + pq \\
&= rsx^2 + (sq + pr)x + pq \qquad \text{for some integers } r, s, p, \text{ and } q
\end{aligned}
$$

so that

$$ax^2 + bx + c = rsx^2 + (sq + pr)x + pq$$

The corresponding coefficients are indicated in the diagram above.

Therefore, to factor a quadratic trinomial of the form $ax^2 + bx + c$ we must be able to find integers s, p, r, and q such that

$$a = rs \qquad b = sq + pr \qquad \text{and} \qquad c = pq$$

For example, to factor the quadratic trinomial $3x^2 + x - 2$, we wish to find some integers r, s, p, and q such that

$$3x^2 + x - 2 = (sx + p)(rx + q) = rsx^2 + (sq + pr)x + pq$$

We start with the fact that $rs = 3$ and $pq = -2$. Since there is no simple rule for finding r, s, p, and q, we must resort to trial and error. The following possibilities exist:

$$3 = 1 \times 3 \quad \text{and} \quad -2 = 1 \times (-2) \quad \text{or} \quad -2 = (-1) \times 2$$

We do not have to try the negative factors of 3, namely, $3 = (-1) \times (-3)$ because, if there is a factorization, it will follow from the positive factors of 3.

Since $[3 \times 1] + [(-2) \times 1] = 1$, the coefficient of the middle term, we try the possibility $(3x - 2)(x + 1)$, which yields

$$(3x - 2)(x + 1) = 3x^2 + x - 2$$

Therefore, the correct factorization of the above trinomial is

$$3x^2 + x - 2 = (3x - 2)(x + 1)$$

Notice that one way to determine if binomial factors $sx + p$ and $rx + q$ of $3x^2 + x - 2$ exist is to list all the possible combinations of factors that satisfy the requirements $rs = 3$ and $pq = -2$ and to check these possibilities to see if one does satisfy the requirement $sq + pr = 1$. Thus, to factor $3x^2 + x - 2$, we test the following possibilities:

Possible combinations	Products
$(3x + 1)(x - 2)$	$3x^2 - 5x - 2$
$(3x - 1)(x + 2)$	$3x^2 + 5x - 2$
$(3x + 2)(x - 1)$	$3x^2 - x - 2$
$(3x - 2)(x + 1)$	$3x^2 + x - 2$

From the possibilities above we see that the factorization of $3x^2 + x - 2$ is $(3x - 2)(x + 1)$.

With sufficient practice, one is usually able to factor a quadratic trinomial either by using trial and error or by checking possible combinations of factors mentally without actually listing all possibilities. The following examples illustrate these procedures.

EXAMPLES

Factor the following trinomials.

1    $x^2 + 6x + 8$

SOLUTION.   To factor the trinomial, the clues are provided as follows:
a)   Since the coefficients of the expression are all positive integers, the terms whose product is x^2 are x and x, so that

$$x^2 + 6x + 8 = (x + \underline{\quad})(x + \underline{\quad})$$

b)   The product of the constant terms is 8 and their sum is 6. Therefore, both constant terms must be positive. The only factorizations of 8 into positive integers are 4×2 and 8×1. Since $4 + 2 = 6$ gives the coefficient of the middle term, we try the possibility $(x + 2)(x + 4)$, which yields $(x + 2)(x + 4) = x^2 + 6x + 8$.
c)   Therefore, the correct factorization is

$$x^2 + 6x + 8 = (x + 2)(x + 4)$$

2    $x^2 - 9x + 20$

SOLUTION

a)   The terms whose product is x^2 are x and x, so that

$$x^2 - 9x + 20 = (x + \underline{\quad})(x + \underline{\quad})$$

b)   The product of the constant terms is 20 and their sum is -9. Therefore, both constant terms must be negative. The only factorizations of 20 into negative integers are

$$(-1) \times (-20) \quad \text{and} \quad (-2) \times (-10) \quad \text{and} \quad (-4) \times (-5)$$

and $(-4) + (-5) = -9$, which is the coefficient of the middle term.
c)   Therefore, we try the possibility $(x - 5)(x - 4)$, which yields $(x - 5)(x - 4) = x^2 - 9x + 20$. The correct factorization is

$$x^2 - 9x + 20 = (x - 5)(x - 4)$$

3    $2x^2 + 5x - 3$

SOLUTION

a)   The terms whose product is $2x^2$ are x and $2x$, so that

$$2x^2 + 5x - 3 = (x + \underline{\quad})(2x + \underline{\quad})$$

b) The product of the constant terms is -3. Therefore, the terms must have opposite signs. The only factorizations of -3 are

$$(-3) \times (1) \quad \text{and} \quad (-1) \times (3)$$

The coefficient of the middle term is found by trying different possible combinations of the factors of 2 and -3. By trial and error, $[1 \times (-1)] + [2 \times 3] = 5$ gives the coefficient of the middle term. Therefore, we try the possibility $(x + 3)(2x - 1)$, which yields

$$(x + 3)(2x - 1) = 2x^2 + 5x - 3$$

c) The correct factorization is

$$2x^2 + 5x - 3 = (x + 3)(2x - 1)$$

4    $6x^2 - 11x - 10$

SOLUTION

a) The terms whose product is $6x^2$ are $6x$ and x, or $2x$ and $3x$, so that

$$6x^2 - 11x - 10 = (6x + \underline{\hspace{1cm}})(x + \underline{\hspace{1cm}})$$

or

$$6x^2 - 11x - 10 = (3x + \underline{\hspace{1cm}})(2x + \underline{\hspace{1cm}})$$

b) The product of the constant terms is -10 and their sum is -11. Therefore, both terms must have opposite signs. The only factorizations of -10 are

$$(-10) \times (1) \quad \text{and} \quad (-1) \times (10) \quad \text{and}$$
$$(-2) \times (5) \quad \text{and} \quad (-5) \times 2$$

Trying different possible combinations of the factors of 6 and -10, and using trial and error, we have $[3 \times (-5)] + [2 \times 2] = -11$, which gives the coefficient of the middle term. Therefore, we try $(3x + 2)(2x - 5)$, which yields

$$(3x + 2)(2x - 5) = 6x^2 - 11x - 10$$

c) The correct factorization is

$$6x^2 - 11x - 10 = (3x + 2)(2x - 5)$$

PROBLEM SET 2.7

In problems 1–40, factor the quadratic trinomial expressions.

1 $x^2 + 4x + 3$ $(x+1)(x+3)$ **2** $x^2 + 5x + 6$

3 $x^2 - 3x + 2$ **4** $x^2 - 3x - 4$

5 $x^2 + 15x + 56$ **6** $x^2 + 3x - 10$

7 $x^2 - 2x - 15$ **8** $y^2 - 3yx - 28x^2$

9 $x^2 - 16x + 63$ **10** $x^2 - 3x - 40$ $(x-8)(x+5)$

11 $x^2 + 11xy + 30y^2$ **12** $x^2 - 9xy - 10y^2$

13 $x^2 - 7x - 18$ **14** $x^2 - 17x + 30$

15 $x^2 + 2xy - 120y^2$ **16** $x^2 - 13xy + 30y^2$

17 $12 - x^2 - 4x$ **18** $40 - 3x - x^2$

19 $-5x + 36 - x^2$ **20** $16 - x^2 - 6x$ $(x-8)(-x+2)$

21 $2x^2 + 7x + 3$ **22** $2x^2 + x - 6$

23 $3x^2 + 5x - 2$ **24** $2y^2 + 9y - 5$

25 $5x^2 - 11x + 2$ → **26** $4x^2 - 35xy - 9y^2$

27 $3x^2 + 7xy + 2y^2$ **28** $10x^2 - 19x + 6$

29 $6x^2 + 13x + 6$ **30** $6y^2 - y - 7$ $(6y-7)(y+1)$

31 $6x^2 + 5xy - 6y^2$ **32** $6x^2 - 7xy - 5y^2$

33 $12x^2 + 17x - 5$ **34** $6x^2 - 7xy - 3y^2$

35 $56x^2 - 83x + 30$ **36** $42x^2 + x - 30$

37 $12 - 2x^2 - 5x$ →**38** $18x^2 + 101x + 90$

39 $6xy + 5y^2 - 8x^2$ **40** $24x^2 - 67xy + 8y^2$

In problems 41–50, use common factors and the factoring of quadratic trinomials to factor the polynomials.

41 $5x^3 - 55x^2 + 140x$ **42** $x^2yz^2 - xyz^2 - 12yz^2$

43 $128st^3 - 32s^2t^2 + 2s^3t$ **44** $bx^2c + 7bcx + 12bc$

45 $x^2y^2 + 10xy^2 + 21y^2$ **46** $a^2x^2z^2 + 5a^2xz^2 - 14a^2z^2$

47 $4m^2n^2 + 24m^2n - 28m^2$ **48** $7hkx^2 + 21hkx + 14hk$

49 $wx^2y - 9wxy + 14wy$ **50** $pq^2x^2y - 2pq^2xy - 15pq^2y$

2.8 Division of Polynomials

Recall from Section 2.1 the division property of exponents: that for $a \neq 0$, $a^m/a^n = a^{m-n}$ if m is greater than n; $a^m/a^n = 1$ if $m = n$; and $a^m/a^n = 1/a^{n-m}$ if m is less than n. To divide a monomial expression by another monomial expression, we need only to apply this property. For example, to divide $4x^5y^3$ by $2x^2y$, we have

$$\frac{4x^5y^3}{2x^2y} = 2x^{5-2}y^{3-1} = 2x^3y^2$$

Similarly, we divide x^3yz^2 by xyz^5 as follows:

$$\frac{x^3yz^2}{xyz^5} = (x^{3-1})(1)\left(\frac{1}{z^{5-2}}\right) = \frac{x^2}{z^3}$$

However, to divide a polynomial by a monomial, we use the definition of division $[a/b = a(1/b)$, for $b \neq 0]$ and the distributive property. For instance, to divide $x^3 + 2x^2 + x$ by x, we have

$$\frac{x^3 + 2x^2 + x}{x} = (x^3 + 2x^2 + x)\left(\frac{1}{x}\right) \qquad \text{(definition of division)}$$

$$= x^3\left(\frac{1}{x}\right) + 2x^2\left(\frac{1}{x}\right) + x\left(\frac{1}{x}\right) \qquad \text{(distributive property)}$$

$$= \frac{x^3}{x} + \frac{2x^2}{x} + \frac{x}{x} \qquad \text{(definition of division)}$$

$$= x^2 + 2x + 1$$

Notice in the example above that the division of a polynomial by a monomial is accomplished by dividing each term of the polynomial by the monomial divisor. Additional examples will be given to illustrate this procedure.

EXAMPLES

1 Divide $4x^3y^2 - 8x^2y^3$ by $2x^2y$.

SOLUTION

$$\frac{4x^3y^2 - 8x^2y^3}{2x^2y} = \frac{4x^3y^2}{2x^2y} - \frac{8x^2y^3}{2x^2y}$$

$$= 2xy - 4y^2$$

2   Divide $x^3 + 2x^2 - 3x + 7$ by x^2.

SOLUTION

$$\frac{x^3 + 2x^2 - 3x + 7}{x^2} = \frac{x^3}{x^2} + \frac{2x^2}{x^2} - \frac{3x}{x^2} + \frac{7}{x^2}$$

$$= x + 2 - \frac{3}{x} + \frac{7}{x^2}$$

Note that in Example 2 the result is not a polynomial because of the nature of the two terms $-3/x$ and $7/x^2$. This confirms the fact that the quotient of two polynomials may not be a polynomial.

One method for dividing polynomials in which the divisor is not a monomial is a method similar to the long-division method of arithmetic. For example, $6{,}741 \div 21$ is performed by long division as follows:

$$
\begin{array}{r}
321 \\
21\,\overline{)6741} \\
\underline{63} \qquad [= 21 \times 3] \\
44 \\
\underline{42} \qquad [= 21 \times 2] \\
21 \\
\underline{21} \qquad [= 21 \times 1] \\
0
\end{array}
$$

We can also perform this division by changing the dividend 6,741 and the divisor 21 to their expanded forms first, and then dividing as follows:

$$
\begin{array}{r}
3 \cdot 10^2 + 2 \cdot 10\ + 1 \\
2 \cdot 10 + 1\,\overline{)6 \cdot 10^3 + 7 \cdot 10^2 + 4 \cdot 10 + 1} \\
\underline{6 \cdot 10^3 + 3 \cdot 10^2} \qquad [= (2 \cdot 10 + 1)(3 \cdot 10^2)] \\
4 \cdot 10^2 + 4 \cdot 10 \\
\underline{4 \cdot 10^2 + 2 \cdot 10} \qquad [= (2 \cdot 10 + 1)(2 \cdot 10)] \\
2 \cdot 10 + 1 \\
\underline{2 \cdot 10 + 1} \qquad [= (2 \cdot 10 + 1)(1)] \\
0
\end{array}
$$

Note that the expanded form of the numbers in the example above represents polynomials, with each term having a known base of 10. Let us change this problem to a similar one by changing the known base of 10 to one of base x. We then have the long division of $6x^3 + 7x^2 + 4x + 1$ by $2x + 1$:

$$
\begin{array}{r}
3x^2 + 2x + 1 \\
2x + 1 \overline{)\, 6x^3 + 7x^2 + 4x + 1} \\
6x^3 + 3x^2 \\
\hline
4x^2 + 4x \\
4x^2 + 2x \\
\hline
2x + 1 \\
2x + 1 \\
\hline
0
\end{array}
\qquad
\begin{array}{l}
[= (2x + 1)(3x^2)] \\
\\
\\
[= (2x + 1)(2x)] \\
\\
[= (2x + 1)(1)]
\end{array}
$$

In general, to divide two polynomials in one variable, we arrange both polynomials in descending powers of the variable and follow the procedures above.

EXAMPLES

3   Divide $x^4 + 3x^3 + 3x^2 - 2x - 3$ by $x + 1$.

SOLUTION

$$
\begin{array}{r}
x^3 + 2x^2 + x - 3 \\
x + 1 \overline{)\, x^4 + 3x^3 + 3x^2 - 2x - 3} \\
x^4 + x^3 \\
\hline
2x^3 + 3x^2 \\
2x^3 + 2x^2 \\
\hline
x^2 - 2x \\
x^2 + x \\
\hline
-3x - 3 \\
-3x - 3 \\
\hline
0
\end{array}
$$

4   Divide $x^3 + 3x - 2x^2 + x^4 - 1$ by $1 + x^2 - x$.

SOLUTION.   Arranging both polynomials in descending powers of x, we have

$$(x^4 + x^3 - 2x^2 + 3x - 1) \div (x^2 - x + 1)$$

so that

$$
\begin{array}{r}
x^2 + 2x - 1 \\
x^2 - x + 1 \overline{)\, x^4 + x^3 - 2x^2 + 3x - 1} \\
x^4 - x^3 + x^2 \\
\hline
2x^3 - 3x^2 + 3x \\
2x^3 - 2x^2 + 2x \\
\hline
- x^2 + x - 1 \\
- x^2 + x - 1 \\
\hline
0
\end{array}
$$

We can apply this method to division of polynomials involving more than one variable by first arranging the dividend and divisor in descending powers of one of the variables, and then dividing as illustrated above.

EXAMPLE

5   Divide $x^4 - y^4 + 3xy^3 - 3x^3y$ by $x + y$.

SOLUTION.   After arranging in descending powers of x, we have

$$(x^4 - 3x^3y + 0x^2y^2 + 3xy^3 - y^4) \div (x + y)$$

so that

$$
\begin{array}{r}
x^3 - 4x^2y + 4xy^2 - y^3 \\
x + y \overline{)\,x^4 - 3x^3y + 0x^2y^2 + 3xy^3 - y^4} \\
\underline{x^4 + x^3y} \\
-4x^3y + 0x^2y^2 \\
\underline{-4x^3y - 4x^2y^2} \\
4x^2y^2 + 3xy^3 \\
\underline{4x^2y^2 + 4xy^3} \\
- xy^3 - y^4 \\
\underline{- xy^3 - y^4} \\
0
\end{array}
$$

Recall from arithmetic that for two numbers a and b

$$\frac{a}{b} = c \qquad \text{if } a = bc \qquad \text{for } b \neq 0$$

For instance, $\frac{12}{3} = 4$ since $12 = 3 \times 4$. If the denominator is not an exact divisor of the numerator, the rule is

$$\frac{a}{b} = q + \frac{r}{b} \quad \text{if } a = bq + r$$

For example, $\frac{23}{4} = 5 + \frac{3}{4}$ since $23 = 4 \times 5 + 3$.

The same rules apply to the division of polynomials. If $P, D, Q,$ and R represent the dividend, divisor, quotient, and remainder, respectively, where $P, D, Q,$ and R are polynomial expressions, then

$$\frac{\text{dividend}}{\text{divisor}} = \text{quotient} + \frac{\text{remainder}}{\text{divisor}}$$

That is,

$$\frac{P}{D} = Q + \frac{R}{D} \quad \text{or} \quad P = QD + R$$

We can use this rule for checking polynomial division in the same way that we checked division in arithmetic.

EXAMPLE

6   Find the quotient $(x^3 + 5x^2 + 6x + 3) \div (x^2 + x + 1)$ and check the result.

SOLUTION

$$
\begin{array}{r}
x + 4 \\
x^2 + x + 1 \,\overline{\smash{\big)}\, x^3 + 5x^2 + 6x + 3} \\
\underline{x^3 + x^2 + x} \\
4x^2 + 5x + 3 \\
\underline{4x^2 + 4x + 4} \\
x - 1
\end{array}
$$

Here P represents the polynomial expression $x^3 + 5x^2 + 6x + 3$, D represents the polynomial $x^2 + x + 1$, Q represents the polynomial $x + 4$, and R represents the polynomial $x - 1$. Thus,

$$\frac{x^3 + 5x^2 + 6x + 3}{x^2 + x + 1} = x + 4 + \frac{x - 1}{x^2 + x + 1}$$

The result is checked as follows:

$$x^3 + 5x^2 + 6x + 3 \overset{?}{=} (x^2 + x + 1)(x + 4) + (x - 1)$$
$$(x^2 + x + 1)(x + 4) + x - 1 = (x^3 + 5x^2 + 5x + 4) + (x - 1)$$
$$= x^3 + 5x^2 + 6x + 3$$

which confirms that the solution is correct.

PROBLEM SET 2.8

In problems 1–11, divide as noted.

1   $6x^5$ by $2x^2$

2   $10y^7$ by $3y^3$

3   $12x^6y^7$ by $-4x^4y^9$

4   $-24x^5y^3$ by $-8x^2y$

5   $14x^3yz^4$ by $7x^5yz^2$

6   $30xy^3z^5$ by $-15xy^7z^8$

7   $9xy^2 - 6x^2$ by $3x$

8   $15x^2y^5 - 25x^4y^3$ by $-5xy^2$

9 $4x^2y^3 - 16xy^3 + 4xy$ by $2xy$ **10** $4(x + y)^4 + 12(x + y)^3$
$-8(x + y)$ by $2(x + y)$

11 $6x^3y^5 + 3x^2y^4 - 12xy + 9$ by $3x^2y^3$

12   Use the definition that $a/b = c$ if $a = bc$, for $b \neq 0$, to check the divisions in problems 1–11.

In problems 13–28, divide by the long-division method.

13 $x^2 - 7x + 10$ by $x - 5$

14 $2y^2 - 5y - 6$ by $2y - 1$

15 $4 - 13x + 3x^2$ by $x - 4$

16 $1 - 4x + 4x^2$ by $2x - 1$

17 $3x^3 - 5x^2 + 2x + 2x^4 - 1$ by $1 + x$

18 $5x^5 - 2x^3 + 1$ by $x - 1$

19 $3x^2y - 2xy^2 - 8y^3 + x^3$ by $x + 2y$

20 $2x^3 + xy^2 + x^2y + 4y^3$ by $x + y$

21 $x^2y - 6x^3 - 6y^3 - 12xy^2$ by $2x - 3y$

22 $5x^3 - 2x^2 + 3x - 4$ by $x^2 - 2x + 1$

23 $x^3 - y^3$ by $x - y$

24 $x^5 + y^5$ by $x + y$

25 $x^3 - 3x^2 + 16x + 52$ by $x^2 - 5x + 26$

26 $x^3 + 2x - 4x^2 + 1$ by $x^2 - 1 - 3x$

27 $x^5 + 2x^2 - 24 + 3x^4$ by $x^3 - 2x^2 + x^4 + 6x - 12$

28 $3x^4 + 2x^2 - 28$ by $3x^3 + 6x^2 + 14x + 28$

In problems 29–35, divide as noted and express the result in the form $P = QD + R$. Use this form to check the division.

29 $x^2 - 5x + 8$ by $x - 2$

30 $x^2 + 7x + 10$ by $x + 3$

31 $x^3 + 3x^2 - 5 - 2x$ by $2 + x$

32 $2x^4 + 3x^3 + 2x - 5x^2 - 1$ by $1 + x$

33 $x^3 + 2y^3 - x^2y - xy^2$ by $x + y$

34 $2x^4 + 3x^3y - 5x^2y^2 + 4xy^3 - 7y^4$ by $2x - y$

35 $x^5 - y^5$ by $x - y$

REVIEW PROBLEM SET

In problems 1–16, use the properties of exponents to simplify each expression.

1 $x^3 x^8$

2 $(-x)^3(-x)^2$

3 $x^2 x^7 x^3$

4 $(-y)^2(-y)^3(-y)^7$

5 $(y^3)^8$

6 $(-x^4)^6$

7 $(xy^2)^3$

8 $(-x^2 y^3)^4$

9 $(-3x^3 y)^3$

10 $\left(\dfrac{2x^3}{y^2}\right)^4$

11 $\left(\dfrac{2x}{y^3}\right)^4$

12 $\dfrac{x^5}{x^5}$

13 $\dfrac{x^{12}}{x^9}$

14 $\dfrac{3y^2}{9y^5}$

15 $\dfrac{(3x^2 y^4)^3}{(3x^3 y)^2}$

16 $\dfrac{(-2xy^2 z^4)^2}{(2x^2 yz^3)^3}$

In problems 17–24, determine the degree and numerical coefficients of each polynomial.

17 $4x^2 - 39x + 100$

18 $x^4 - 3x^2 + 6x + 5$

19 $81x^3 - 2x^2 + 25$

20 $x^3 - 7x^2 + 10x - 1$

21 $4x + 36x^2 + 14$

22 $x^2 y^3 - 2xy^3 + y^4$

23 $x^4 - 3$

24 $x^2 yz + 3xy^2 - 7yz$

In problems 25–30, evaluate each polynomial for $x = -2$.

25 $21x^2 + 5$

26 $5x^3 + 2x^2 - 7$

27 $4x^4 - 11x^3 + 19$

28 $x^4 + x^3 + 4$

29 $2x^2 - 3x + 4$

30 $3x^4 - 17$

In problems 31–36, perform the additions or subtractions.

31 $(2x^2 + 3x - 4) + (x^2 - 5x + 7)$

32 $(4x^3 + 3x^2 - 5) + (-5x^2 + 2x + 3)$

33 $(3x^2 + 7x + 8) - (2x^2 + 3x + 2)$

34 $(-x^3 + 2x^2 + 3x - 7) - (-3x^3 + x^2 - 5x - 2)$

35 $(5x^2 - 3x + 2) + (2x^2 + 5x - 7) - (3x^2 - 4x - 1)$

36 $(7x^3 - 3x^2y + y^3) - (5x^3 - 4xy^2 + 2y^3) - (x^3 - y^3)$

In problems 37–48, find the products.

37 $3x^2(2x^3 - 4x)$

38 $-4y^3(-5y^2 - 2y + 1)$

39 $-2xy^3(-3x^2y + 5xy^2 - y)$

40 $3xy^2z^3(7xy - 2x^2z + 5y)$

41 $(x - y)(x^2 - 2xy + y^2)$

42 $(x^2 - x + 1)(2x^2 - 3x + 2)$

43 $(x + 7)(x + 3)$

44 $(3x + 7)(2x + 5)$

45 $(2x - 3)(x + 5)$

46 $(3y - 9x)(4y + 9x)$

47 $(4x - 5y)(x - 3y)$

48 $(7 - 5xy)(4 - 3xy)$

In problems 49–58, use special products to find the products.

49 $(x - 8)(x + 8)$

50 $(x^2 + 4)(x^2 - 4)$

51 $(3x - 2)^2$

52 $(2y + 5z)^2$

53 $(x + 6y)^2$

54 $(2 - y)(4 + 2y + y^2)$

55 $(x + 5)(x^2 - 5x + 25)$

56 $(x^2 - y^2)^2$

57 $(1 + y)^3$

58 $(x^2 - y^2)^3$

In problems 59–64, factor the polynomials by finding common factors.

59 $7x^2y - 21xy^3$

60 $8xy^3 + 16x^2y$

61 $26x^3y^2 + 39x^5y^4 - 52x^2y^3$

62 $25xyz^2 - 50xy^2z^2 + 75x^2y^2z^2$

63 $2y(y + z) - 4x(z + y)$

64 $7y(x - z) + 14y^2(z - x)$

In problems 65–74, use special products to factor the polynomials.

65 $x^2 + 16x + 64$

66 $9x^2 - 24xy + 16y^2$

67 $y^2 - 121$

68 $y^3 - 216$

69 $x^3 + 64$

70 $25y^2 - 81z^2$

71 $y^3 - 3y^2 + 3y - 1$

72 $x^3y^3 + 3x^2y^2 + 3xy + 1$

73 $16x^4 - 81$

74 $x^8 - 256$

In problems 75–80, use the method of grouping to factor the expressions.

75 $3x - 3y + xz - yz$ **76** $x^4 - y^4 - z^4 - 2y^2z^2$

77 $x^2 - a^2 + 2xy - 2ab + y^2 - b^2$ **78** $x^5 + 3x^4 - x - 3$

79 $x^4 + x^2 + 1$ **80** $x^4 - 3x^2 + 9$

In problems 81–90, factor the trinomials.

81 $x^2 + 2xy - 3y^2$ **82** $x^2 + 7x - 18$

83 $x^2 - 5x - 36$ **84** $x^2 - xy - 12y^2$

85 $3x^2 + 17x + 10$ **86** $6x^2 - 29x + 35$

87 $2x^2 - x - 6$ **88** $33x^2 + 14xy - 40y^2$

89 $20x^2 - 31xy + 12y^2$ **90** $68 - 31y - 15y^2$

In problems 91–100, divide as noted.

91 $18x^4$ by $3x^2$

92 $-12x^3y^4$ by $-6xy^3$

93 $32x^5y^3 - 16x^4y^2$ by $4x^3y$

94 $100x^3y^2 - 90x^4y^3 + 70x^5y^4$ by $10x^2y$

95 $5x^2 - 16x + 3$ by $x - 3$

96 $2y^2 + 17y + 21$ by $3 + 2y$

97 $x^3 - 6x^2 + 12x - 8$ by $x^2 - 4x + 4$

98 $x^2 + y^2 + z^2 + 2xy - 2xz - 2yz$ by $x + y - z$

99 $x^{15} + 1$ by $x^5 + 1$

100 $x^6 - y^6$ by $x - y$

CHAPTER 3

Algebra of Fractions

In Chapter 1 we assumed the basic knowledge of the rules for the operations with rational numbers. Here we shall extend this discussion to "rational expressions" which are quotients of polynomials. We will find that the techniques for combining rational expressions are essentially the same as those for rational numbers. From now on we shall use the terms "fraction" or "rational expression" to mean the same thing and shall use them interchangeably.

3.1 Rational Expressions

In Section 1.2 we indicated that a rational number can be represented in the form of a/b, where a and b are integers and $b \neq 0$. The numeral a/b is also called a *fraction* whose numerator is a and whose denominator is b.

If the numerator and the denominator of a fraction are polynomial expressions, the fraction is called a *rational expression*. That is, a rational expression can be expressed in the form P/Q, where P and Q are polynomials and $Q \neq 0$. Examples of rational expressions are $1/x$, $(x + 1)/(5x - 2)$, $5/(t^2 - 7)$, $(3x^2 + 4x - 1)/1$, and $(3x^2 + x)/(5x + 1)$.

In order to find specific values for rational expressions, we use the substitution property. For example, considered the rational expression $x/(x - 2)$:

if $x = 3$ we have $\dfrac{3}{3 - 2} = \dfrac{3}{1} = 3$

if $x = -1$ we have $\dfrac{-1}{-1 - 2} = \dfrac{-1}{-3} = \dfrac{1}{3}$

However, if $x = 2$, we obtain $2/(2 - 2) = 2/0$, which does not exist as a real number. Thus, a rational expression represents a real number for any specific value substituted for the variable except for those values which cause the denominator to take on the value zero. In those cases we say that the rational expression is *not defined*. For instance, consider the rational expression

$$\frac{x^2 - 2x + 1}{x^2 - 3x + 2} = \frac{(x - 1)^2}{(x - 1)(x - 2)}$$

Substituting 1 for x in this expression, we have $\frac{0}{0}$; hence this expression is not defined for $x = 1$. Also, if we substitute $x = 2$, we have $\frac{1}{0}$; hence, the expression is not defined for $x = 2$. We shall assume throughout this text that the variables in any fraction may not be assigned values that will result in a division by zero.

Since much of our work in algebra involves the substitution property, we need a method for determining "equivalent" rational expressions. We say that two rational expressions (or fractions) are *equivalent* if they give equal real numbers for each assignment of values to their variables, except when such an assignment gives an undefined value to either (or both) expressions. We can point out, however, that in deference to long-established convention, the fact that two rational expressions are equivalent will be denoted by writing an *equals sign* between them. Accordingly, we can develop a procedure for determining equivalent rational expressions. Such a procedure is based on the following property.

PROPERTY

The fractions a/b and c/d, with $b \neq 0$ and $d \neq 0$, are equivalent if and only if $ad = bc$. In symbols we have

$$\frac{a}{b} = \frac{c}{d} \qquad \text{if and only if} \qquad ad = bc$$

To prove this property, we first prove that if $a/b = c/d$, then $ad = bc$. Since $a/b = a(1/b)$ and $c/d = c(1/d)$, we have

$$a\left(\frac{1}{b}\right) = c\left(\frac{1}{d}\right)$$

Multiplying both sides by bd, we have

$$(bd)a\left(\frac{1}{b}\right) = (bd)c\left(\frac{1}{d}\right)$$

or

$$db\left(\frac{1}{b}\right)a = (bd)\left(\frac{1}{d}\right)c \qquad \text{(Commutative property)}$$

Using the fact that $b(1/b) = 1$ and $d(1/d) = 1$, we have

$$da = bc \qquad \text{or} \qquad ad = bc$$

To prove that if $ad = bc$, then $a/b = c/d$, we reverse the preceding steps. Hence, the two statements are combined in the statement

$$\frac{a}{b} = \frac{c}{d} \qquad \text{if and only if} \qquad ad = bc$$

EXAMPLE

1   Indicate which pairs of fractions are equivalent.

a) $\dfrac{2}{3}$ and $\dfrac{4}{6}$ b) $\dfrac{-3}{7}$ and $\dfrac{3}{-7}$ c) $\dfrac{4}{5}$ and $\dfrac{5}{9}$

SOLUTION

a)   Since $2 \times 6 = 3 \times 4 = 12$, then $\dfrac{2}{3} = \dfrac{4}{6}$.

b)   Since $(-3) \times (-7) = 7 \times 3 = 21$, then $\dfrac{-3}{7} = \dfrac{3}{-7}$.

c)   Since $4 \times 9 \ne 5 \times 5$, then $\dfrac{4}{5} \ne \dfrac{5}{9}$.

The property of equivalence demonstrated above also holds true for rational expressions. That is, if P/Q and R/S are rational expressions, with $Q \ne 0$ and $S \ne 0$, then $P/Q = R/S$ if and only if $PS = QR$. The following example will illustrate the application of this property.

EXAMPLE

2   Indicate which pairs of rational expressions are equivalent.

a) $\dfrac{3x}{6x^2}$ and $\dfrac{1}{2x}$

b) $\dfrac{x+3}{x^2+5x+6}$ and $\dfrac{1}{x+2}$

c) $\dfrac{5}{x}$ and $\dfrac{3}{x}$

SOLUTION

a)   Since $(3x)(2x) = (6x^2)(1)$, then $\dfrac{3x}{6x^2} = \dfrac{1}{2x}$.

b)   Since $(x+3)(x+2) = (x^2+5x+6)(1)$, then $\dfrac{x+3}{x^2+5x+6} = \dfrac{1}{x+2}$.

c)   Since $5x \ne 3x$, then $\dfrac{5}{x} \ne \dfrac{3}{x}$.

To change a fraction into an equivalent fraction, we use the following property.

PROPERTY   FUNDAMENTAL PRINCIPLE OF FRACTIONS

If a/b is a fraction and k is a real number, where $b \neq 0$ and $k \neq 0$, then

$$\frac{a}{b} = \frac{ak}{bk}$$

The proof of the fundamental principle of fractions follows from the statement

$$\frac{a}{b} = \frac{ak}{bk} \qquad \text{if and only if} \qquad (ak) \cdot b = (bk) \cdot a$$

This can be shown as follows:

$$
\begin{aligned}
(ak)b &= a(kb) &&\text{(associative property of multiplication)} \\
&= (kb)a &&\text{(commutative property of multiplication)} \\
&= (bk)a &&\text{(commutative property of multiplication)}
\end{aligned}
$$

EXAMPLES

Use the fundamental principle of fractions to find equivalent rational expressions having the indicated denominators. That is, find the missing expression for each of the following equivalent fractions.

3    $\dfrac{3}{4} = \dfrac{?}{16}$

SOLUTION

$$\frac{3}{4} = \frac{3 \times 4}{4 \times 4} = \frac{12}{16}$$

so that the unknown number is 12.

4    $\dfrac{3a}{b} = \dfrac{?}{ba}$

SOLUTION

$$\frac{3a}{b} = \frac{(3a)(a)}{b(a)} = \frac{3a^2}{ba}$$

so that the unknown expression is $3a^2$.

5    $\dfrac{x+3}{x-2} = \dfrac{?}{x^2-4}$

SOLUTION

$$\frac{x+3}{x-2} = \frac{(x+3)(x+2)}{(x-2)(x+2)} = \frac{x^2+5x+6}{x^2-4}$$

so that the unknown expression is $x^2 + 5x + 6$.

If the numerator and denominator of a fraction are integers and the denominator is not zero, the fraction is said to be *in lowest terms* when the numerator and the denominator have no common factor other than 1. For example, $\frac{2}{5}$, $\frac{5}{7}$, and $\frac{7}{19}$ are in lowest terms, whereas $\frac{4}{6}$ and $\frac{3}{9}$ are not, since each fraction has a common factor in its numerator and denominator. This notion can be extended to rational expressions. For instance, $x/(x+1)$, $x/(x^2+5)$, and $(x^2+1)/(x^3+5x+7)$ are in lowest terms, since the numerator and the denominator of each have no common factors, whereas $3xy/x^2z$ is not, since x is a common factor in both the numerator and the denominator of the fraction. The procedure of expressing a given fraction as an equivalent fraction in lowest terms is called *reducing the fraction to lowest terms*. This is illustrated as follows.

EXAMPLES

Reduce each of the following fractions to lowest terms.

6    $\dfrac{36}{44}$

SOLUTION

$$\frac{36}{44} = \frac{9 \times 4}{11 \times 4} = \frac{9}{11}$$

7    $\dfrac{28x^3y}{21xy^2}$

SOLUTION

$$\frac{28x^3y}{21xy^2} = \frac{4x^2(7xy)}{3y(7xy)} = \frac{4x^2}{3y}$$

8    $\dfrac{30x^2y^2}{25x^2y - 20xy^2}$

SOLUTION

$$\frac{30x^2y^2}{25x^2y - 20xy^2} = \frac{5xy(6xy)}{5xy(5x - 4y)} = \frac{6xy}{5x - 4y}$$

9    $\dfrac{x^2 - y^2}{x^2 + xy}$

SOLUTION

$$\frac{x^2 - y^2}{x^2 + xy} = \frac{(x - y)(x + y)}{x(x + y)} = \frac{x - y}{x}$$

10    $\dfrac{x^2 + 4x - 21}{x^2 - x - 6}$

SOLUTION

$$\frac{x^2 + 4x - 21}{x^2 - x - 6} = \frac{(x + 7)(x - 3)}{(x + 2)(x - 3)} = \frac{x + 7}{x + 2}$$

PROBLEM SET 3.1

In problems 1–10, determine all values for which the rational expressions are not defined.

1    $\dfrac{3}{x}$

2    $\dfrac{3x - 2}{5x^2}$

3    $\dfrac{3x + 5}{x^2}$

4    $\dfrac{x^2 + 5x + 1}{x(x - 4)}$

5    $\dfrac{x + 1}{4x - 12}$

6    $\dfrac{x^2 - 5}{x(x - 1)(x + 1)}$

7    $\dfrac{x^2 + 5x + 2}{(x - 3)(x + 3)}$

8    $\dfrac{x}{(x - 7)(x + 6)}$

9    $\dfrac{x^3 + 5x^2 + 7}{(x - 1)(x + 2)(x - 3)}$

10    $\dfrac{7x^5}{(5 + x)(3 - x)(1 - x)}$

In problems 11–22, use the property $a/b = c/d$ if and only if $ad = bc$ to indicate which pairs of fractions are equivalent.

11    $\dfrac{14}{3}, \dfrac{2}{21}$

12    $\dfrac{-5}{13}, \dfrac{5}{-13}$

13 $\dfrac{-2}{9}, \dfrac{2}{-9}$

14 $\dfrac{-3}{-29}, \dfrac{6}{58}$

15 $\dfrac{3}{x}, \dfrac{18}{6x}$

16 $\dfrac{a+b}{x}, \dfrac{7a+b}{7x}$

17 $\dfrac{a+b}{1}, \dfrac{ca+cb}{cb}$

18 $\dfrac{x^2+2x+1}{x+1}, \dfrac{3x+3}{3}$

19 $\dfrac{x^2-9}{x-3}, x+3$

20 $\dfrac{x+y}{1}, \dfrac{x^3+y^3}{x^2-xy+y^2}$

21 $\dfrac{x+2}{x-2}, \dfrac{-2-x}{2-x}$

22 $\dfrac{x^4-y^4}{x^2-y^2}, \dfrac{x^2+y^2}{1}$

In problems 23–34, use the fundamental principle of fractions to determine the missing expressions.

23 $\dfrac{15}{18} = \dfrac{?}{36}$

24 $\dfrac{65}{26} = \dfrac{130}{?}$

25 $\dfrac{5x^3y}{7xy^2} = \dfrac{?}{21x^4y^3}$

26 $\dfrac{5^2}{t} = \dfrac{?}{rst^2}$

27 $\dfrac{x-y}{3} = \dfrac{5(x-y)^2}{?}$

28 $\dfrac{7a}{9a^2(a-b)} = \dfrac{7a^4(a-b)}{?}$

29 $\dfrac{11}{3x(x-1)(x+5)} = \dfrac{?}{6x^2(x-1)^2(x+5)^2}$

⇢ 30 $\dfrac{x^2-y^2}{x^2+xy} = \dfrac{?}{(x^2-y^2)x}$

31 $\dfrac{x+3}{x^2-9} = \dfrac{x^2+3x}{?}$

32 $\dfrac{x}{x^2+3x+9} = \dfrac{?}{x^3-27}$

33 $\dfrac{8}{2x^2-7x+6} = \dfrac{?}{(x^2-4)(2x-3)}$

34 $\dfrac{xy}{x^2-xy+y^2} = \dfrac{?}{x^3+y^3}$

In problems 35–66, use the fundamental principle of fractions to reduce each fraction to an equivalent fraction in its lowest terms.

35 $\dfrac{9}{21}$

36 $\dfrac{15}{25}$

37 $\dfrac{26}{91}$

38 $\dfrac{-35}{56}$

39 $\dfrac{x^2y}{xy^3}$

40 $\dfrac{-a^2bc}{-ab^2c}$

41 $\dfrac{3x^3y(a-b)}{15x^4y^2(a-b)}$

42 $\dfrac{2x(-y)(-z)^2}{6x^3y^3(-z)^3}$

43 $\dfrac{15x^2y^5c^7}{45x^3y^3c^6}$

44 $\dfrac{9xy}{24x^3y^2-36xy}$

45 $\dfrac{x^2+x}{x^2-x}$

46 $\dfrac{x^3-x}{x^2+2x-3}$

47 $\dfrac{x^2+2x-3}{x^2+5x+6}$

48 $\dfrac{x^2+4x}{x^2+6x+8}$

49 $\dfrac{4x^2-9}{6x^2-9x}$

50 $\dfrac{x^2-1}{x^3-1}$

51 $\dfrac{x^2+x-12}{x^2+4x-21}$

52 $\dfrac{x+x^2-y-xy}{x^2-2xy+y^2}$

53 $\dfrac{x^2-9x+18}{3x^2-5x-12}$

54 $\dfrac{x^2-4+2y-xy}{2xy-4y-x^2+4}$

55 $\dfrac{x^2+xy-2y^2}{(x+2y)(x-y)^2}$

56 $\dfrac{7x^2-5xy}{49x^3-25xy^2}$

57 $\dfrac{x^2-y^2}{y^2+xy-2x^2}$

58 $\dfrac{x^2-8x-9}{x^2-14x+45}$

59 $\dfrac{x^2-4x-32}{x^2-10x+16}$

60 $\dfrac{x^3+1}{x^3+x^2}$

61 $\dfrac{3x^2-12x+12}{24-6x^2}$

62 $\dfrac{a^2+ac-ab-bc}{a^2-b^2}$

63 $\dfrac{6x^4-48x}{5x^5-80x}$

64 $\dfrac{ab-bc+ad-dc}{ab+bc+ad+dc}$

65 $\dfrac{xz+xw-yz-yw}{xy+xz-y^2-yz}$

66 $\dfrac{x^3-2x^2+5x-10}{3x^5+15x^3-x^2-5}$

In problems 67 and 68, use the fundamental principle of fractions to show that each statement is true.

67 $\dfrac{a}{b}=\dfrac{-a}{-b}$

68 $\dfrac{-a}{b}=\dfrac{a}{-b}$

3.2 Multiplication and Division of Fractions

Multiplying and dividing fractions is really simpler than adding and subtracting them, so we shall study these operations first.

Multiplication of Fractions

We recall from arithmetic that when we multiply two fractions such as $\frac{2}{5}$ and $\frac{3}{7}$, we get

$$\frac{2}{5} \times \frac{3}{7} = \frac{2 \times 3}{5 \times 7} = \frac{6}{35}$$

That is, if a/b and c/d are fractions, with $b \neq 0$ and $d \neq 0$, then $(a/b) \cdot (c/d) = ac/bd$.

The rule for multiplying rational expressions is the same as that for multiplying rational numbers. That is, if P/Q and R/S are rational expressions, with $Q \neq 0$ and $S \neq 0$, then

$$\frac{P}{Q} \cdot \frac{R}{S} = \frac{P \cdot R}{Q \cdot S}$$

To state this rule in words: To multiply two rational expressions, we obtain a rational expression whose numerator is the product of the numerators and whose denominator is the product of the denominators. The following examples will illustrate this rule.

EXAMPLES

Find the following products and reduce them to lowest terms.

1 $\dfrac{7}{8} \times \dfrac{5}{6}$

SOLUTION

$$\frac{7}{8} \times \frac{5}{6} = \frac{7 \times 5}{8 \times 6} = \frac{35}{48}$$

2 $\dfrac{a^3 b^2}{a^2 b} \cdot \dfrac{ab^3}{a^3 b^2}$

SOLUTION

$$\frac{a^3 b^2}{a^2 b} \cdot \frac{ab^3}{a^3 b^2} = \frac{(a^3 b^2) \cdot (ab^3)}{(a^2 b) \cdot (a^3 b^2)} = \frac{a^4 b^5}{a^5 b^3} = \frac{b^2}{a}$$

3   $\dfrac{x}{x-1} \cdot \dfrac{x^2-1}{x^3}$

SOLUTION

$$\frac{x}{x-1} \cdot \frac{x^2-1}{x^3} = \frac{x(x^2-1)}{x^3(x-1)} = \frac{x(x+1)(x-1)}{x \cdot x^2(x-1)} = \frac{x+1}{x^2}$$

4   $\dfrac{x^2+xy-2y^2}{xy} \cdot \dfrac{5xy^2}{x+2y} \cdot \dfrac{1}{5(x-y)}$

SOLUTION

$$\frac{x^2+xy-2y^2}{xy} \cdot \frac{5xy^2}{x+2y} \cdot \frac{1}{5(x-y)} = \frac{(x^2+xy-2y^2)(5xy^2)(1)}{(xy)(x+2y)(5)(x-y)}$$

$$= \frac{5xy(x+2y)(x-y)y}{5xy(x+2y)(x-y)} = y$$

5   $\dfrac{2x-2}{x^2+2x} \cdot \dfrac{x^2+4x+4}{2x^2+2x-4}$

SOLUTION

$$\frac{2x-2}{x^2+2x} \cdot \frac{x^2+4x+4}{2x^2+2x-4} = \frac{(2x-2)(x^2+4x+4)}{(x^2+2x)(2x^2+2x-4)}$$

$$= \frac{2(x-1)(x+2)^2}{x(x+2)(2)(x+2)(x-1)} = \frac{1}{x}$$

Division of Fractions

Recall from arithmetic that when we divide fractions such as $\frac{3}{4}$ by $\frac{2}{3}$, we have

$$\frac{3}{4} \div \frac{2}{3} = \frac{3}{4} \times \frac{3}{2} = \frac{9}{8}$$

That is, if a/b and c/d are fractions, with $b \neq 0$ and $c/d \neq 0$, then

$$\frac{a}{b} \div \frac{c}{d} = \frac{a}{b} \cdot \frac{d}{c} = \frac{ad}{bc}$$

This rule, which applies to rational expressions, is stated as follows:

If P, Q, R, and S are polynomials with $Q \neq 0$ and $R/S \neq 0$, then

$$\frac{P}{Q} \div \frac{R}{S} = \frac{P}{Q} \cdot \frac{S}{R} = \frac{PS}{QR}$$

This rule is stated in words as follows: To divide one fraction by another, multiply the first fraction by the reciprocal of the second fraction. To prove this rule, let P/Q and R/S be rational expressions with $Q \neq 0$ and $R/S \neq 0$. Suppose that $x = P/Q$ and $y = R/S$. Then x and y are rational expressions and $x \div y = x \cdot 1/y$, so that

$$\frac{P}{Q} \div \frac{R}{S} = x \div y = x \cdot \frac{1}{y} = \frac{P}{Q} \cdot \frac{1}{\frac{R}{S}} \qquad \text{(Substitution)}$$

$$= \frac{P}{Q} \cdot \frac{S}{R} \qquad \text{(Multiplicative inverse)}$$

$$= \frac{PS}{QR}$$

EXAMPLES

Find the following quotients and reduce them to lowest terms.

6    $\dfrac{34}{57} \div \dfrac{51}{95}$

SOLUTION

$$\frac{34}{57} \div \frac{51}{95} = \frac{34}{57} \times \frac{95}{51} = \frac{34 \times 95}{57 \times 51} = \frac{2 \times 17 \times 5 \times 19}{3 \times 19 \times 3 \times 17} = \frac{10}{9}$$

7    $\dfrac{3a^2}{5b} \div \dfrac{2a^3}{6b^3}$

SOLUTION

$$\frac{3a^2}{5b} \div \frac{2a^3}{6b^3} = \frac{3a^2}{5b} \cdot \frac{6b^3}{2a^3} = \frac{(3a^2)(6b^3)}{(5b)(2a^3)} = \frac{18a^2b^3}{10a^3b} = \frac{9b^2}{5a}$$

8    $\dfrac{a+b}{c} \div \dfrac{(a+b)^2}{c^4}$

SOLUTION

$$\frac{a+b}{c} \div \frac{(a+b)^2}{c^4} = \frac{a+b}{c} \cdot \frac{c^4}{(a+b)^2} = \frac{(a+b)c^4}{c(a+b)^2} = \frac{c^3}{a+b}$$

9   $\dfrac{x^2 + 5x + 6}{x^2 - 4} \div \dfrac{x^2 + 4x + 4}{x^2 - 4x + 4}$

SOLUTION

$$\frac{x^2 + 5x + 6}{x^2 - 4} \div \frac{x^2 + 4x + 4}{x^2 - 4x + 4} = \frac{x^2 + 5x + 6}{x^2 - 4} \cdot \frac{x^2 - 4x + 4}{x^2 + 4x + 4}$$

$$= \frac{(x^2 + 5x + 6)(x^2 - 4x + 4)}{(x^2 - 4)(x^2 + 4x + 4)}$$

$$= \frac{(x + 2)(x + 3)(x - 2)(x - 2)}{(x - 2)(x + 2)(x + 2)(x + 2)}$$

$$= \frac{(x + 3)(x - 2)}{(x + 2)(x + 2)}$$

$$= \frac{x^2 + x - 6}{x^2 + 4x + 4}$$

10   $\dfrac{5x + 10y}{x - y} \div \dfrac{x^2 + 2xy}{x^2 - y^2}$

SOLUTION

$$\frac{5x + 10y}{x - y} \div \frac{x^2 + 2xy}{x^2 - y^2} = \frac{5x + 10y}{x - y} \cdot \frac{x^2 - y^2}{x^2 + 2xy}$$

$$= \frac{5(x + 2y)}{x - y} \cdot \frac{(x - y)(x + y)}{x(x + 2y)}$$

$$= \frac{5(x + 2y)(x - y)(x + y)}{(x - y)(x)(x + 2y)}$$

$$= \frac{5(x + y)}{x}$$

11   $\dfrac{x^2 - x}{y^2 - y} \cdot \dfrac{y^2x - yx}{x - 1} \div \dfrac{x^2}{x - 1}$

SOLUTION

$$\frac{x^2 - x}{y^2 - y} \cdot \frac{y^2x - yx}{x - 1} \div \frac{x^2}{x - 1} = \frac{x^2 - x}{y^2 - y} \cdot \frac{y^2x - yx}{x - 1} \cdot \frac{x - 1}{x^2}$$

$$= \frac{x(x - 1)}{y(y - 1)} \cdot \frac{xy(y - 1)}{x - 1} \cdot \frac{x - 1}{x^2}$$

$$= x - 1$$

12   $\dfrac{x^2 - y^2}{3x + 3y} \div \left[\dfrac{2xy}{9x - 9y} \cdot \dfrac{(x - y)^2}{x^2y^2} \right]$

SOLUTION. In working this example, it is necessary to perform the operation inside the bracket first, so that

$$\frac{x^2 - y^2}{3x + 3y} \div \left[\frac{2xy}{9x - 9y} \cdot \frac{(x - y)^2}{x^2 y^2} \right] = \frac{x^2 - y^2}{3x + 3y} \div \frac{2xy(x - y)^2}{(9x - 9y)x^2 y^2}$$

$$= \frac{x^2 - y^2}{3x + 3y} \cdot \frac{(9x - 9y)x^2 y^2}{2xy(x - y)^2}$$

$$= \frac{(x - y)(x + y)}{3(x + y)} \cdot \frac{9(x - y)x^2 y^2}{2xy(x - y)^2}$$

$$= \frac{3xy}{2}$$

PROBLEM SET 3.2

In problems 1–22, find the products and reduce them to lowest terms.

1 $\dfrac{7}{13} \times \dfrac{26}{51}$

2 $\dfrac{-10}{17} \times \dfrac{-95}{25}$

3 $\dfrac{-7x^2}{8y} \cdot \dfrac{2y^2}{5x^3}$

4 $\dfrac{3xy^3}{7x^2 y} \cdot \dfrac{14x^5}{9xy^6}$

5 $\dfrac{5x^2 y}{3x^2 y} \cdot \dfrac{6y}{10x^2}$

6 $\dfrac{15xyz}{16a^2} \cdot \dfrac{12a^3}{25x^2 yz^2}$

7 $\dfrac{x + 4}{x^2} \cdot \dfrac{5x}{3x + 12}$

8 $\dfrac{x^2 + xy}{xy} \cdot \dfrac{5y}{x^2 - y^2}$

9 $\dfrac{3x + 6}{5x + 5} \cdot \dfrac{x + 1}{x^2 - 6x - 16}$

10 $\dfrac{x + 2}{x^2 + 8x - 9} \cdot \dfrac{2x + 18}{x^2 - 4}$

11 $\dfrac{a^2 - 1}{a + 1} \cdot \dfrac{7a^2 - 5a - 2}{a^2 - 2a + 1}$

12 $\dfrac{a^2 - 4}{a^2 - 1} \cdot \dfrac{a - 1}{6a + 12}$

13 $\dfrac{x^2 - 144}{x + 4} \cdot \dfrac{x^2 - 16}{x - 12}$

14 $\dfrac{a^2 - 9b^2}{a^2 - b^2} \cdot \dfrac{5a - 5b}{a^2 + 6ab + 9b^2}$

15 $\dfrac{x^2 - 1}{x - 3} \cdot \dfrac{3x^2 - 8x - 3}{x^2 - 10x + 9}$

16 $\dfrac{x^2 - 7x + 10}{x^2 - 10x + 25} \cdot \dfrac{x - 5}{x + 12}$

17 $\dfrac{(x - 5)(x + 5)}{(x - 1)(3x + 2)} \cdot \dfrac{3x^2 - x - 2}{x^2 - 10x + 25}$

18 $\dfrac{a^2 b^2 - 9}{x^2 - x - 2} \cdot \dfrac{x^3 - x^2 - 2x}{a^2 b^2 - 6ab + 9}$

19 $\dfrac{a^2 - 4b^2}{(a + b)^2} \cdot \dfrac{a^2 - b^2}{a^2 - 4ab + 4b^2}$

20 $\dfrac{x^2 - 10x + 25}{x^2 - 100} \cdot \dfrac{x^2 + 12x + 20}{x^2 - 7x + 10}$

21 $\dfrac{a^3 - b^3}{5a + 5b} \cdot \dfrac{a^2 + 2ab + b^2}{a^2 + ab + b^2}$

22 $\dfrac{8x^3 + y^3}{3x - 5y} \cdot \dfrac{9x^2 - 25y^2}{4x^2 - 2xy + y^2}$

In problems 23–42, find the quotients and reduce them to lowest terms.

23 $\dfrac{7}{15} \div \dfrac{-14}{25}$

24 $\dfrac{-4}{11} \div \dfrac{-12}{33}$

25 $\dfrac{4x}{y} \div \dfrac{3x}{5y}$

26 $\dfrac{5xy^3}{9a^2b} \div \dfrac{25x^3y}{18a^3b}$

27 $\dfrac{4x+8}{x^3} \div \dfrac{x+2}{x^4}$

28 $\dfrac{x^2-y^2}{x-y} \div \dfrac{1}{x+y}$

29 $\dfrac{7x-14}{11x-33} \div \dfrac{2x-4}{3x-9}$

30 $\dfrac{9a^2-1}{a+1} \div \dfrac{3a+1}{2(a+1)}$

31 $\dfrac{x^3+3x}{2x-1} \div \dfrac{x^2+3}{x+1}$

32 $\dfrac{2a}{a^3b^2-a^2b^3} \div \dfrac{4a}{a^2-b^2}$

33 $\dfrac{a^2+4a+4}{a-2} \div \dfrac{a^2-4}{a+2}$

34 $\dfrac{x^3-4x^2+3x}{x+2} \div (x^2-3x)$

35 $\dfrac{3x+3y}{x^2+6x+9} \div \dfrac{x^2-y^2}{x^2-9}$

36 $\dfrac{x^2-3}{x^3-4x} \div \dfrac{x^4-9}{x^2-4}$

37 $\dfrac{y^3-x^3}{y^3} \div \dfrac{x-y}{xy}$

38 $\dfrac{a^2+8a+16}{a^2-8a+16} \div \dfrac{a^3+4a^2}{a^2-16}$

39 $\dfrac{x^2-11x+10}{9x^2-25} \div \dfrac{x^2-8x-20}{12x^2+20x}$

40 $\dfrac{x^4-16}{x^4+3x^3} \div \dfrac{x^3+4x}{(x+3)^2}$

41 $\dfrac{3x-2}{x^3+3x^2+2x} \div \dfrac{9x^2-4}{x^2+10x+16}$

42 $\dfrac{a^3-b^3}{(a+b)(a^2-ab+b^2)} \div \dfrac{(a-b)(a^2+ab+b^2)}{a^3+b^3}$

In problems 43–52, perform the indicated operations and reduce the results to lowest terms.

→ 43 $\dfrac{3a^2b^2}{7x^2y^3} \cdot \dfrac{5xy^4}{6ab^5} \div \dfrac{ab^3}{21x^4y^3}$

44 $\dfrac{51x^3y^3}{5a^2b^2} \cdot \dfrac{xy^5}{3x^2y} \div \dfrac{17x^4y^6}{25ab^4}$

45 $\dfrac{x-1}{x^2-1} \cdot \dfrac{2x+2}{x^2-4} \div \dfrac{3x+3}{x^2+4x+4}$

46 $\dfrac{x-3}{x^2+2x-3} \cdot \dfrac{x^2-2x+1}{x^2-2x-3} \div \dfrac{x^2-9}{x^2-1}$

$\rightarrow$ **47** $\quad \dfrac{x-1}{21-4a-a^2} \cdot \dfrac{x-2}{x-x^3} \div \dfrac{2-x}{a^2+6a-7}$

$\rightarrow$ **48** $\quad \left(\dfrac{x^2-1}{x^2}\right)^2 \cdot \dfrac{2}{x-1} \div \dfrac{x^2+2x+1}{x}$

$\rightarrow$ **49** $\quad \dfrac{7}{5x^2(x+3)} \div \left[\dfrac{x^2-5x+6}{8x^2} \cdot \dfrac{24}{5(x^2-9)}\right]$

50 $\quad \dfrac{x^3-16x}{x^2-3x-10} \div \left(\dfrac{x^3+x^2-12x}{x^2+5x+6} \cdot \dfrac{x^2-4}{x^2-9}\right)$

51 $\quad \dfrac{x^2-x-6}{x^2-5x+6} \div \left(\dfrac{x^2-8x+7}{x^2-4} \cdot \dfrac{x^2+4x+4}{x^2-6x-7}\right)$

52 $\quad \dfrac{x^2+3x-28}{x^2-x-20} \div \left(\dfrac{x^2-5x+4}{x^2+5x+6} \cdot \dfrac{x^2-3x-18}{x^2-6x+5}\right)$

3.3 Addition and Subtraction of Fractions

Fractions are either real numbers or algebraic expressions representing real numbers. Hence, we can develop rules for adding and subtracting fractions by using the properties of real numbers displayed in Chapter 1. First we shall consider adding and subtracting "like fractions," that is, fractions with the same denominator.

Addition and Subtraction of Like Fractions

Recall that $a/b = a \cdot (1/b)$ for real numbers a and b with $b \neq 0$. This is also true for polynomials; that is, if P and Q are polynomials, then $P/Q = P \cdot (1/Q)$, where $Q \neq 0$. We can apply this definition and the distributive property of real numbers to add the like fractions P/Q and R/Q with $Q \neq 0$,

$$\frac{P}{Q} + \frac{R}{Q} = P \cdot \frac{1}{Q} + R \cdot \frac{1}{Q} = (P+R) \cdot \frac{1}{Q} = \frac{P+R}{Q}$$

This result can be stated as follows: To add like fractions, add the numerators to find the numerator of the sum, and retain the common denominator as the denominator of the sum.

EXAMPLES

Determine each of the following sums.

1 $\quad \dfrac{3}{7} + \dfrac{2}{7}$

SOLUTION

$$\frac{3}{7} + \frac{2}{7} = \frac{3+2}{7} = \frac{5}{7}$$

2    $\dfrac{7}{x} + \dfrac{13}{x}$

SOLUTION

$$\frac{7}{x} + \frac{13}{x} = \frac{7+13}{x} = \frac{20}{x}$$

3    $\dfrac{x}{x^2+1} + \dfrac{2x+1}{x^2+1}$

SOLUTION

$$\frac{x}{x^2+1} + \frac{2x+1}{x^2+1} = \frac{x+2x+1}{x^2+1} = \frac{3x+1}{x^2+1}$$

4    $\dfrac{-x^2}{5-x} + \dfrac{25}{5-x}$

SOLUTION

$$\frac{-x^2}{5-x} + \frac{25}{5-x} = \frac{25-x^2}{5-x} = \frac{(5-x)(5+x)}{5-x} = 5+x$$

We have already defined the subtraction of real numbers in terms of addition by $a - b = a + (-b)$, where $-b$, the additive inverse of b, has the property that $b + (-b) = 0$. Now we shall use this definition to develop a rule for subtracting fractions.

First, the *additive inverse* of a fraction a/b is denoted by $-(a/b)$ and is defined to have the property

$$\frac{a}{b} + \left(-\frac{a}{b}\right) = 0$$

If the inverse $-(a/b)$ exists, it is unique, that is, $-(a/b)$ is the only inverse. Note that

$$-\left(\frac{a}{b}\right) = \frac{-a}{b}$$

since $-\left(\dfrac{a}{b}\right) = (-1)\left(\dfrac{a}{b}\right) = \dfrac{(-1)a}{b} = \dfrac{-a}{b}$.

Thus, to subtract the rational numbers a/b and c/b, with $b \neq 0$, we have

$$\frac{a}{b} - \frac{c}{b} = \frac{a}{b} + \left(-\frac{c}{b}\right) = \frac{a}{b} + \frac{-c}{b} = \frac{a-c}{b}$$

We can follow a similar procedure to find the subtraction of two like rational expressions.

If P/Q and R/Q are rational expressions, with $Q \neq 0$, then

$$\frac{P}{Q} - \frac{R}{Q} = \frac{P}{Q} + \left(\frac{-R}{Q}\right) = \frac{P-R}{Q}$$

This is stated in words as follows: To subtract like fractions, subtract the numerators to find the numerator of the difference, and retain the common denominator as the denominator of the difference.

EXAMPLES

Perform the following subtractions and reduce to lowest terms.

5   $\dfrac{7}{9} - \dfrac{2}{9}$

SOLUTION

$$\frac{7}{9} - \frac{2}{9} = \frac{7-2}{9} = \frac{5}{9}$$

6   $\dfrac{15}{x} - \dfrac{8}{x}$

SOLUTION

$$\frac{15}{x} - \frac{8}{x} = \frac{15-8}{x} = \frac{7}{x}$$

7   $\dfrac{x^2}{x-1} - \dfrac{1}{x-1}$

SOLUTION

$$\frac{x^2}{x-1} - \frac{1}{x-1} = \frac{x^2-1}{x-1} = \frac{(x-1)(x+1)}{x-1} = x+1$$

Addition and Subtraction of Unlike Fractions

Let us now consider the sum of *unlike fractions* (fractions with different denominators) such as $x/5$ and $x/6$. We can write

$$\frac{x}{5} + \frac{x}{6} = \frac{6x}{30} + \frac{5x}{30}$$

so that

$$\frac{x}{5} + \frac{x}{6} = \frac{6x + 5x}{30} = \frac{11x}{30}$$

This example can be generalized by the following rule. If P/Q and R/S are rational expressions, with $Q \neq 0$, $S \neq 0$, then

$$\frac{P}{Q} + \frac{R}{S} = \frac{PS + QR}{QS}$$

We prove this rule as follows:

$$\frac{P}{Q} + \frac{R}{S} = \frac{PS}{QS} + \frac{RQ}{SQ} \qquad \text{(fundamental principle of fractions)}$$

$$= \frac{PS}{QS} + \frac{QR}{QS} \qquad \text{(commutative property for multiplication)}$$

$$= \frac{PS + QR}{QS}$$

We dispense with the word statements of the rules of unlike fractions since the formulas involved clarify the procedure.

EXAMPLES

Perform the following additions.

8    $\dfrac{y}{14} + \dfrac{y}{5}$

SOLUTION

$$\frac{y}{14} + \frac{y}{5} = \frac{5y + 14y}{14(5)} = \frac{19y}{70}$$

9    $\dfrac{3}{x - 1} + \dfrac{x}{x + 1}$

SOLUTION

$$\frac{3}{x-1} + \frac{x}{x+1} = \frac{3(x+1) + x(x-1)}{(x-1)(x+1)}$$

$$= \frac{3x+3+x^2-x}{(x-1)(x+1)}$$

$$= \frac{x^2+2x+3}{(x-1)(x+1)}$$

The subtraction of unlike rational expressions is accomplished by using the following rule: If P/Q and R/S are rational expressions, with $Q \neq 0$, $S \neq 0$, then

$$\frac{P}{Q} - \frac{R}{S} = \frac{PS - QR}{QS}$$

To prove this rule, we have

$$\frac{P}{Q} - \frac{R}{S} = \frac{P}{Q} + \left(-\frac{R}{S}\right) = \frac{P}{Q} + \left(\frac{-R}{S}\right) = \frac{PS + (-QR)}{QS} = \frac{PS - QR}{QS}$$

EXAMPLES

Perform the following subtractions.

10   $\dfrac{5}{x} - \dfrac{2}{3}$

SOLUTION

$$\frac{5}{x} - \frac{2}{3} = \frac{5(3) - 2x}{3x} = \frac{15 - 2x}{3x}$$

11   $\dfrac{8}{a-2} - \dfrac{4}{a+2}$

SOLUTION

$$\frac{8}{a-2} - \frac{4}{a+2} = \frac{8(a+2) - 4(a-2)}{(a-2)(a+2)}$$

$$= \frac{8a + 16 - 4a + 8}{a^2 - 4} = \frac{4a + 24}{a^2 - 4}$$

12   $\dfrac{x+1}{x-1} - \dfrac{x-1}{x+1}$

SOLUTION

$$\frac{x+1}{x-1} - \frac{x-1}{x+1} = \frac{(x+1)(x+1) - (x-1)(x-1)}{(x-1)(x+1)}$$

$$= \frac{(x^2 + 2x + 1) - (x^2 - 2x + 1)}{(x-1)(x+1)}$$

$$= \frac{x^2 + 2x + 1 - x^2 + 2x - 1}{x^2 - 1} = \frac{4x}{x^2 - 1}$$

There are occasions when a direct application of the addition or subtraction rule as derived above is not the most convenient approach. For example,

$$\frac{x}{x^2 - 1} + \frac{2}{x^2 + 2x + 1} = \frac{x(x^2 + 2x + 1) + 2(x^2 - 1)}{(x^2 - 1)(x^2 + 2x + 1)}$$

$$= \frac{x(x+1)(x+1) + 2(x-1)(x+1)}{(x-1)(x+1)(x+1)(x+1)}$$

$$= \frac{(x+1)[x(x+1) + 2(x-1)]}{(x+1)[(x-1)(x+1)(x+1)]}$$

$$= \frac{x^2 + x + 2x - 2}{(x-1)(x+1)(x+1)}$$

$$= \frac{x^2 + 3x - 2}{(x-1)(x+1)(x+1)}$$

This same problem could also be solved as follows:

$$\frac{x}{x^2 - 1} + \frac{2}{x^2 + 2x + 1} = \frac{x}{(x-1)(x+1)} + \frac{2}{(x+1)(x+1)}$$

$$= \frac{x(x+1)}{(x-1)(x+1)(x+1)}$$

$$+ \frac{2(x-1)}{(x-1)(x+1)(x+1)}$$

$$= \frac{x(x+1) + 2(x-1)}{(x-1)(x+1)(x+1)}$$

$$= \frac{x^2 + x + 2x - 2}{(x-1)(x+1)(x+1)}$$

$$= \frac{x^2 + 3x - 2}{(x-1)(x+1)(x+1)}$$

In the first solution, the factor $x + 1$, which was included in the denominator, was subsequently removed, since this inclusion is unnecessary. In the second solution we converted the fractions to equivalent fractions with a common denominator containing fewer factors than the common denominator of the first solution. In this case we found a polynomial with the smallest number of factors that both denominators "divide into evenly." Such a polynomial expression is called the *least common denominator* and is abbreviated by the symbol L.C.D. To find the L.C.D. of a set of fractions, we adopt the following steps:

1   Factor each denominator into a product of prime factors or into prime factors and -1.

2   List each prime factor with the largest exponent it has in any factored denominators; then find the product of the factors so listed.

This procedure is illustrated by the following examples.

EXAMPLES

Express each fraction in an equivalent form having the L.C.D. of the given set as its denominator.

13   $\dfrac{5x}{7y^3z^2}, \dfrac{9y^2}{14x^3z^3},$ and $\dfrac{11z}{21y^5x^6}$

SOLUTION.   First we find the L.C.D. of the denominators.

$$7y^3z^2 = 7y^3z^2$$

$$14x^3z^3 = 2(7)x^3z^3$$

$$21y^5x^6 = 3(7)y^5x^6$$

The L.C.D. of these expressions is $2(3)7x^6y^5z^3 = 42x^6y^5z^3$. Using the fundamental principle of fractions, we have

$$\frac{5x}{7y^3z^2} = \frac{5x(6x^6y^2z)}{7y^3z^2(6x^6y^2z)} = \frac{30x^7y^2z}{42x^6y^5z^3}$$

$$\frac{9y^2}{14x^3z^3} = \frac{9y^2(3x^3y^5)}{14x^3z^3(3x^3y^5)} = \frac{27x^3y^7}{42x^6y^5z^3}$$

$$\frac{11z}{21y^5x^6} = \frac{11z(2z^3)}{21y^5x^6(2z^3)} = \frac{22z^4}{42x^6y^5z^3}$$

14    $\dfrac{x+2}{x^2+2x-3}$ and $\dfrac{x-6}{x^2+3x-4}$

SOLUTION

$$x^2 + 2x - 3 = (x+3)(x-1)$$

$$x^2 + 3x - 4 = (x+4)(x-1)$$

The L.C.D. of these expressions is $(x+3)(x+4)(x-1)$. Using the fundamental principle of fractions, we have

$$\frac{x+2}{x^2+2x-3} = \frac{x+2}{(x+3)(x-1)} = \frac{(x+2)(x+4)}{(x+3)(x-1)(x+4)}$$

$$\frac{x-6}{x^2+3x-4} = \frac{x-6}{(x+4)(x-1)} = \frac{(x-6)(x+3)}{(x+3)(x-1)(x+4)}$$

The use of the L.C.D. in adding or subtracting fractions usually simplifies the computations. This is illustrated in the following examples.

EXAMPLES

Perform the operations and simplify.

15    $\dfrac{6}{x^2-2x-8} + \dfrac{x}{x+2}$

SOLUTION. $x^2 - 2x - 8 = (x+2)(x-4)$, so that the L.C.D. of the given fractions is $(x+2)(x-4)$. Thus,

$$\frac{6}{x^2-2x-8} + \frac{x}{x+2} = \frac{6}{(x+2)(x-4)} + \frac{x}{x+2}$$

$$= \frac{6}{(x+2)(x-4)} + \frac{x(x-4)}{(x+2)(x-4)}$$

$$= \frac{6+x(x-4)}{(x+2)(x-4)} = \frac{6+x^2-4x}{(x+2)(x-4)}$$

16    $\dfrac{a}{a-x} - \dfrac{x}{a+x} - \dfrac{a^2+x^2}{a^2-x^2}$

SOLUTION.   Since $a^2 - x^2 = (a-x)(a+x)$, the L.C.D. is $(a-x)(a+x)$. Therefore,

$$\frac{a}{a-x} - \frac{x}{a+x} - \frac{a^2+x^2}{a^2-x^2} = \frac{a}{a-x} - \frac{x}{a+x} - \frac{a^2+x^2}{(a-x)(a+x)}$$

$$= \frac{a(a+x)}{(a-x)(a+x)} - \frac{x(a-x)}{(a+x)(a-x)}$$

$$- \frac{a^2+x^2}{(a-x)(a+x)}$$

$$= \frac{a(a+x) - x(a-x) - (a^2+x^2)}{(a-x)(a+x)}$$

$$= \frac{a^2 + ax - xa + x^2 - a^2 - x^2}{(a-x)(a+x)}$$

$$= \frac{0}{(a-x)(a+x)} = 0$$

17   $\dfrac{x}{x^2+6x+5} - \dfrac{2}{x^2+4x-5} + \dfrac{3}{x^2-1}$

SOLUTION. The denominators in factored form are $(x+1)(x+5)$, $(x-1)(x+5)$, and $(x-1)(x+1)$, so that the L.C.D. is $(x+1)(x+5)(x-1)$. Therefore,

$$\frac{x}{x^2+6x+5} - \frac{2}{x^2+4x-5} + \frac{3}{x^2-1}$$

$$= \frac{x}{(x+1)(x+5)} - \frac{2}{(x-1)(x+5)} + \frac{3}{(x-1)(x+1)}$$

$$= \frac{x(x-1)}{(x+1)(x+5)(x-1)} - \frac{2(x+1)}{(x+1)(x+5)(x-1)}$$

$$+ \frac{3(x+5)}{(x+1)(x+5)(x-1)}$$

$$= \frac{x(x-1) - 2(x+1) + 3(x+5)}{(x+1)(x+5)(x-1)} = \frac{x^2 - x - 2x - 2 + 3x + 15}{(x+1)(x+5)(x-1)}$$

$$= \frac{x^2+13}{(x+1)(x+5)(x-1)}$$

PROBLEM SET 3.3

In problems 1–20, perform the indicated operations of the like fractions and reduce them to lowest terms.

1   $\dfrac{5}{11} + \dfrac{3}{11}$

2   $\dfrac{7}{12} + \dfrac{5}{12}$

3 $\dfrac{6}{x} + \dfrac{18}{x}$

4 $\dfrac{7}{y^2} + \dfrac{3}{y^2}$

5 $\dfrac{7}{13} - \dfrac{3}{13}$

6 $\dfrac{71}{y} - \dfrac{35}{y}$

7 $\dfrac{x-1}{7} + \dfrac{x+1}{7}$

8 $\dfrac{x+2}{9} - \dfrac{x+2}{9}$

9 $\dfrac{5x}{3x+1} - \dfrac{2x}{3x+1}$

10 $\dfrac{3x^2+5x}{x} - \dfrac{3x-3x^2}{x}$

11 $\dfrac{x+x^2}{9x} + \dfrac{3x-x^2}{9x}$

12 $\dfrac{7xy-8x^2y}{x^2-y^2} - \dfrac{3x^2y-6xy}{x^2-y^2}$

13 $\dfrac{x^3-27}{x-3} - \dfrac{x^3+27}{x-3}$

14 $\dfrac{a^3+3a^2}{5a-2} - \dfrac{7a^3-8a^2}{5a-2}$

15 $\dfrac{a^2-b^2}{7ab} - \dfrac{a^2+b^2}{7ab}$

16 $\dfrac{x^2-xy-2y^2}{x+y} + \dfrac{x^2-2xy-3y^2}{x+y}$

17 $\dfrac{x^2+5x+16}{x+1} - \dfrac{3x^2-2x+7}{x+1}$

18 $\dfrac{x^3-5x^2+8}{x^2-9} - \dfrac{6x^2-10}{x^2-9}$

19 $\dfrac{5}{x} + \dfrac{3}{x} - \dfrac{8}{x}$

20 $\dfrac{5}{x} + \dfrac{7}{x} + \dfrac{x-12}{x}$

In problems 21–40, perform the indicated operations of the unlike fractions and reduce them to lowest terms.

21 $\dfrac{5}{8} + \dfrac{3}{7}$

22 $\dfrac{11}{4} + \dfrac{11}{8}$

23 $\dfrac{7}{6} - \dfrac{3}{10}$

24 $\dfrac{11}{11} - \dfrac{5}{9}$

25 $\dfrac{2x}{3} + \dfrac{9x}{10}$

26 $\dfrac{3x}{5} + \dfrac{2x}{15}$

27 $\dfrac{7y}{8} - \dfrac{5y}{6}$

28 $\dfrac{5a}{7} - \dfrac{3a}{8}$

29 $\dfrac{3}{x} + \dfrac{2}{x^2}$

30 $\dfrac{9}{10y} + \dfrac{3}{5y}$

31 $\dfrac{7-3x}{8} - \dfrac{2-5x}{6}$

32 $\dfrac{1}{a^2} - \dfrac{3}{ab}$

33 $\dfrac{a}{4b^2} - \dfrac{b}{5a^2b}$

34 $\dfrac{b}{2a^2c} - \dfrac{c}{5b^2a}$

35 $\dfrac{9}{y-5} + \dfrac{6}{y-3}$

36 $\dfrac{1}{x+y} - \dfrac{1}{x-y}$

37 $\dfrac{7}{x-3} - \dfrac{5}{x+3}$

38 $\dfrac{2a}{a-3} + \dfrac{3}{a+7}$

39 $\dfrac{5}{x-7} + \dfrac{3}{x-5}$

40 $\dfrac{2}{x-3} - \dfrac{1}{x-4}$

In problems 41–52, express each fraction in an equivalent form having the L.C.D. of the given set as its denominator.

41 $\dfrac{7}{12}, \dfrac{1}{10}, \dfrac{2}{45}$

42 $\dfrac{5}{8}, \dfrac{3}{4}, \dfrac{8}{6}$

43 $\dfrac{3x}{84z}, \dfrac{5}{12xz}, \dfrac{7z}{6xy}$

44 $\dfrac{x-2}{5xy^3}, \dfrac{4y+1}{25x^3y}, \dfrac{3}{75x^4y^2}$

45 $\dfrac{5}{x+3}, \dfrac{4}{x^2+3x}$

46 $\dfrac{7}{x-4}, \dfrac{5}{x^2-16}$

47 $\dfrac{5}{12x^2}, \dfrac{7}{6x^2-24}, \dfrac{1}{12x-24}$

48 $\dfrac{2}{x+1}, \dfrac{2}{x^2-1}, \dfrac{4}{x^2+2x+1}$

49 $\dfrac{8}{x^2-2x}, \dfrac{4}{x^2-4x+4}, \dfrac{x}{x^2-x-2}$

50 $\dfrac{x}{x^2+3x+9}, \dfrac{x-2}{x^2-3x}, \dfrac{1}{x^3-27}$

51 $\dfrac{2x-3}{x-5}, \dfrac{x+1}{x+5}, \dfrac{x-1}{x^2-25}$

52 $\dfrac{2}{x+4}, \dfrac{x-3}{x^2-4x+16}, \dfrac{x^2}{x^3+64}$

In problems 53–70, perform the operations and reduce the results to lowest terms.

53 $\dfrac{x}{x^2-25} + \dfrac{1}{x+5}$

54 $\dfrac{x}{x-1} - \dfrac{1}{x^2-x}$

55 $\dfrac{x}{x^2-9} - \dfrac{x-1}{x^2-5x+6}$

56 $\dfrac{x}{x^2+5x-6} + \dfrac{3}{x+6}$

57 $\dfrac{x-5}{x^2-5x-6} + \dfrac{x+4}{x^2-6x}$

58 $\dfrac{2}{a^2-4} + \dfrac{7}{a^2-4a-12}$

59 $\dfrac{x-2}{x^2+10x+16} + \dfrac{x+1}{x^2+9x+14}$

60 $\dfrac{a}{a^2-2ab+b^2} - \dfrac{b}{a^2-b^2}$

61 $\dfrac{1}{x^2+7x+10} - \dfrac{1}{x^2+x-20}$

62 $\dfrac{1}{3x^2-x-2} - \dfrac{1}{2x^2-x-1}$

63 $\dfrac{x}{x+2} - \dfrac{x}{x-2} - \dfrac{x^2}{x^2-4}$

64 $\dfrac{x-3}{x+3} - \dfrac{x+3}{3-x} + \dfrac{x^2}{9-x^2}$

65   $\dfrac{2x^2 - x}{3x^2 - 27} - \dfrac{x - 3}{3x - 9} + \dfrac{6x^2}{9 - x^2}$

66   $\dfrac{4}{a^2 - 4} + \dfrac{2}{a^2 - 4a + 4} + \dfrac{1}{a^2 + 4a + 4}$

67   $\dfrac{x - 1}{x^2 - x - 6} + \dfrac{x + 2}{x^2 - 6x + 9} + \dfrac{x - 2}{x^2 - x - 6}$

68   $\dfrac{x}{x^2 y - y^3} - \dfrac{y}{x^2 + xy} + \dfrac{4}{x^2 y - y^3}$

69   $\dfrac{2x^2}{x^4 - 1} + \dfrac{1}{x^2 + 1} - \dfrac{1}{x^2 - 1}$

70   $\dfrac{1}{a^2 + 6a + 9} - \dfrac{3}{a^2 + 2a - 3} + \dfrac{2}{a^2 + 4a + 3}$

3.4 Complex Fractions

We have considered fractions of the form P/Q, $Q \neq 0$, where P and Q are polynomial expressions. In this section we shall consider fractions in which the numerator or denominator, or both, contain fractions. Fractions of this type are called *complex fractions*. For example, the fractions

$$\frac{1}{\dfrac{2}{3}}, \quad \frac{\dfrac{2}{3}}{\dfrac{7}{5}}, \quad \frac{3}{\dfrac{x}{1-x}}, \quad \frac{\dfrac{2}{3}}{1 + \dfrac{1}{x}} \quad \text{and} \quad \frac{x - \dfrac{1}{x}}{\dfrac{2}{1+x} - \dfrac{x}{1-x}}$$

are complex fractions.

To simplify a complex fraction is to express it as a single fraction in lowest terms. This can be accomplished by using the following approaches.

1   Express the fraction as a quotient of simplified fractions by performing additions and subtractions in the numerator and the denominator of the complex fraction; then perform the division indicated in the resulting fraction.

2   Multiply the numerator and the denominator of the complex fraction by the least common denominator (L.C.D.) of all the fractions they contain, by making use of the fundamental principle of fractions.

EXAMPLES

Express each of the following fractions as a single fraction in lowest terms.

1 $\dfrac{\dfrac{7}{11}}{\dfrac{5}{44}}$

SOLUTION

First approach

$$\frac{\dfrac{7}{11}}{\dfrac{5}{44}} = \frac{7}{11} \div \frac{5}{44} = \frac{7}{11} \times \frac{44}{5}$$

$$= \frac{28}{5}$$

Second approach

The L.C.D. of the fractions $\frac{7}{11}$ and $\frac{5}{44}$ is 44, so that

$$\frac{\dfrac{7}{11}}{\dfrac{5}{44}} = \frac{\dfrac{7}{11} \times 44}{\dfrac{5}{44} \times 44} = \frac{7 \times 4}{5 \times 1} = \frac{28}{5}$$

2 $\dfrac{\dfrac{9}{xy^2}}{\dfrac{12}{x^2y}}$

SOLUTION

First approach

$$\frac{\dfrac{9}{xy^2}}{\dfrac{12}{x^2y}} = \frac{9}{xy^2} \div \frac{12}{x^2y} = \frac{9}{xy^2} \cdot \frac{x^2y}{12}$$

$$= \frac{3x}{4y}$$

Second approach

The L.C.D. of $9/xy^2$ and $12/x^2y$ is x^2y^2, so that

$$\frac{\dfrac{9}{xy^2}}{\dfrac{12}{x^2y}} = \frac{\left(\dfrac{9}{xy^2}\right) \cdot x^2y^2}{\left(\dfrac{12}{x^2y}\right) \cdot x^2y^2}$$

$$= \frac{9x}{12y} = \frac{3x}{4y}$$

3 $\dfrac{8 + \dfrac{4}{x}}{\dfrac{2x + 1}{12}}$

SOLUTION

First approach

We first express the numerator and the denominator separately as single fractions, so that

$$8 + \frac{4}{x} = \frac{8}{1} + \frac{4}{x} = \frac{8x + 4}{x}$$

Then we proceed as follows:

$$\frac{8 + \dfrac{4}{x}}{\dfrac{2x + 1}{12}} = \frac{\dfrac{8x + 4}{x}}{\dfrac{2x + 1}{12}}$$

$$= \frac{8x + 4}{x} \div \frac{2x + 1}{12}$$

$$= \frac{8x + 4}{x} \cdot \frac{12}{2x + 1}$$

$$= \frac{4(2x + 1)}{x} \cdot \frac{12}{2x + 1} = \frac{48}{x}$$

Second approach

The L.C.D. of $8 + 4/x$ and $(2x + 1)/12$ is $12x$, so that

$$\frac{8 + \dfrac{4}{x}}{\dfrac{2x + 1}{12}} = \frac{\left(8 + \dfrac{4}{x}\right) \cdot 12x}{\left(\dfrac{2x + 1}{12}\right) \cdot 12x} = \frac{96x + 48}{2x^2 + x}$$

$$= \frac{48(2x + 1)}{x(2x + 1)}$$

$$= \frac{48}{x}$$

In these examples, we used both approaches to simplify complex fractions. The following examples are illustrated by either approach.

EXAMPLES

Express each of the following complex fractions as a single fraction in lowest terms.

4 $$\dfrac{\dfrac{1}{x}+\dfrac{1}{y}}{\dfrac{y}{x}-\dfrac{x}{y}}$$

SOLUTION. The L.C.D. of $(1/x) + (1/y)$ and $(y/x) - (x/y)$ is xy, so that

$$\frac{\dfrac{1}{x}+\dfrac{1}{y}}{\dfrac{y}{x}-\dfrac{x}{y}}=\frac{\left(\dfrac{1}{x}+\dfrac{1}{y}\right)\cdot xy}{\left(\dfrac{y}{x}-\dfrac{x}{y}\right)\cdot xy}=\frac{y+x}{y^{2}-x^{2}}=\frac{y+x}{(y+x)(y-x)}=\frac{1}{y-x}$$

5 $$\dfrac{x-\dfrac{1}{y}}{1-\dfrac{x}{\dfrac{1}{y}}}$$

SOLUTION. We first simplify the term $-x/(1/y)$ in the denominator, so that

$$\frac{-x}{\dfrac{1}{y}}=-x\div\frac{1}{y}=-x\left(\frac{y}{1}\right)=-xy$$

Then we proceed as follows:

$$\frac{x-\dfrac{1}{y}}{1-\dfrac{x}{\dfrac{1}{y}}}=\frac{x-\dfrac{1}{y}}{1-xy}=\frac{\dfrac{xy-1}{y}}{1-xy}=\frac{xy-1}{y}\div(1-xy)$$

$$=\frac{xy-1}{y}\cdot\frac{1}{1-xy}=\frac{-(1-xy)}{y}\cdot\frac{1}{1-xy}$$

$$=\frac{-1}{y}$$

6 $$\dfrac{\dfrac{1}{x^{2}-y^{2}}}{\dfrac{1}{x+y}+\dfrac{1}{x-y}}$$

SOLUTION

$$\frac{\dfrac{1}{x^2 - y^2}}{\dfrac{1}{x+y} + \dfrac{1}{x-y}} = \frac{\dfrac{1}{x^2 - y^2}}{\dfrac{x - y + x + y}{x^2 - y^2}} = \frac{\dfrac{1}{x^2 - y^2}}{\dfrac{2x}{x^2 - y^2}}$$

$$= \frac{1}{x^2 - y^2} \div \frac{2x}{x^2 - y^2} = \frac{1}{x^2 - y^2} \cdot \frac{x^2 - y^2}{2x}$$

$$= \frac{1}{2x}$$

PROBLEM SET 3.4

In problems 1–26, express each complex fraction as a single fraction in lowest terms.

1 $\dfrac{\dfrac{2}{3}}{\dfrac{4}{5}}$

2 $\dfrac{\dfrac{17}{6}}{\dfrac{34}{9}}$

3 $\dfrac{\dfrac{113x}{24y}}{\dfrac{12x}{9y^2}}$

4 $\dfrac{\dfrac{a^2b^2}{8}}{\dfrac{25ab^3}{16}}$

5 $\dfrac{3 + \dfrac{1}{5}}{3 - \dfrac{1}{5}}$

6 $\dfrac{3 + \dfrac{1}{x}}{3 - \dfrac{1}{x}}$

7 $\dfrac{\dfrac{1}{x}}{1 - \dfrac{1}{x}}$

8 $\dfrac{1 - \dfrac{x}{3}}{1 - \dfrac{x}{5}}$

9 $\dfrac{\dfrac{1}{1+x}}{1 + \dfrac{1}{1+x}}$

10 $\dfrac{\dfrac{1}{x+1} - 1}{3 - \dfrac{1}{x+1}}$

11 $\dfrac{x + \dfrac{3}{x}}{1 + \dfrac{3}{x^2}}$

12 $\dfrac{\dfrac{3}{y} - \dfrac{x}{y}}{\dfrac{3}{y} - 1}$

13 $\dfrac{\dfrac{x}{y} - \dfrac{y}{x}}{\dfrac{x}{y} + \dfrac{y}{x}}$

14 $\dfrac{\dfrac{x}{y} + 1 + \dfrac{y}{x}}{\dfrac{x^3 - y^3}{xy}}$

15 $\dfrac{\dfrac{a}{b} - 1 + \dfrac{b}{a}}{\dfrac{a^3 + b^3}{a^2 b + ab^2}}$

16 $\dfrac{\dfrac{a}{b} - 2 + \dfrac{b}{a}}{\dfrac{a}{b} + 2 + \dfrac{b}{a}}$

17 $\dfrac{\dfrac{1}{x + y} - \dfrac{1}{x - y}}{\dfrac{4}{x^2 - y^2}}$

18 $\dfrac{\dfrac{x - 3}{x - 2} + \dfrac{x + 1}{x + 2}}{\dfrac{x^2 - 3x + 6}{x^2 - 4}}$

19 $\dfrac{\dfrac{3}{1 - x} + \dfrac{x}{x - 1}}{\dfrac{1}{1 - x}}$

20 $\dfrac{3 + \dfrac{5x + 1}{x^2 - 1}}{1 + \dfrac{1}{x - 1}}$

21 $\dfrac{1 - x + \dfrac{2x^2}{1 + x}}{1 - \dfrac{1}{1 + x}}$

22 $\dfrac{y + \dfrac{x}{x - y}}{x + \dfrac{y}{y - x}}$

23 $\dfrac{\dfrac{x - 1}{x + 1} - \dfrac{x + 1}{x - 1}}{\dfrac{x - 1}{x + 1} + \dfrac{x + 1}{x - 1}}$

24 $\dfrac{\dfrac{a - b}{a + b} + \dfrac{a + b}{a - b}}{\dfrac{a - b}{a + b} - \dfrac{a + b}{a - b}}$

25 $\dfrac{1 - \dfrac{x}{x - y}}{1 + \dfrac{x}{x - y}}$

26 $\dfrac{\left(\dfrac{a^2}{b} - \dfrac{b^2}{a} \right) \dfrac{a}{a^2 + ab + b^2}}{\dfrac{a^2 - 2ab + b^2}{b^2}}$

REVIEW PROBLEM SET

In problems 1–16, reduce each fraction to lowest terms.

1 $\dfrac{17}{85}$

2 $\dfrac{-26}{78}$

3 $\dfrac{-6xy}{9yz}$

4 $\dfrac{36xy^2}{54x^8 y^3}$

5 $\dfrac{96a^2 bx^2 y^3}{84a^2 b^2 x^3 y^2}$

6 $\dfrac{-46x(a + b)^3}{69y(a + b)^2}$

7 $\dfrac{(a-b)(x+y)}{(b-a)(x-y)}$

8 $\dfrac{(x-2y)(y-3x)}{(x^2-4y^2)(3x-y)}$

9 $\dfrac{x^2-7x+12}{x^2+3x-18}$

10 $\dfrac{2a^2-2b^2}{5a^2+10ab+5b^2}$

11 $\dfrac{x^2-9}{x^2-2x-3}$

12 $\dfrac{a^3-8b^3}{a^2-4ab+4b^2}$

13 $\dfrac{x^2-7x+12}{x^2-2x-8}$

14 $\dfrac{12x^3-27xy^2}{20x^2+60xy+45y^2}$

15 $\dfrac{x^2+xz-xy-yz}{3x^2-3y^2}$

16 $\dfrac{a^3x-b^3x}{ca^2x-cb^2x}$

In problems 17–20, use the fundamental principle of fractions to determine the missing expression.

17 $\dfrac{x}{y}=\dfrac{?}{y^3}$

18 $\dfrac{a}{a-b}=\dfrac{?}{b^2-a^2}$

19 $\dfrac{a}{a-b}=\dfrac{?}{(a-b)^2}$

→20 $\dfrac{6a^2b}{18ab^2(x-y)}=\dfrac{?}{3b}$

$6ab(x-y)$

In problems 21–50, perform the indicated operations.

21 $\dfrac{a^4x^2y^3}{b^2az^5}\cdot\dfrac{ab^2x^2z^2}{ay^3}$

22 $\dfrac{-36x^2}{y^2}\cdot\dfrac{5y^4}{27x^3}$

23 $\dfrac{b-1}{b-5}\cdot\dfrac{b^2-6b+5}{b-1}$

24 $\dfrac{x^2-6x}{x-6}\cdot\dfrac{x+3}{x}$

25 $\dfrac{9-y^2}{x^3-x}\cdot\dfrac{x-1}{y+3}$

26 $\dfrac{a^2+8a+16}{a^2-9}\cdot\dfrac{a-3}{a+4}$

27 $\dfrac{a^2-b^2}{(a+b)^2}\cdot\dfrac{3a+3b}{6a}$

28 $\dfrac{2x-4}{9}\cdot\dfrac{6}{5x-10}$

29 $\dfrac{(x+y)^2}{a+b}\cdot\dfrac{(a+b)^2}{x+y}$

30 $\dfrac{(x+y)^2}{a+b}\cdot\dfrac{x^3+y^3}{(x+y)^3}$

31 $\dfrac{x^2+5x+6}{x^2+x-2}\cdot\dfrac{x^2+3x-4}{x^2+7x+12}$

32 $\dfrac{x^2-16}{x^2-4x}\cdot\dfrac{x-4}{x+4}$

33 $\dfrac{ab^2-b^3}{a^3+a^2b}\cdot\dfrac{a^2-ab-2b^2}{a^2-2ab+b^2}$

34 $\dfrac{x^4-y^4}{(x-y)^2}\cdot\dfrac{x-y}{x^2+y^2}$

35 $\dfrac{-14x^2}{10b^2} \div \dfrac{21x^2}{15b^2}$

36 $\dfrac{-5a}{12yz^2} \div \dfrac{15a^2}{18y^2z^3}$

37 $\dfrac{10x^2}{(x+y)^2} \div \dfrac{5x}{x+y}$

38 $\dfrac{(d-b)(a+b)}{(x-y)(x+y)} \div \dfrac{d-b}{x+y}$

39 $\dfrac{c^2-b^2}{x^2-y^2} \div \dfrac{c+b}{x-y}$

40 $\dfrac{x^2-9}{x^2-4} \div \dfrac{x+3}{x-2}$

41 $\dfrac{x+7}{x+2} \div \dfrac{x^2-49}{x^2-4}$

42 $\dfrac{y^3+1}{x^2-4y^2} \div \dfrac{y^2-y+1}{x-2y}$

43 $\dfrac{x^2-x-2}{x^2-x-6} \div \dfrac{x^2-2x}{2x+x^2}$

44 $\dfrac{x^2+5x+6}{x^2+x-2} \div \dfrac{x^2+7x+12}{x^2+3x-4}$

45 $\dfrac{(x-y)^2}{a^2-b^2} \div \dfrac{x^2-y^2}{(a+b)^2}$

46 $\dfrac{x^2+3xy+2y^2}{x^2-2xy} \div \dfrac{x^2+4xy+3y^2}{x^2+xy-6y^2}$

47 $\dfrac{x^2-y^2}{x^3-3x^2y+2xy^2} \cdot \dfrac{xy-2y^2}{y^2+xy} \div \dfrac{(x-y)^2}{x(x-y)}$

48 $\left(1 - \dfrac{a+b}{a^2+b^2}\right) \div \dfrac{(a-b)^2}{a^4-b^4}$

$\Rightarrow$ **49** $\dfrac{b^3-8b^2+16b}{10b^2+3b-1} \div \left(\dfrac{b^3-10b^2+24b}{9b^2-1} \cdot \dfrac{3b^2+19b+6}{2b^2-11b-6}\right)$

50 $\dfrac{x}{x-y} \cdot \dfrac{x+y}{y} \cdot \dfrac{x^2-2xy+y^2}{y^3-x^3} \div \dfrac{x^3+y^3}{x^4+x^2y^2+y^4}$

In problems 51–74, perform the indicated operations.

51 $\dfrac{5}{7x} + \dfrac{9}{7x}$

52 $\dfrac{2}{x+4} + \dfrac{3}{x+4}$

53 $\dfrac{2a+b}{8} + \dfrac{3a-4b}{8}$

54 $\dfrac{3x-5}{9} + \dfrac{2x+7}{6}$

55 $\dfrac{x+2}{4} - \dfrac{x-5}{5}$

56 $\dfrac{3}{x-2} - \dfrac{7}{x+2}$

57 $\dfrac{2}{3x} - \dfrac{1}{4x}$

58 $\dfrac{a}{x^3y} + \dfrac{b}{x^2y^2}$

59 $\dfrac{5}{xy} - \dfrac{6}{yz} + \dfrac{7}{xz}$

60 $\dfrac{3}{2xy} - \dfrac{4}{3yz} + \dfrac{5}{4xz}$

61 $\dfrac{5}{x} - \dfrac{2}{x^3} + \dfrac{3}{x^2}$

62 $\dfrac{y+2}{y+4} - \dfrac{y-1}{y+6}$

63 $\dfrac{1}{x+3} + \dfrac{1}{x-2}$

64 $\dfrac{5a}{b^2} - \dfrac{3a}{b} + \dfrac{7a}{b^3}$

65 $\dfrac{5x}{7x+2y} - \dfrac{6y}{3x+4y}$

66 $\dfrac{1}{x-y} + \dfrac{1}{x+y}$

67 $\dfrac{x}{xy-y^2} - \dfrac{y}{x^2-xy}$

68 $\dfrac{1}{x^2-y^2} + \dfrac{1}{x^2+2xy+y^2}$

69 $\dfrac{3}{x^2-7x+12} + \dfrac{2}{x^2-5x+4}$

70 $\dfrac{5}{x^2+8x+15} - \dfrac{4}{x^2+2x-3}$

71 $\dfrac{2}{x-2} - \dfrac{3}{x+3} + \dfrac{4}{x-4}$

72 $\dfrac{1}{(x-y)^2} + \dfrac{x}{x-y} + 1$

73 $\dfrac{3}{x+2} - \dfrac{5}{x-7} + 4$

74 $\dfrac{3x}{x+5} - \dfrac{2x+1}{x+3} - 5$

In problems 75–90, simplify each complex fraction.

75 $\dfrac{\dfrac{1}{2}}{\dfrac{-3}{2}}$

76 $\dfrac{\dfrac{6x}{y}}{\dfrac{18x^2}{y^3}}$

77 $\dfrac{\dfrac{34a}{4x^2y}}{\dfrac{17a^3}{2xy}}$

78 $\dfrac{\dfrac{-25x^2}{(3yz)^2}}{\dfrac{100x^3}{18yz^3}}$

79 $\dfrac{\dfrac{-10x^2}{(x+y)^2}}{\dfrac{5x}{x+y}}$

80 $\dfrac{\dfrac{5}{2a+3b}}{\dfrac{10}{4a^2-9b^2}}$

81 $\dfrac{\dfrac{2}{3}+\dfrac{1}{6}}{1-\dfrac{3}{4}}$

82 $\dfrac{\dfrac{x}{y}+5}{\dfrac{x}{y}-5}$

83 $\dfrac{2+\dfrac{2}{y-1}}{\dfrac{2}{y-1}}$

84 $\dfrac{x+\dfrac{y}{2}}{x-\dfrac{y}{2}}$

85 $\dfrac{1-\dfrac{1}{x}}{x-2+\dfrac{1}{x}}$

86 $\dfrac{\dfrac{x^2+y^2}{y}-x}{\dfrac{1}{y}-\dfrac{1}{x}}$

87 $\dfrac{\dfrac{y}{y^2 - 1} - \dfrac{1}{y + 1}}{\dfrac{y}{y - 1} + \dfrac{1}{y + 1}}$

88 $\dfrac{\dfrac{1}{x} - \dfrac{1}{y}}{\dfrac{x^2 - y^2}{xy}}$

89 $\dfrac{\dfrac{1}{2x - 3} - \dfrac{1}{2x + 3}}{\dfrac{x}{4x^2 - 9}}$

90 $\dfrac{\dfrac{x - 3}{x - 2} + \dfrac{x + 1}{x + 2}}{\dfrac{x^2 - 3x + 6}{x^2 - 4}}$

CHAPTER 4

First-Degree Equations and Inequalities in One Variable

In this chapter we apply the techniques learned in the preceding chapters to solving equations and inequalities of the first degree. In addition, we shall discuss the solution of fractional equations, literal equations, absolute value equations, and absolute value inequalities. A variety of applications are also considered in the chapter.

4.1 Equations

An *equation* is a mathematical statement that expresses the relation of equality between two expressions representing real numbers. For example, $15 - 2 = 13$ is an equation. Also, if x represents a real number, $3x + 6 = 14$ is an equation. This latter equation is stated in words as follows: "Three times the number represented by x added to 6 is equal to 14." Generalizing the above example, we can say that an equation is an abbreviated form of a statement of equality that could be written in sentence form. Other examples of equations are $x^2 - 5x + 6 = 0$ and $x/(x-1) - 2/(x+1) = 1$. Any value of the variable x that satisfies the equation (makes it a true statement) is called a *solution* of the equation. Thus, to solve an equation is to find *all* values of the variable for which the equation holds true. For example, in the equation $3x + 6 = 14$, $x = \frac{8}{3}$ is the solution, since $\frac{8}{3}$ satisfies the equation; that is, $3(\frac{8}{3}) + 6 = 8 + 6 = 14$.

The solution or solutions of an equation form a set, and this set is called the *solution set* of the equation. The solution set of the equation $3x + 6 = 14$ is designated by $\{\frac{8}{3}\}$.

An equation may have one, more than one, or no solutions. If there are no solutions, its solution set will be designated by the empty set $\varnothing$. In this chapter we shall be concerned primarily with those equations in one variable that have one solution.

EXAMPLE

1 Determine if any of the numbers, 3, -1, and 5 are solutions of the equation $3x - 2 = 13$.

SOLUTION.   Substituting $x = 3$ in the equation $3x - 2 = 13$, we have $3(3) - 2 = 9 - 2 = 7$. Since $7 \neq 13$, 3 is not a solution.

For $x = -1$, we have $3(-1) - 2 = -3 - 2 = -5$. Since $-5 \neq 13$, $x = -1$ is not a solution.

For $x = 5$, we obtain $3(5) - 2 = 15 - 2 = 13$. Therefore, $x = 15$ is a solution.

Equations are classified into two general types, called conditional equations and identical equations. *Conditional equations* are those equations which are true only for certain values of the variable. For example, if x is a real number and $x + 1 = 3$, we have placed a condition on the variable x; that is, $x = 2$. Other examples of conditional equations are $7x + 3 = 16$ and $x^2 - 3x + 2 = 0$.

Identities are those equations that are true for all possible values of the variable. For example, $x^2 - 1 = (x - 1)(x + 1)$ is an identity, since the equation is true regardless of what values are assigned to x. Other examples of identities were discussed in Chapter 1. For instance, the properties $ab = ba$, $a(bc) = (ab)c$, and $a(b + c) = ab + ac$ are examples of identities, since they are true for all choices for the variables.

Two equations are *equivalent* if all values, and no other, that satisfy one of those equations also satisfy the second equation. In other words, two equations that involve the same variable are equivalent if and only if they have the same solution sets. For example, the equations $5x + 2 = 7$, $5x = 5$, and $x = 1$ are equivalent since they each have the solution set $\{1\}$. To solve equations, we shall make use of some of the properties of real numbers discussed in previous chapters; in particular, we make use of the substitution properties of addition and multiplication, which may be stated as follows:

1   ADDITION PROPERTY

The addition of the same number to both sides of an equation produces an equivalent equation. That is, if P, Q, and R represent real numbers, the equations $P = Q$ and $P + R = Q + R$ are equivalent.

2   MULTIPLICATION PROPERTY

The multiplication of both sides of an equation by the same nonzero number produces an equivalent equation. That is, if P, Q, and R represent real numbers with $R \neq 0$, the equations $P = Q$ and $PR = QR$ are equivalent.

First-Degree Equations

Equations such as $2x - 1 = 5$, $3(x + 2) = 7 - 2(x + 1)$, and $\frac{1}{3}x + 3 = 5 - x$ are called *first-degree equations* (or *linear equations*). We shall use the properties above to solve first-degree equations.

EXAMPLE

2   Find the value of x for which the equation $3x - 1 = 11$ is true.

SOLUTION

$$3x - 1 = 11$$
$$3x - 1 + 1 = 11 + 1 \qquad \text{(addition property)}$$
$$3x = 12$$
$$\tfrac{1}{3}(3x) = \tfrac{1}{3}(12) \qquad \text{(multiplication property)}$$
$$x = 4$$

In this example, the addition and the multiplication properties provide a method for writing the equations $3x = 12$ and $x = 4$, which are equivalent to the equation $3x - 1 = 11$. Such a procedure of transforming the given equation $3x - 1 = 11$ to the equivalent equation $x = 4$ is called solving the equation. Thus, in solving the equation $3x - 1 = 11$, we had to "remove" the numbers 3 and -1 from the side containing x. To do so, we first added 1, the additive inverse of -1, to both sides of the equation, because $-1 + 1 = 0$. Next, we multiplied both sides of the equation by $\frac{1}{3}$, the "multiplicative inverse" of 3, because $\frac{1}{3}(3) = 1$. The above procedure of solving first-degree equations is generalized as follows: Employ the addition and multiplication properties; in applying the addition property, we add the additive inverse of the term to be removed, and in applying the multiplication property, we multiply by the multiplicative inverse of the number to be removed.

EXAMPLES

Find the solution set of each of the following first-degree equations and indicate the property applied to produce each new equivalent equation.

3   $\frac{1}{2}x + 2 = 5$

SOLUTION

$$\tfrac{1}{2}x + 2 = 5$$
$$\tfrac{1}{2}x + 2 + (-2) = 5 + (-2) \qquad \text{(addition property)}$$

so that

$$\tfrac{1}{2}x = 3$$
$$2(\tfrac{1}{2}x) = 2(3) \qquad \text{(multiplication property)}$$

Therefore, $x = 6$.

CHECK:

$$\tfrac{1}{2}(6) + 2 \overset{?}{=} 5$$
$$3 + 2 \overset{?}{=} 5$$
$$5 = 5$$

The solution set is $\{6\}$.

4   $3(x - 1) = 5 - x$

SOLUTION

$$\begin{aligned}
3(x - 1) &= 5 - x \\
3x - 3 &= 5 - x &\text{(distributive property)} \\
3x - 3 + 3 &= 5 - x + 3 &\text{(addition property)} \\
3x &= 8 - x \\
3x + x &= 8 - x + x &\text{(addition property)}
\end{aligned}$$

so that

$$4x = 8$$
$$\tfrac{1}{4}(4x) = \tfrac{1}{4}(8) \qquad \text{(multiplication property)}$$

Therefore, $x = 2$. The solution set is $\{2\}$.

Fractional Equations

Equations of the form $(x + 3)/4 = 6$ and $(x - 1)/2x = 3$ are called *fractional equations*. The procedure employed in solving fractional equations is as follows:

1   Determine the L.C.D. of the fractions.

2   Multiply both sides of the equation by the L.C.D. in order to clear the equation of fractions.

3   Solve the resulting equation by the methods previously illustrated.

It should be pointed out that checking the proposed solutions is often essential when working with fractional equations. Such practice will help in rejecting the solution that makes the denominator of the fraction equal to zero.

EXAMPLES

Find the solution set for each of the following equations.

5    $\dfrac{x}{4} + \dfrac{2x}{3} = 5$

SOLUTION.   Multiply both sides of the equation by 12, the L.C.D. of both fractions, so that

$$12\left(\frac{x}{4} + \frac{2x}{3}\right) = 12(5)$$

$$12\left(\frac{x}{4}\right) + 12\left(\frac{2x}{3}\right) = 12(5) \qquad \text{(distributive property)}$$

or $3x + 8x = 60$, so that $11x = 60$ or $x = \frac{60}{11}$.

CHECK:

$$\frac{\frac{60}{11}}{4} + \frac{2\left(\frac{60}{11}\right)}{3} \overset{?}{=} 5$$

$$\tfrac{60}{44} + \tfrac{120}{33} \overset{?}{=} 5$$

$$\tfrac{180}{132} + \tfrac{480}{132} \overset{?}{=} 5$$

$$\tfrac{660}{132} \overset{?}{=} 5$$

$$5 = 5$$

The solution set is $\left\{\dfrac{60}{11}\right\}$.

6    $\dfrac{x-2}{3} - \dfrac{x-3}{5} = \dfrac{13}{15}$

SOLUTION.   The L.C.D. of the fractions in the equation is 15. Multiplying both sides of the equation by 15, we have

$$(15)\left(\frac{x-2}{3} - \frac{x-3}{5}\right) = (15)\frac{13}{15}$$

$$15\left(\frac{x-2}{3}\right) - 15\left(\frac{x-3}{5}\right) = (15)\frac{13}{15}$$

so that

$$5(x-2) - 3(x-3) = 13$$
$$5x - 10 - 3x + 9 = 13$$

or

$$2x - 1 = 13$$
$$2x = 14$$

Thus, $x = 7$.

CHECK:

$$\frac{7-2}{3} - \frac{7-3}{5} \stackrel{?}{=} \frac{13}{15}$$

$$\frac{5}{3} - \frac{4}{5} \stackrel{?}{=} \frac{13}{15}$$

$$\frac{25}{15} - \frac{12}{15} \stackrel{?}{=} \frac{13}{15}$$

$$\frac{13}{15} = \frac{13}{15}$$

The solution set is $\{7\}$.

7     $\dfrac{1}{x} + \dfrac{1}{2} = \dfrac{5}{6x} + \dfrac{1}{3}$

SOLUTION.   Multiplying both sides of the equation by the L.C.D. $6x$, we have

$$6x\left(\frac{1}{x} + \frac{1}{2}\right) = 6x\left(\frac{5}{6x} + \frac{1}{3}\right)$$

$$6x\left(\frac{1}{x}\right) + 6x\left(\frac{1}{2}\right) = 6x\left(\frac{5}{6x}\right) + 6x\left(\frac{1}{3}\right)$$

That is,

$$6 + 3x = 5 + 2x \quad \text{or} \quad 3x - 2x = 5 - 6$$

Therefore, $x = -1$ and the solution set is $\{-1\}$.

8    $\dfrac{1}{x} + \dfrac{1}{x-1} = \dfrac{5}{x-1}$

SOLUTION. The equation $1/x + 1/(x-1) = 5/(x-1)$ is equivalent to the equation

$$\frac{1}{x} = \frac{5}{x-1} - \frac{1}{x-1} \qquad \text{(addition property)}$$

or

$$\frac{1}{x} = \frac{4}{x-1}$$

Multiplying both sides of the equation by the L.C.D. $x(x-1)$, we have

$$x(x-1)\left(\frac{1}{x}\right) = x(x-1)\left(\frac{4}{x-1}\right)$$

$$x - 1 = 4x$$

so that $-3x = 1$ or $x = -\tfrac{1}{3}$. Notice that the denominators x and $x-1$ do not equal zero for $x = -\tfrac{1}{3}$. Hence, the solution set is $\{-\tfrac{1}{3}\}$.

9    $\dfrac{5}{x-5} + 6 = \dfrac{x}{x-5}$

SOLUTION. Multiplying both sides of the equation by $x-5$, we have

$$(x-5)\left(\frac{5}{x-5} + 6\right) = (x-5)\left(\frac{x}{x-5}\right)$$

so that

$$5 + 6(x-5) = x$$
$$5 + 6x - 30 = x$$
$$6x - 25 = x$$
$$5x = 25$$
$$x = 5$$

To check the proposed solution $x = 5$, we run into trouble, since division by zero is impossible. Therefore, we must reject $x = 5$ as a solution and say that the solution set is $\varnothing$.

PROBLEM SET 4.1

In problems 1–10, determine which of the given numbers is a solution of the given equation.

1 $3x + 1 = 7$, for $-1, 0, 1, 2$

2 $2x - 1 = x - 1$, for $0, 1, 2$

3 $2(x + 1) - 3 = 1$, for $-1, 0, 1$

4 $3(x - 2) = 5 - (x - 1)$, for $1, 2, 3$

5 $\dfrac{x + 1}{3} = 2x - 3$, for $0, 1, 2$

6 $\dfrac{2x - 1}{3} + 1 = x + 1$, for $-1, 0, 1$

7 $\dfrac{x}{x - 1} + \dfrac{x + 1}{x} = \dfrac{x + 5}{x}$, for $0, 1, 2, 3$

8 $\dfrac{x + 3}{x} + \dfrac{x + 5}{x + 1} = 7$, for $-1, 0, 1, 2$

9 $x^2 - 5x + 6 = 0$, for $0, 1, 2, 3, 4$

10 $x^2 + x - 12 = 0$, for $-4, -3, 0, 3, 4$

In problems 11–30, solve the first-degree equations, check the solutions, and write the solution sets.

11 $2x - 1 = 9$

12 $\frac{1}{2}x + 3 = 7$

13 $5x + 2 = x + 10$

14 $2x - 1 = 5x + 8$

15 $5(x - 1) = 3x - 7$

16 $8x = 10 + 2(x + 1)$

17 $3x - 2(x + 1) = 2(x - 1)$

18 $8(5x - 1) + 36 = -3(x + 5)$

19 $1 - 2(5 - 2x) = 26 - 3x$

20 $2(1 - 2x) = 3(2x - 4) + 94$

21 $7x - 3(9 - 5x) = 4x - 9x$

22 $7(x - 3) = 4(x + 5) - 47$

23 $6(x - 10) + 3(2x - 7) = -45$

24 $8 = 3x - 8(3 - 2x) - 63$

25 $13 - 5(2 - x) - 18 = 0$

26 $16 - 9(3 - x) + 4x = 15$

27 $34 - 3x = 8(7 - x) + 23$

28 $3(x - 2) + 5(x + 1) = 4(x - 1)$

29 $5 + 8(x + 2) = 23 - 2(2x - 5)$

30 $11 - 7(1 - 2x) = 9(x + 1)$

In problems 31–50, solve the fractional equations, check the solutions, and write the solution sets.

31 $\dfrac{5x}{4} - 1 = \dfrac{3x}{4} + \dfrac{1}{2}$

32 $x + \dfrac{16}{3} = \dfrac{3x}{2} + \dfrac{25}{6}$

33 $x - 1 = \dfrac{2x}{5} - \dfrac{7}{5}$

34 $\dfrac{9x}{11} + \dfrac{3}{11} = 2$

35 $\dfrac{3x - 6}{4} - \dfrac{6 + x}{6} + \dfrac{2x}{3} = 5$

36 $\dfrac{3x - 2}{3} + \dfrac{x - 3}{2} = \dfrac{5}{6}$

37 $\dfrac{x + 9}{4} = 2 + \dfrac{3}{14}(2x - 3)$

38 $\dfrac{x - 14}{5} + 4 = \dfrac{x + 16}{10}$

39 $\dfrac{1}{2x} + \dfrac{5}{8} = \dfrac{3}{x}$

40 $\dfrac{5}{x} + \dfrac{3}{8} = \dfrac{7}{16}$

41 $\dfrac{1}{x} + \dfrac{2}{x} = 3 - \dfrac{3}{x}$

42 $\dfrac{2}{3x} + \dfrac{1}{6x} = \dfrac{1}{4}$

43 $\dfrac{8}{x - 3} = \dfrac{12}{x + 3}$

44 $\dfrac{x}{x - 1} - \dfrac{3}{x + 1} = 1$

45 $\dfrac{2x}{x - 2} = \dfrac{4}{x - 2} - 1$

46 $\dfrac{9}{5x - 3} = \dfrac{5}{3x + 7}$

47 $\dfrac{x - 1}{x + 1} + x = 2 + x$

48 $\dfrac{3(x - 1)}{x + 7} = -\dfrac{3}{5}$

49 $\dfrac{x}{x + 1} + 2 = \dfrac{3x}{x + 2}$

50 $\dfrac{1}{x - 1} + \dfrac{x}{(x + 4)(x - 1)} = \dfrac{1}{x + 4} + \dfrac{1}{(x + 4)(x - 1)}$

4.2 Literal Equations and Formulas

Equations that involve one or more literal numbers besides the one whose value is required are called *literal equations*. Examples of such equations are $ax + b = d$, $mx + q = 3c$, and $2l + 2w = p$. Formulas from science, engineering, business, and other disciplines are literal equations.

The steps used in solving such equations are identical with those employed in Section 4.1; we treat the equation as though it was one in a single variable (the variable to be solved for) and the other literal numbers are taken as constants. The following examples will illustrate this procedure.

EXAMPLES

Find the solution set for each of the following equations by solving for x.

1    $\dfrac{3x}{2} - a = x + 3a$, where a is constant.

SOLUTION. $\dfrac{3x}{2} - a = x + 3a$, so that

$$3x - 2a = 2x + 6a \qquad \text{(multiplication property)}$$

or

$$3x - 2x = 6a + 2a \qquad \text{(addition property)}$$
$$x = 8a$$

Therefore, the solution set is $\{8a\}$.

2    $a(x - b) = cx + d$, where a, b, c, and d are constants.

SOLUTION

$$a(x - b) = cx + d$$
$$ax - ab = cx + d \qquad \text{(distributive property)}$$
$$ax - ab - cx = d \qquad \text{(addition property)}$$
$$ax - cx = d + ab \qquad \text{(addition property)}$$
$$(a - c)x = d + ab \qquad \text{(distributive property)}$$
$$\left(\dfrac{1}{a - c}\right)(a - c)x = \left(\dfrac{1}{a - c}\right)(d + ab) \qquad \text{(multiplication property)}$$

so that

$$x = \dfrac{d + ab}{a - c}$$

Therefore, the solution set is $\left\{\dfrac{d + ab}{a - c}\right\}$.

3    $\dfrac{a}{x} = \dfrac{5}{x} + 1$, where a is constant.

SOLUTION

$$\frac{a}{x} = \frac{5}{x} + 1$$

$$x\left(\frac{a}{x}\right) = x\left(\frac{5}{x} + 1\right) \qquad \text{(multiplication property)}$$

$$a = 5 + x$$

$$a - 5 = x \qquad \text{(addition property)}$$

Therefore, the solution set is $\{a - 5\}$.

We know from geometry that the area A of a rectangle (see Appendix B, Section B.1) is given by the formula $A = lw$. Now suppose that we want to find the length of a rectangle with known area and width. We can do so by solving the formula $A = lw$ for l. Thus, we have

$$A = lw \qquad \text{or} \qquad \frac{1}{w}(A) = \frac{1}{w}(lw)$$

so that

$$\frac{A}{w} = l$$

The following examples further illustrate how we can solve formulas for specific variables.

EXAMPLES

4 Solve the formula $F = \frac{9}{5}C + 32$ for C, and then find C when $F = 98$. Also, find C when $F = 212$.

SOLUTION

$$F = \frac{9}{5}C + 32$$

$$F - 32 = \frac{9}{5}C + 32 - 32 \qquad \text{(addition property)}$$

$$F - 32 = \frac{9}{5}C$$

$$\frac{5}{9}(F - 32) = \frac{5}{9}\left(\frac{9}{5}C\right) \qquad \text{(multiplication property)}$$

Hence,

$$C = \frac{5}{9}(F - 32)$$

When $F = 98$, we have

$$C = \tfrac{5}{9}(98 - 32) = \tfrac{5}{9}(66) = 36\tfrac{2}{3}$$

When $F = 212$, we have

$$C = \tfrac{5}{9}(212 - 32) = \tfrac{5}{9}(180) = 100$$

5   The area A of a trapezoid is represented by the formula $A = \tfrac{1}{2}(a+b)h$, where h is the height, a is one base, and b is the other base. Solve for h in terms of a, b, and A, and then find h if $A = 165$ square inches, $a = 18$ inches, and $b = 12$ inches.

SOLUTION. $A = \tfrac{1}{2}(a+b)h$, so that $2A = (a+b)h$. Thus, $h = 2A/(a+b)$. Then, substituting the given values, we have

$$h = \frac{2(165)}{18 + 12} = 11 \text{ inches}$$

6   If the original cost of a property is C dollars and it is depreciating linearly over N years, its undepreciated value V at the end of n years is given by the formula $V = C - (C/N)n$.
a)   Solve this formula for C.
b)   Solve this formula for n.

SOLUTION

a) $V = C - \left(\dfrac{C}{N}\right)n$

so that

$$V = C\left(1 - \frac{1}{N}n\right) = C\left(1 - \frac{n}{N}\right) = C\left(\frac{N-n}{N}\right)$$

Thus,

$$C = V\left(\frac{N}{N-n}\right) = \frac{VN}{N-n}$$

b) $V = C - \left(\dfrac{C}{N}\right)n$

so that

$$\left(\frac{C}{N}\right)n = C - V$$

Thus,

$$n = \frac{N}{C}(C - V) = N - \frac{NV}{C}$$

PROBLEM SET 4.2

In problems 1–20, solve the literal equations for the indicated variable and write the solution sets.

1 $ax = b$, for x

2 $\frac{x}{a} = b$, for x

3 $x + a = b$, for x

4 $x - a = b$, for x

5 $ax + b = c$, for x

6 $ax - b = c$, for x

7 $\frac{ay}{b} = c + \frac{d}{b}$, for y

8 $\frac{-5}{a} + 3b = 16 + \frac{9}{a}$, for a

9 $\frac{3}{x} - \frac{4}{b} = \frac{5}{3b}$, for x

10 $\frac{ax}{7} - \frac{bc}{3} = \frac{2x}{3b^2}$, for x

11 $\frac{a + b}{x} + \frac{a - b}{x} = 2a$, for x

12 $\frac{b - x}{2} = 2a - b - 3x$, for x

13 $a(b - y) = b(y - a)$, for y

14 $\frac{b}{x} - \frac{a}{3} = \frac{b - a}{3x}$, for x

15 $\frac{x - a}{b} = \frac{x - b}{a}$, for x

16 $a(y - a) = ab + 2b(y - b)$, for y

17 $3(a - 2b) + 4(b + a) = 5$, for a

18 $\frac{x + a}{2x - b} = \frac{x + b}{2x - a}$, for x

19 $\frac{3}{a - x} + \frac{a}{a + x} = \frac{1}{a^2 - x^2}$, for x

20 $\frac{1}{y} + \frac{2}{y + a} = \frac{3}{y - a}$, for y

In problems 21–36, solve each formula for the indicated variable.

21 $V = lwh$, for l

22 $A = \frac{1}{2}bh$, for h

23 $s = \frac{1}{2}gt^2$, for g

24 $F = mx + b$, for m

25 $E = I(R + r)$, for R

26 $s = gt^2 + ut$, for u

27 $C = 2\pi r$, for r

28 $s = \frac{n}{2}[2a + (n - 1)d]$, for d

29 $V = \pi r^2 h$, for h

30 $S = 2\pi rh$, for h

31 $P = 2l + 2w$, for w

32 $s = \frac{a - rl}{l - r}$, for r

33 $L = a + (m - 1)d$, for m **34** $\dfrac{1}{C} = \dfrac{1}{a} + \dfrac{1}{b} + \dfrac{1}{c}$, for c

35 $I = Prt$, for t ◁**36** $pv = k\left(1 + \dfrac{t}{m}\right)$, for t

37 The volume V of a square prism is given by the formula $V = lw$
Solve the formula for l and then find l when $V = 36$ cubic inches an
$w = 2$ inches.

38 The volume V of a right circular cone is given by the formula $V = \frac{1}{3}\pi r^2$
Solve the formula for h and then find h when $r = 6$ inches and $V = 48$
cubic inches.

39 If P dollars is invested at a simple interest rate of r percent annually
the amount A dollars accumulated after n years is given by th
formula $A = P(1 + nr)$. Solve this formula for n.

40 A rocket is fired vertically upward at an initial velocity v feet pe
second; its distance s feet above the ground t seconds after bein
fired is given by the formula $s = vt - 16t^2$. Solve this formula for

◁**41** If two electrical resistances R_1 and R_2 are connected in parallel, th
resulting resistance R is given by the formula $1/R = (1/R_1) + (1/R_2$
Find R_1 in terms of R and R_2.

42 If a taxicab fare is 80 cents for the first mile and 50 cents for eac
additional mile, the cost C for n miles is represented by the formul
$C = 80 + 50(n - 1)$. If a trip cost \$3.75, how many miles were traveled

4.3 Applications of First-Degree Equations

In this section we consider some applications (word problems c
story problems) of first-degree equations in solving problems that dea
with finding plane areas from geometry, converting temperatur
from Fahrenheit degrees to centigrade degrees in physics, computin
interest in business, and solving many other everyday types of appl
cations. Since most story problems are not solved by trial-and-errc
methods, the following steps are suggested as a guide:

1 Read the problem carefully to determine what facts are give
and precisely what is to be found.

2 Draw diagrams if possible to help interpret the given informa
tion and the nature of the solution.

3 Express the given information and the solution in mathematic
symbols and/or numbers.

4 Express the problem in the form of an equation.

5 Solve the equation.

6 Check the solution to see if it satisfies the equation and to see if it makes sense with regard to the story problem.

The procedure is illustrated in the following examples.

EXAMPLES

1 Find a number such that when two times the number is increased by 5, the result is 11.

SOLUTION. We first note that the solution is a number. Let x be the symbol for this unknown number. Next, identify the given statement of equality and translate it into a mathematical expression: "When two times the number is increased by 5, the result is 11," which translates into the equation $2x + 5 = 11$. Solving this equation, we get

$$2x + 5 = 11$$
$$2x = 6$$
$$x = 3$$

CHECK:

$$2(3) + 5 \overset{?}{=} 11$$
$$6 + 5 \overset{?}{=} 11$$
$$11 = 11$$

The solution is correct.

2 The sum of the ages of Joe and Jane is 35. Joe is 5 years older than twice Jane's age. How old are they?

SOLUTION. Let x years be Jane's age; then $2x + 5$ years is Joe's age, and $2x + 5 + x$ years is the sum of their ages. Therefore, we have $2x + 5 + x = 35$ or $3x = 30$, so that $x = 10$. Therefore, Jane's age is 10 years and Joe's age is 25 years.

3 A piggybank holds nickels, dimes, and quarters. If it contains four times as many nickels as dimes, and two less quarters than dimes, and if the total value of its contents is \$2.25, how many dimes are there in the piggybank?

SOLUTION. Let x = number of dimes in the piggybank. Then $4x$ = number of nickels and $x - 2$ = number of quarters. Since the total value of the coins is \$2.25, the equation is

$$0.10x + 0.05(4x) + 0.25(x - 2) = 2.25$$
$$10x + 5(4x) + 25(x - 2) = 225$$
$$10x + 20x + 25x - 50 = 225$$
$$55x - 50 = 225$$
$$55x = 275$$
$$x = 5$$

Hence, there are 5 dimes in the piggybank.

4   A water tank can be filled by an intake pipe in 4 hours and can be emptied by a drain pipe in 5 hours. How long would it take to fill the tank with both pipes open?

SOLUTION.   Let t be the number of hours required to fill the tank with both pipes open. The rate of flow for the intake pipe is one-fourth of a tank per hour, and the rate of flow for the drain pipe is one-fifth of a tank per hour. Hence,

$$\left(\frac{1}{4}\right)t = \frac{t}{4} = \text{fraction of a tank that the intake pipe can fill in } t \text{ hours}$$

$$\left(\frac{1}{5}\right)t = \frac{t}{5} = \text{fraction of a tank that the drain pipe can empty in } t \text{ hours}$$

Since both pipes are open and one tank is to be filled, we have the equation $(t/4) - (t/5) = 1$ or $5t - 4t = 20$, so that $t = 20$ hours.

Geometric Applications

Problems involving geometric figures usually require a formula such as $A = lw$, that is, the area of a rectangle is the product of its length and width. Other formulas from geometry are listed in Appendix B for reference.

EXAMPLE

5   Find the dimensions of the floor of a rectangular classroom whose perimeter is 130 feet and whose length is 5 feet more than its width.

SOLUTION.   The unknowns in this problem are the length and width of the classroom. Suppose that $x = $ the width in feet, so that $x + 5$ = the length in feet (Figure 4.1). We have the known quantity A = perimeter = 130 feet and the formula for the perimeter of a rec tangle is $P = 2l + 2w$, so that $2x + 2(x + 5) = 130$. Solving this equation, we get $2x + 2x + 10 = 130$ or $4x = 120$, so that $x = 30$ and $x + 5 = 35$. Thus, the width is 30 feet and the length is 35 feet.

Figure 4.1

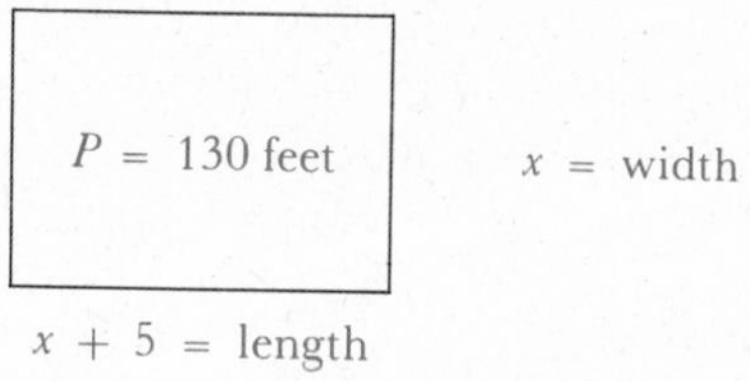

$P = 130$ feet $x = $ width

$x + 5 = $ length

Investment Applications

Investment problems are those involving income from various sources such as stocks and bonds, mutual funds, and savings accounts offered by banks and credit unions. In this section we deal with investments that offer simple interest income. Investments that offer compound interest will be considered later in Chapter 8.

The income from a simple interest investment is given by the formula $I = Prt$, where I represents the income in dollars earned on an investment of P dollars at an annual interest rate of r percent for t years.

EXAMPLE

6 A woman invested $24,000 in the stock market. She bought two kinds of stock. One kind pays 8 percent annual dividends and the other kind pays 11 percent annual dividends. If the total annual income from both stocks is $2,340, how much did she invest in the 8 percent stock?

SOLUTION.   Let $\$x = $ amount invested at 8 percent, so $\$(24{,}000 - x)$ = amount invested at 11 percent. The income I_1 from the 8 percent stock is $I_1 = x(8 \text{ percent})(1) = 0.08x$, and the income I_2 from the 11 percent stock is $I_2 = (24{,}000 - x)(11 \text{ percent})(1) = 0.11(24{,}000 - x)$. The total income from both stocks is $I_1 + I_2$, or $\$2{,}340$, so we have the equation $0.08x + 0.11(24{,}000 - x) = 2{,}340$. Solving this equation, we get

$$8x + 11(24{,}000 - x) = 234{,}000$$
$$8x + 264{,}000 - 11x = 234{,}000$$
$$-3x = -30{,}000$$
$$x = 10{,}000$$

Hence, the woman invested $10,000 in the 8 percent stock.

Mixture Applications

Many applications from nursing, medicine, pharmacy, and industry are classified as "mixture problems." These applications involve mixing solutions or substances containing different percentages of ingredients to obtain a solution or a substance containing a specified percent of the ingredients. The following example illustrates this type of application.

EXAMPLE

7   A nurse has a medicine containing 25 percent alcohol. How much medicine containing no alcohol should she add to 120 cubic centimeters of the 25 percent mixture so that the final mixture will contain only 20 percent alcohol?

SOLUTION.   The amount of alcohol contained in the two medicines to be mixed must equal the amount of alcohol in the final mixture. That is,

$$\begin{pmatrix} \text{amount of} \\ \text{alcohol in the} \\ 25\% \text{ medicine} \end{pmatrix} + \begin{pmatrix} \text{amount of} \\ \text{alcohol in the} \\ 0\% \text{ medicine} \end{pmatrix} = \begin{pmatrix} \text{amount of} \\ \text{alcohol in the} \\ 20\% \text{ medicine} \end{pmatrix}$$

Let

$$x = \text{amount of medicine in cubic centimeters containing no alcohol (0\% medicine)}$$

$$x + 120 = \text{amount of final mixture in cubic centimeters (20\% medicine)}$$

so that

$$(25\% \text{ of } 120) + (0\% \text{ of } x) = 20\% \text{ of } (x + 120)$$

or $$0.25(120) = 0.20(x + 120)$$

Solving this equation, we get

$$25(120) = 20(x + 120)$$
$$3{,}000 = 20x + 2{,}400$$
$$600 = 20x$$
$$30 = x$$

Hence, the nurse must add 30 cubic centimeters.

Physics Applications—Uniform Motion

We know from physics that if an object travels at a uniform rate of speed, the distance traveled during a specified time interval is given by

$$\text{distance} = \text{rate} \times \text{time}$$

For example, an automobile traveling at a uniform rate of 50 miles per hour for 2 hours will go 50(2), or 100 miles. If we let $d =$ distance, $r =$ rate of speed, and $t =$ time, we have the formula $d = rt$. The following example illustrates this application.

EXAMPLE

8   A girl rides her bicycle 5 miles from her home to the school bus stop at the rate of 8 miles per hour. She arrives in time to catch the school bus, which travels at 25 miles per hour. If the girl spent $1\frac{1}{2}$ hours traveling from home to school, how far did she travel on the school bus?

SOLUTION.   Let $x =$ distance in miles traveled on bus. We draw a line diagram to help solve the problem.

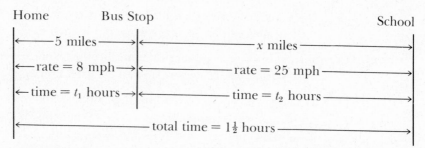

We see from the diagram that the total time for the trip is $t_1 + t_2$ $= 1\frac{1}{2} = \frac{3}{2}$, where t_1 hours represents the time to ride the bicycle and t_2 hours represents the time on the bus. Using the formula $d = rt$, we have $t = d/r$, so that $t_1 = 5/8$ and $t_2 = x/25$. Thus, the equation is $(5/8) + (x/25) = 3/2$. Solving this equation, we get

$$200\left(\frac{5}{8} + \frac{x}{25}\right) = (\tfrac{3}{2})200$$

$$125 + 8x = 300$$

$$8x = 175$$

$$x = 21\tfrac{7}{8}$$

Hence, the girl traveled $21\frac{7}{8}$ miles on the school bus.

PROBLEM SET 4.3

Solve the following word problems and check your solution.

1   Three more than twice a certain number is 57. Find the number.

2   Find two numbers whose sum is 18 if one number is 8 larger than the other.

3   Find two consecutive even integers such that seven times the first exceeds five times the second by 54.

4   Find three consecutive even integers such that the first plus twice the second plus four times the third equals 174.

5   One-fourth of a number is 3 greater than one-sixth of it. Find the number.

6   Two-thirds of a number plus five-sixths of the same number is equal to 42. What is the number?

7   Gus is 10 years older than his brother, and 6 years from now he will be twice as old as his brother is then. How old is each now?

8   Linda's mother is three times as old as Linda, and 14 years from now she will be twice as old as Linda is then. How old is each now?

9   Roy is 3 years older than his brother, and 4 years from now the sum of their ages will be 33 years. How old is each now?

10   Doreen is 5 years younger than her brother, and 3 years ago the sum of their ages was 23 years. How old is each now?

11   A payphone slot receives quarters, dimes, and nickels. When the phone box was emptied, there was $6.50 in coins. If there are 4 more dimes than quarters and three times as many nickels as dimes, find the number of coins of each kind.

12   In a collection of nickels and quarters there are four times as many nickels as quarters, and the value of the collection is $3.60. Find the number of each kind of coin.

13   A parking meter slot receives dimes and nickels. Emptying the box produced 70 coins worth $4.85. How many nickels and how many dimes were there?

14   A person withdrew $75 in $1 bills and $5 bills from the bank. There were 3 more $1 bills than $5 bills. How many were there of each kind?

15   A vending machine receives dimes and quarters. In emptying the coin box of this machine, there were found coins worth $12.50. If the number of dimes is 20 more than the number of quarters, find the number of each kind of coin.

16 A charter jet has 210 passengers signed up to make a trip at a fixed price for the flight. If 50 more people sign up for the trip, each person will save $10 on the cost of transportation. What is the cost per passenger if 260 people make the trip?

17 One pipe can fill a tank in 18 minutes and another pipe can fill it in 24 minutes. The drain pipe can empty the tank in 15 minutes. With all pipes open, how long will it take to fill the tank?

18 John can mow a lawn in 1 hour and 20 minutes and Tom can mow the same lawn in 2 hours. How long would it take John and Tom together to mow the lawn?

19 A student had marks of 72 and 84 on his last two tests and he wishes to raise his average to 85. What must he get on a third test to accomplish this?

20 Find the dimensions of a rectangle if its width is 8 centimeters less than its length and half its perimeter is 24 centimeters.

21 The length of a rectangle is 7 inches greater than its width. If its length is increased by 2 inches and its width is decreased by 3 inches, its area is decreased by 37 square inches. Find its dimensions.

22 The sides of two squares differ by 6 centimeters and their areas differ by 468 square centimeters. Find the lengths of the sides of the squares.

23 The length of a rectangle is four times its width. What are its dimensions if its perimeter is 150 meters?

24 The length of a rectangle is 17 less than three times its width, and its perimeter is 238 centimeters. Find its dimensions.

25 The length and width of a square are increased by 6 feet and 8 feet, respectively, thereby increasing its area by 188 square feet. Find a side of the square.

26 A woman has $8,000 invested at 5 percent and $2,000 invested at 7 percent. How much must she invest at 8 percent to make an average of 6 percent of her investment?

27 In a partnership investment, the agreement was that the general partner income from the business should be 20 percent greater than the limited partner income. If the annual income of the investment was $8,800, how much did each receive?

28 Linda borrows a certain sum at $5\frac{1}{2}$ percent interest and a sum that is $5,000 greater at 7 percent. How much does she borrow if the interest for one year is $1,850?

29 A certain amount of money is invested at 6 percent. Another amount of $6,000 is invested at 5 percent. The total income (from the two investments) is at the average rate of $5\frac{1}{4}$ percent of the entire investment. How much is invested at 6 percent?

30 A man has a $1,200 exemption from income tax but pays 20 percent tax on the remainder of his income. If his total taxes paid after the exemption is $5,440, find his total income.

31 A retail store invested $25,000 in two kinds of toys. For one year, it made a profit of 15 percent from the first kind but lost 5 percent from the second kind. If the one-year income from the two investments was a return of 8 percent on the entire amount invested, how much had it invested in each kind of toy?

32 A grocer has 100 pounds of candy selling at $1.80 per pound. How many pounds of a different candy worth $3.00 per pound should he mix with the 100 pounds in order to have a mixture worth $2.40 per pound?

33 Of 80 pounds of saltwater, 12 percent is salt. How many pounds of pure water must be added in order to have a solution which will contain $6\frac{1}{4}$ percent salt?

34 A chemist has a gallon containing 4 percent solution of a certain chemical. How much water should she add in order that the resulting mixture will be a 1 percent solution?

35 How much water must be added to 4 quarts of an acid solution which is 12 percent full strength to get a mixture that is 10 percent full strength?

36 A man started for a town 75 miles away at the rate of 30 miles per hour. After traveling part of the way, road construction forced him to reduce his speed to 22 miles per hour for the remainder of the trip. If the entire trip took 2 hours and 38 minutes, how far did he travel at the reduced speed?

37 If a freight train traveling at 30 miles per hour is 300 miles ahead of an express train traveling at 55 miles per hour, how long will it take the express train to catch up with the freight train?

38 An airplane travels 1,620 kilometers in the same time a train travels 180 kilometers. If the airplane goes 480 kilometers per hour faster than the train, find the rate of each.

4.4 Inequalities

An *inequality* is a statement that an expression representing some real number is greater than (or less than) another expression representing a real number.

Recall from Chapter 1 that the real numbers were located on the real line as follows: The numbers located to the right of the origin are the positive real numbers and they are arranged in order of increasing magnitude; and the numbers located to the left of the origin are the negative real numbers and they are arranged in order of decreasing magnitude. Accordingly, we agree to extend this "ordering" to the entire real line in the following manner: Any number whose corresponding point on the number line lies to the right of the corresponding point of a second number is said to be *greater than* (denoted by $>$) the second number. The second number is also said to be *less than* (denoted by $<$) the first number. For example, $7 > -3$ or $-3 < 7$, since the point that corresponds to the number 7 lies to the right of the point that corresponds to the number -3 (Figure 4.2).

Figure 4.2

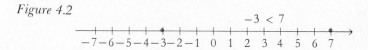

$$-3 < 7$$

$-7 -6 -5 -4 -3 -2 -1 \quad 0 \quad 1 \quad 2 \quad 3 \quad 4 \quad 5 \quad 6 \quad 7$

More formally, we have the following definition.

DEFINITION 1   INEQUALITIES

Let a and b be any two real numbers. We say that *a is less than b* ($a < b$), or, equivalently, *b is greater than a* ($b > a$) if $b - a$ is a positive number.

For example, $3 < 7$ (or $7 > 3$), since the difference $7 - 3$ is 4, a positive number. Also, $-5 < -3$ (or $-3 > -5$), since $-3 - (-5) = 2$.

Properties of Inequalities

We shall now investigate the properties of inequalities. Suppose that we are given two real numbers a and b. If these two numbers are located on the number line, only one of three possible situations can occur (Figure 4.3).

Figure 4.3

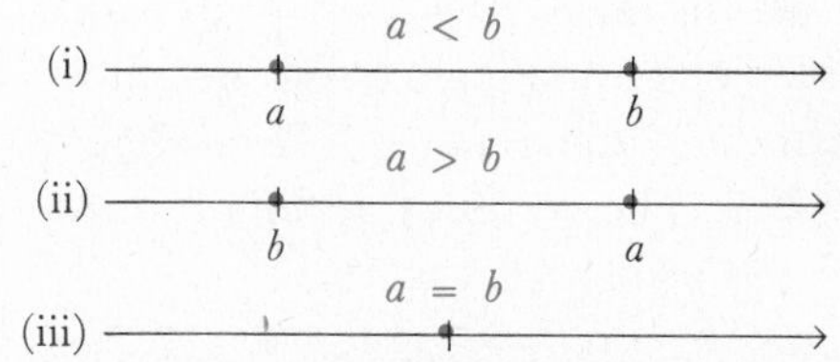

(i) $\quad a < b$

$\qquad a \qquad\qquad b$

(ii) $\quad a > b$

$\qquad b \qquad\qquad a$

(iii) $\quad a = b$

These three possible situations suggest the following property, which is stated as follows.

PROPERTY 1 TRICHOTOMY LAW

If a and b are real numbers, then one and only one of the following situations can hold:

$$a < b \qquad a > b \qquad \text{or} \qquad a = b$$

We can also indicate that a number is positive or negative by use of the inequality relation. If the number a is positive, then we write $a > 0$ (Figure 4.4), that is, a lies to the right of 0. Similarly, if the number b is negative, then we write $b < 0$ (Figure 4.4), that is, b lies to the left of zero.

Figure 4.4

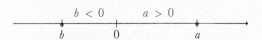

If $a < b$ (a lies to the left of b) and if $b < c$ (b lies to the left of c), then a must lie to the left of c; that is, $a < c$ (Figure 4.5).

Figure 4.5

For example, since $3 < 5$ and $5 < 7$, then $3 < 7$. This notion can be stated as follows.

PROPERTY 2 TRANSITIVE LAW OF INEQUALITIES

If a, b, and c are real numbers such that $a < b$ and $b < c$, then $a < c$.

Let us now consider the effect of adding the same number to both sides of an inequality. For instance, suppose that 3 is added to both sides of the inequality $2 < 4$. We have $2 + 3 < 4 + 3$, or, equivalently, $5 < 7$. The result of adding 3 to both sides of the inequality is illustrated in Figure 4.6.

Figure 4.6

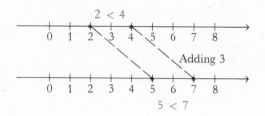

This example can be generalized as follows.

PROPERTY 3   ADDITION LAW OF INEQUALITIES

If $a < b$ and c is any real number, then $a + c < b + c$.

Since subtraction is defined in terms of addition, we can use Property 3 to show the fact that "if the same number is subtracted from both sides of an inequality, the order is preserved." That is, if a and b are real numbers, and $a < b$, then $a + (-c) < b + (-c)$. Using the definition of subtraction, we have

$$a + (-c) = a - c \quad \text{and} \quad b + (-c) = b - c$$

so that if $a < b$ and c is any real number, then $a - c < b - c$.

Consider the effect of multiplying both sides of an inequality by any real number. We shall see that the result of such multiplication can either preserve or reverse the original inequality, depending on whether we multiply by a positive number or a negative number. For example, if both sides of the inequality $3 < 5$ are multiplied by 2, the result is $6 < 10$, whereas if $3 < 5$ is multiplied on both sides by -2, we have $-6 > -10$. That is, when multiplying both sides of an inequality by a positive number, we preserve the order of inequality; and when multiplying by a negative number, we reverse the order of inequality. This can be generalized as follows.

PROPERTY 4   MULTIPLICATION LAW OF INEQUALITIES

a)   If $a < b$ and $c > 0$, then $ac < bc$.

b)   If $a < b$ and $c < 0$, then $ac > bc$.

It is important to note that the examples above, which we used to motivate this property, do not suffice as a proof; they merely serve as illustrations of the property. We shall now consider a proof of part (a) of Property 4, and leave part (b) for the problem sets.

*Reversal of
sign of ineq.*

PROOF OF (a). $a < b$ means $b - a = p, p > 0$ (why?), so that $c(b-a)=cp$ where $cp > 0$, since $c > 0$ and $p > 0$ (the product of two positive numbers is a positive number). Hence, $bc - ac = cp$, which is positive, so that, by the definition of inequality, we have $ac < bc$.

Since division by a nonzero real number can be expressed as multiplication, we can conclude from Property 4 that division of both sides of an inequality by a positive number does not change the order of the inequality, whereas division on both sides of an inequality by a negative number reverses the order; that is,

a)   If $a < b$ and $c > 0$, then $a/c < b/c$.

b)   If $a < b$ and $c < 0$, then $a/c > b/c$.

For example, since $3 < 4$ and if $c = 5$, then $\frac{3}{5} < \frac{4}{5}$; however, if $c = -5$ then $-\frac{3}{5} > -\frac{4}{5}$.

In summary, we have the following addition and multiplication properties of inequalities:

1   If $a < b$ and c any number, then $a + c < b + c$.

2   If $a < b$ and $c > 0$, then $ac < bc$.

3   If $a < b$ and $c < 0$, then $ac > bc$.

We have similar properties for the equality relation (Chapter 1) that is:

1   If $a = b$ and c is any number, then $a + c = b + c$.

2   If $a = b$ and c is any number, then $ac = bc$.

We can combine the two sets of properties into one set that includes both the relation of equality and inequality. (We interpret the symbol $\leq$ to mean *less than or equal*. That is, if $a \leq b$, then either $a < b$ or $a = b$. Both relations cannot be true simultaneously; one or the other relation is true.) Hence, we have the following result:

1   If $a \leq b$ and c is any number, then $a + c \leq b + c$.

2   If $a \leq b$ and $c > 0$, then $ac \leq bc$.

3   If $a \leq b$ and $c < 0$, then $ac \geq bc$.

EXAMPLES

Illustrate each of the following sets on a number line.

1   $\{x \mid x < 2\}$

SOLUTION. Notice that the point 2 is excluded from this set (Figure 4.7), as indicated by the use of the symbol).

Figure 4.7

2   $\{x \,|\, x \geq 2\}$

SOLUTION. Here, the point 2 is included in the set (Figure 4.8), as indicated by the use of the symbol [.

Figure 4.8

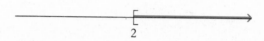

3   $\{x \,|\, x < 3\} \cap \{x \,|\, x \geq -1\}$

SOLUTION. It should be noted that the point -1 is included in this set, while the point 3 is excluded. This set can also be indicated as $\{x \,|\, -1 \leq x < 3\}$ or even more briefly as $-1 \leq x < 3$ (Figure 4.9).

Figure 4.9

4   $\{x \,|\, x < -2\} \cup \{x \,|\, x > 3\}$

SOLUTION. Notice that the point -2 is excluded, and the point 3 is also excluded (Figure 4.10).

Figure 4.10

First-Degree Inequalities

Inequalities such as $3x + 2 \geq 5$ and $(-3x + 5)/2 < 3$ are called *first-degree inequalities* or *linear inequalities*.

The *solution set* of a first-degree inequality is the set of all numbers which makes the inequality a true statement. *Equivalent inequalities* are those inequalities that have the same solution set. Inequalities that are true for all permissible values of the variable involved, such as $x^2 + 4 > 0$, where x is any real number, are called *unconditional* or *absolute* inequalities. Inequalities that are true only for certain values of the variables involved are called *conditional inequalities*. For example,

$3x - 4 > 5$ is only true for $x > 3$. The process of solving first-degree inequalities follows the same pattern that was previously used in solving first-degree equations. In this case we apply the properties above to the given inequality to produce a series of equivalent inequalities until an inequality of one of the forms $x < a$, $x > a$, $x \leq a$, or $x \geq a$ is obtained. For example, to find the solution set of the first-degree inequality $5x + 6 > 8x + 2$: First, we subtract $8x$ from both sides of the inequality, so that $-8x + 5x + 6 > -8x + 8x + 2$, or $-3x + 6 > 2$. Next, we add -6 to both sides of the inequality to get $-3x + 6 - 6 > 2 - 6$ or $-3x > -4$. Dividing both sides by -3 and reversing the order of the inequality, we have $x < \frac{4}{3}$. Thus, the solution set is $\{x \mid x < \frac{4}{3}\}$.

EXAMPLES

Find the solution set of each of the following first-degree inequalities and represent the solution on the real line.

5    $3x - 2 < 7$

SOLUTION

$$3x - 2 < 7$$
$$3x - 2 + 2 < 7 + 2 \qquad \text{(addition property)}$$

We have $3x < 9$, so that

$$\tfrac{1}{3}(3x) < \tfrac{1}{3}(9) \qquad \text{(multiplication property)}$$

Therefore, $x < 3$. Hence, the solution set is $\{x \mid x < 3\}$ (Figure 4.11).

Figure 4.11

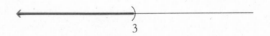

3

6    $x + 2 > 7x - 1$

SOLUTION

$$x + 2 > 7x - 1$$

Adding $-7x$ to both sides of the inequality, we get

$$-6x + 2 > -1$$

Adding -2 to both sides, we have

$$-6x > -3$$

Multiplying each side by $-\frac{1}{6}$ and reversing the order of the inequality, we have $x < \frac{1}{2}$. The solution set is $\{x \,|\, x < \frac{1}{2}\}$ (Figure 4.12).

Figure 4.12

7   $3 - x \leq -4$

SOLUTION

$$3 - x \leq -4$$

Adding -3 to each side, we get

$$-x \leq -7$$

Multiplying both sides by -1 and reversing the order of inequality, we get $x \geq 7$ (Figure 4.13). The solution set is $\{x \,|\, x \geq 7\}$.

Figure 4.13

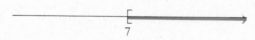

8   $\dfrac{3x + 7}{5} < 2x - 1$

SOLUTION

$$\frac{3x + 7}{5} < 2x - 1$$

Multiplying both sides by 5, we get

$$3x + 7 < 10x - 5 \qquad \text{or} \qquad 3x + 12 < 10x$$

Adding $-3x$ to both sides, we have

$$12 < 7x \qquad \text{or} \qquad \tfrac{12}{7} < x \qquad \text{or} \qquad x > \tfrac{12}{7}$$

The solution set is $\{x \,|\, x > \frac{12}{7}\}$ (Figure 4.14).

Figure 4.14

Some word problems require an inequality rather than an equation to express the stated conditions. The method of solving such word problems is the same as that presented in Section 4.3.

EXAMPLE

9   Judy saves $100 monthly in a teacher's credit union. How many months will it take her to save at least $3,865 in order to pay cash for a new car?

SOLUTION.   Let $x =$ the number of months that it takes Judy to save at least $3,865. Then $100x$ represents the amount she can save in x months. The problem is to determine a positive integer x so that

$$100x \geq 3,865 \quad \text{or} \quad x \geq \frac{3,865}{100} \quad \text{or} \quad x \geq 38.65$$

Hence, the positive integer x must be 39 or greater. That is, it will take her 39 or more months to save at least $3,865. The solution set is $\{x \mid x \geq 39\}$.

PROBLEM SET 4.4

In problems 1–10, indicate the properties of inequalities which explain each statement.

1   If $x < 10$, then $-2x > -20$.   2   If $a < 5$, then $a - 3 < 2$.

3   If $x < -3$ and $-3 < y$, then $x < y$.   4   If $0 < a < 1$, then $a^3 < a$.

5   If $a < b$, then $4a < 4b$.   6   If a is a real number, then either $a = 3$ or $a < 3$ or $a > 3$.

7   If $7x + 8 < 15$, then $7x < 7$.   8   If $-\frac{x}{3} \leq -5$, then $x \geq 15$.

9   If $6t \geq -5$, then $-18t \leq 15$.   10   If $-5x < 20$, then $x > -4$.

In problems 11–16, illustrate each set on the number line.

11    $\{x \mid x < 1\}$ 12 $\{x \mid x \geq -3\}$

13    $\{x \mid x < -2\} \cup \{x \mid x \geq 2\}$ 14 $\{x \mid x \geq -1\} \cap \{x \mid x \leq 1\}$

15    $\{x \mid x \geq -2\} \cap \{x \mid x \leq 3\}$ 16 $\{x \mid x \geq 3\} \cup \{x \mid x < -3\}$

In problems 17–48, find the solution set of each inequality and illustrate the solution on the number line.

17 $5x < 15$

18 $6 < x - 2$

19 $-3x \geq -12$

20 $x + 2 \leq 0$

21 $5x + 6 < 13$

22 $3 - x \leq 0$

23 $5 - 3x \geq 7$

24 $3x - 4 \geq 6x$

25 $10 - 5x \leq 0$

26 $3x - 2 + x \leq 5x$

27 $4x + 3 \geq 12$

28 $x + 6 \leq 4 - 3x$

29 $5 + x < -x + 3$

30 $3x - 4 > 2x - 9$

31 $-4x > -21 + 3x$

32 $2x + 1.3 > -4.1$

33 $0 \leq 2x + 7 - 9x$

34 $10(x + 1) < x - 4$

35 $2x - 1 \leq 4(3 - x)$

36 $5x \geq -3(x - 2)$

37 $\dfrac{x}{4} \leq 2$

38 $\dfrac{3x}{5} - \dfrac{6}{10} \leq 0$

39 $\dfrac{3x}{2} > -6 - \dfrac{x}{2}$

40 $\dfrac{x}{3} + 2 < \dfrac{x}{4} - 2x$

41 $\dfrac{4x - 3}{3} \geq 5$

42 $\dfrac{x - 5x}{2} \leq -15$

43 $\dfrac{3x - 2}{5} - 4 < 0$

44 $\dfrac{3x - 7}{2} \leq 7$

45 $\dfrac{2x + 3}{3} \leq \dfrac{3x}{4}$

46 $\dfrac{2}{3}(2x - 1) - \dfrac{2x}{5} \leq 4$

47 $\dfrac{3x - 7}{6} - 13 \geq 1 - \dfrac{x}{2}$

48 $\dfrac{3}{5}(3x - 2) - \dfrac{1}{10}(6x + 7) \leq 0$

49 Give numerical examples to illustrate the following properties.
a) If $a < b$ and c is any number, then $a + c < b + c$.
b) If $a < b$ and $c < 0$, then $ac > bc$.
c) If $a < b$ and $b < c$, then $a < c$.

50 a) Show that if $a < b$ and $c > 0$, then $a/c < b/c$.
b) Show that if $a < b$ and $c < 0$, then $a/c > b/c$.

51 If $a^2 < b^2$, does it follow that $a < b$? Use numerical examples to illustrate the answer.

52 Show that if $0 < a < b$, then $a^2 < b^2$.

53 Harry has a $60 stock which pays him an 11 percent dividend every full year. In how many full years will the stock have paid him more than $58 in dividends?

54 David asks for a $7 weekly payroll deduction from his salary. Ho
many weeks will it take to save at least $437 in order to pay cash f
a microwave oven?

55 In a fund-raising gathering, Lila sold five times as many tickets
Joe, but Lila could not have sold more than 35 tickets. How ma
tickets might Joe have sold?

56 A bank teller is entitled to a 2-week vacation for the first year
employment. Thereafter, she is entitled to a 3-week vacation for ea
full year she works. How many full years must she work withc
taking a vacation to entitle her to at least a 30-week vacation?

4.5 Absolute Value Equations and Inequalities

In order to investigate the solutions of absolute value equations a
inequalities, we shall review the notion of absolute value of a r
number. Recall from Section 1.4 that the notation $|x|$, which is re
"the *absolute value* of x," is used to represent the distance betwe
x and 0, where $|x| \geq 0$. For example, $|3| = 3$, $|0| = 0$, and $|-4| =$
(Figure 4.15).

Figure 4.15

This concept can be formalized in the following definition:

DEFINITION 1 ABSOLUTE VALUE

For x a real number, the *absolute value* of x, denoted by $|x|$, is defined

$$|x| = \begin{cases} x & \text{if } x \geq 0 \\ -x & \text{if } x < 0 \end{cases}$$

Hence, from the definition, $|3| = 3$, because $3 > 0$, $|0| = 0$; a
$|-4| = -(-4) = 4$, because $-4 < 0$.

If a and b are real numbers such that $a \leq b$, then the "distance"
between a and b is considered to be the nonnegative number $b -$
so that $d = b - a$, for $a \leq b$ (Figure 4.16).

Figure 4.16

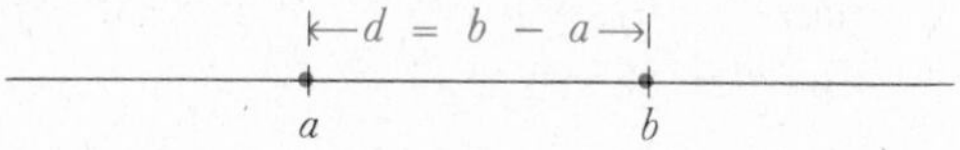

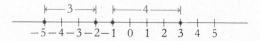

For example, to find the distance d between -1 and 3, we note that $-1 < 3$, so that $d = 3 - (-1) = 4$. Also, since $-5 < -2$, the distance d between -5 and -2 is given by $d = (-2) - (-5) = 3$ (Figure 4.17).

Figure 4.17

$$\overset{\longmapsto 3 \longmapsto \quad \longmapsto 4 \longrightarrow}{\underset{-5\ -4\ -3\ -2\ -1\quad 0\quad 1\quad 2\quad 3\quad 4\quad 5}{\rule{6cm}{0.4pt}}}$$

Using absolute values, we can also express the distance d between two real numbers a and b as $d = |a - b|$ without regard to the relation between a and b.

For example, the distance d between -1 and 3 can be computed as

$$d = |-1 - 3| = |-4| = 4 \quad \text{or} \quad d = |3 - (-1)| = |4| = 4$$

EXAMPLES

Use the formula $d = |a - b|$ to find the distance between the following pairs of numbers.

1   7 and -1

SOLUTION.   Letting $a = 7$ and $b = -1$, we have

$$d = |7 - (-1)| = |8| = 8$$

2   5 and 0

SOLUTION.   Letting $a = 0$ and $b = 5$, we have

$$d = |0 - 5| = |-5| = 5$$

Absolute Value Equations

An equation of the form $|x - a| = b$ is referred to as an *absolute value equation*. This type of equation is not a polynomial equation, hence no degree can be assigned to it. For example, the equation $|x - 1| = 4$ is not a first-degree equation, although it contains the first-degree polynomial $x - 1$. We shall now use the definition of absolute value to solve such equations. The procedure is illustrated in the following examples.

EXAMPLES

Find the solution set of each of the following absolute value equations and illustrate the solution on the number line.

3    $|x| = 3$

SOLUTION.   By definition, we know that $|3| = 3$ and $|-3| = 3$. Therefore, $x = 3$ or $x = -3$; that is, the solution set is $\{-3, 3\}$. The solution can be illustrated on the real line (Figure 4.18).

Figure 4.18

4    $|x - 2| = 5$

SOLUTION. $|x - 2| = 5$ is equivalent to the two equations

$$x - 2 = -5 \qquad\qquad x - 2 = 5$$
$$\text{or}$$
$$x = -3 \qquad\qquad x = 7$$

The solution set is $\{-3, 7\}$ (Figure 4.19).

Figure 4.19

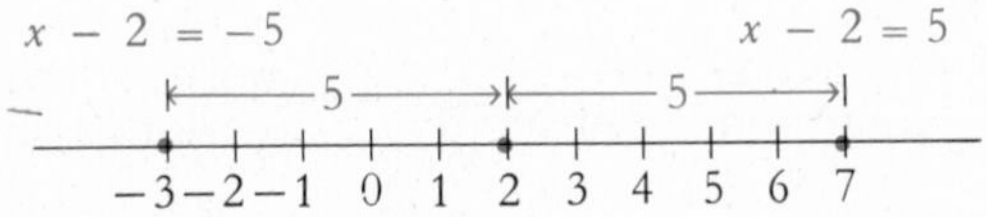

5    $|3 - 2x| = 15$

SOLUTION.   The equation is equivalent to the two equations

$$3 - 2x = -15 \qquad\qquad 3 - 2x = 15$$
$$-2x = -18 \qquad \text{or} \qquad -2x = 12$$
$$x = 9 \qquad\qquad\qquad x = -6$$

The solution set is $\{-6, 9\}$ (Figure 4.20).

Figure 4.20

Note that the absolute value equation in Example 4 is of the form $d = |a - b|$, where $d = 5$, $a = x$, and $b = 2$. Hence, we can interpret this equation as an instruction to find those numbers x on the real line that lie at a distance of 5 units from the number 2.

EXAMPLE

6   Use an absolute value equation to find the two numbers which lie 3 units from 1.

SOLUTION.   In the formula $d = |a - b|$ we have $d = 3$. Let $a = x$ and $b = 1$, so that $|x - 1| = 3$. Solving this equation, we have

$$x - 1 = 3 \qquad\qquad x - 1 = -3$$
$$\text{or}$$
$$x = 4 \qquad\qquad x = -2$$

Hence, the two numbers are -2 and 4 (Figure 4.21).

Figure 4.21

Absolute Value Inequalities

Inequalities involving absolute value notation are called absolute value inequalities. To solve inequalities such as $|x| < 3$, we use the definition of absolute value, so that inequality is equivalent to $x < 3$ for $x \geq 0$, or $-x < 3$ for $x < 0$, which is also equivalent to $-3 < x$. Therefore, $|x| < 3$ can be written as $-3 < x$ and $x < 3$, or simply $-3 < x < 3$. Thus, the solution set is $\{x | -3 < x < 3\}$. This solution set is represented on the number line as shown in Figure 4.22.

Figure 4.22

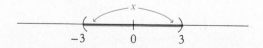

We can also solve the above inequality by using the distance inter
pretation of the notation $|x|$. Since $|x|$ represents the distance of
from the origin, the solution set of $|x| < 3$ is all those number
whose distance from the origin is less than 3 (Figure 4.23). From
Figure 4.23, we can see that all those numbers whose distance from
the origin is less than 3 lie between -3 and 3. Hence, the solution
set of $|x| < 3$ is $\{x|-3 < x < 3\}$.

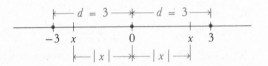

Now consider the absolute value inequality $|x| > 3$. From th
definition of absolute value, this inequality is equivalent to $x > 3$ fo
$x > 0$, or $-x > 3$ for $x < 0$, which is also equivalent to $x < -3$. There
fore, $|x| > 3$ is equivalent to the two inequalities $x < -3$ or $x > 3$
Thus, the solution set is $\{x|x < -3\} \cup \{x|x > 3\}$, which can also b
written as $\{x|x < -3 \text{ or } x > 3\}$. This solution set is represented on
the number line as shown in Figure 4.24. To solve this inequalit
by using the distance interpretation of $|x|$, we note that $|x| >$
represents all those numbers x whose distance from the origin i
greater than 3 (Figure 4.25). From the figure we see that all thos
numbers whose distance from the origin is greater than 3 lie to th
right of 3 ($x > 3$), or to the left of -3 ($x < -3$). Hence, the solution
set of $|x| > 3$ is $\{x|x < -3 \text{ or } x > 3\}$.

Figure 4.24

Figure 4.25

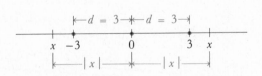

The above results are generalized in the following properties o
absolute value inequalities.

PROPERTY 1

If $|x| < a$, where $a > 0$, then $-a < x < a$ (Figure 4.26).

Figure 4.26

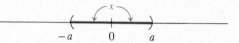

PROPERTY 2

If $|x| > a$, where $a > 0$, then $x < -a$ or $x > a$ (Figure 4.27).

Figure 4.27

EXAMPLES

Find the solution sets of the following absolute value inequalities.

7 $|x| < 5$

SOLUTION. Using Property 1, with $a = 5$, we conclude that $|x| < 5$ is equivalent to $-5 < x < 5$. Hence, the solution set is $\{x \,|\, -5 < x < 5\}$ (Figure 4.28).

Figure 4.28

8 $|x| \geq \frac{7}{2}$

SOLUTION. Using Property 2, and replacing the symbol $>$ by $\geq$, we have that $|x| \geq \frac{7}{2}$ is equivalent to $x \leq -\frac{7}{2}$ or $x \geq \frac{7}{2}$. Hence, the solution set is $\{x \,|\, x \leq -\frac{7}{2}$ or $x \geq \frac{7}{2}\}$ (Figure 4.29).

Figure 4.29

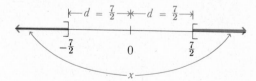

To solve absolute value inequalities such as $|x - 2| < 5$, we first note from the formula $d = |a - b|$ that $|x - 2|$ represents the distance between 2 and any number x. Thus, the inequality $|x - 2| < 5$

represents all those numbers x whose distance from 2 is less than 5 (Figure 4.30). Since 7 lies 5 units to the right of 2 and -3 lies 5 units to the left of 2, we see from the figure that the solution set of $|x-2| < 5$ is $\{x|-3 < x < 7\}$.

Figure 4.30

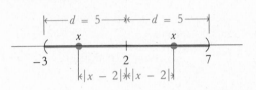

A more direct method of solving the absolute value inequality $|x - 2| < 5$ is to apply Property 1, and replace x by $x - 2$, so that $|x-2| < 5$ is equivalent to $-5 < x - 2 < 5$. Adding 2 to both sides of the two inequalities, we have $-5 + 2 < x - 2 + 2 < 5 + 2$, or $-3 < x < 7$. Therefore, the solution set is $\{x|-3 < x < 7\}$.

EXAMPLES

Use Properties 1 and 2 to find the solution set of each of the following absolute value inequalities. Illustrate the solution on the number line.

9    $|x - 2| < 3$

SOLUTION.   Using Property 1, with $a = 3$ and replacing x by $x - 2$, we have $|x - 2| < 3$ is equivalent to $-3 < x - 2 < 3$, so that

$$-3 + 2 < x - 2 + 2 < 3 + 2 \qquad \text{or} \qquad -1 < x < 5$$

Hence, the solution set is $\{x|-1 < x < 5\}$ (Figure 4.31).

Figure 4.31

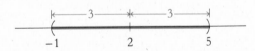

10    $|x + 1| > 4$

SOLUTION.   Using Property 2, with $a = 4$ and replacing x by $x + 1$, we have that $|x + 1| > 4$ is equivalent to $x + 1 < -4$ or $x + 1 > 4$, so that

$$x < -5 \qquad \text{or} \qquad x > 3$$

Hence, the solution set is $\{x|x < -5 \text{ or } x > 3\}$ (Figure 4.32).

Figure 4.32

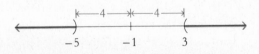

11   $|2x + 7| \geq 11$

SOLUTION. $|2x + 7| \geq 11$ is equivalent to

$$2x + 7 \leq -11 \qquad \text{or} \qquad 2x + 7 \geq 11$$

so that

$$2x \leq -18 \qquad\qquad 2x \geq 4$$
$$\text{or}$$
$$x \leq -9 \qquad\qquad x \geq 2$$

Hence, the solution set is $\{x \mid x \leq -9 \text{ or } x \geq 2\}$ (Figure 4.33).

Figure 4.33

PROBLEM SET 4.5

In problems 1–10, use the formula $d = |a - b|$ to find the distance between the numbers.

1	0 and 12	**2**	0 and 6
3	-4 and 0	**4**	0 and -3
5	4 and 9	**6**	$3\frac{1}{2}$ and 11
7	-2 and -11	**8**	$\frac{7}{2}$ and $-\frac{11}{4}$
9	-7 and 6	**10**	-6.3 and 7.8

In problems 11–24, find the solution set of each absolute value equation.

11	$|x| = 5$	**12**	$|x| = 12$
13	$|2x| = |-4|$	**14**	$|x| = -2$
15	$|x - 2| = 3$	**16**	$|x + 3| = 3$
17	$|x + 5| = 6$	$\rightarrow$**18**	$|3 - x| = 4$
19	$|2x - 3| = 7$	**20**	$|5 - 2x| = 1$
21	$|2 - x| = 5$	**22**	$|-(x - 1)| = 5$
23	$|3 - 2x| = 9$	**24**	$|x - 5| = |-3x + 7|$

25   Use an absolute value equation to find the numbers whose distance from 3 is 7.

26 Use an absolute value equation to find the numbers whose distance from -4 is 8.

27 Use an absolute value equation to find the numbers whose distance from -1 is 1.

28 Use a real line to show that:
   a) If $|x| \leq a$, where $a > 0$, then $-a \leq x \leq a$.
   b) If $|x| \geq a$, where $a > 0$, then $x \leq -a$ or $x \geq a$.

In problems 29–50, find the solution set of each absolute value inequality and represent the solution set on a line graph.

29 $|x| < 5$ **30** $|x| < |-2|$

31 $|x| > 2$ **32** $|-x| > 3$

33 $|x| \leq 4$ **34** $|2x| < 0$

35 $|x| \geq 0$ **36** $|3x| \geq 12$

37 $|x - 1| > 8$ **38** $|5x - 3| > 12$

39 $|x + 2| < 7$ → **40** $|-x + 2| < |-5| = |-x+2| < 5$

41 $|2x + 2| \leq 5$ **42** $\left|\dfrac{x}{2} + 5\right| \geq 3$

43 $|4x - 1| \geq 11$ **44** $|x - \frac{1}{2}| < \frac{5}{2}$

45 $|3 - 2x| < 5$ **46** $\left|4 - \dfrac{x}{3}\right| > 1$

47 $|6 - 7x| \geq 1$ → **48** $\left|\dfrac{1 - x}{3}\right| \leq 2$

49 $|5x + 3| + 2 > 8$ → **50** $|1 - 2x| - 5 \leq 7$

REVIEW PROBLEM SET

In problems 1–10, find the solution set for each first-degree equation.

1 $5x + 2 = 32$ **2** $-4x - 7 = 21$

3 $3x - 4 = x - 12$ **4** $17 - 3x = x + 5$

5 $2(x - 1) = 3 - 3(x - 5)$ **6** $12(x - 2) + 8 = 5(x - 1) + 2x$

7 $x - 7(4 + x) = 5x - 6(3 - 4x)$

8 $5(x + 2) - 3(x - 1) = 4(x + 1) + 15$

9 $2(x - 1) + 3(x - 2) + 4(x - 3) = 0$

10 $x - 2(3 - x) = 2(x + 3) - (x - 2)$

In problems 11–20, find the solution set for each fractional equation.

11 $\dfrac{x}{3} - \dfrac{x}{6} = 1$

12 $\dfrac{5}{x} + \dfrac{3}{8} = \dfrac{7}{16}$

13 $\dfrac{2x - 1}{3} = \dfrac{x}{5} + \dfrac{2}{15}$

14 $\dfrac{x + 4}{3} - \dfrac{x + 2}{2} = \dfrac{x - 2}{5}$

15 $\dfrac{13}{14x} - \dfrac{1}{2x} = \dfrac{1}{7}$

16 $\dfrac{3}{x - 2} = \dfrac{5}{x + 2}$

17 $\dfrac{2}{x} + \dfrac{x - 1}{3x} = \dfrac{2}{5}$

18 $\dfrac{10 - x}{x} + \dfrac{3x + 3}{3x} = 3$

19 $\dfrac{x + 1}{x - 1} - \dfrac{x}{x + 1} = \dfrac{5 - x}{x^2 - 1}$

20 $\dfrac{12}{x^2 - 25} = \dfrac{1}{x + 5} + \dfrac{2}{x - 5}$

In problems 21–30, solve each equation for the indicated variable and write the solution set.

21 $a - bx = c$, for x

22 $x - y = a(y - 1)$, for y

23 $ax = b + cx$, for x

24 $y - a = m(x - b)$, for x

25 $7x^2 - 3y + 5c = y + c$, for y

26 $x^2y - 2xy = 1 - y$, for y

27 $3(ax - 2b) = 4(2b - ax)$, for a

28 $\dfrac{x - 1}{a} + \dfrac{2}{a} = b$, for a

29 $\dfrac{x}{a} - \dfrac{3}{a - b} = 1$, for b

30 $\dfrac{2x - a}{5} + \dfrac{x}{3} - 3a = 1$, for x

In problems 31–40, solve each formula for the indicated variable.

31 $G = \dfrac{m}{r^3}$, for m

32 $E = mc^2$, for m

33 $g = \dfrac{v^2}{R + h}$, for h

34 $S = 2A + Ph$, for A

35 $s = 2\pi rh + 2\pi r^2$, for h

36 $s = \dfrac{n}{2}[2a + (n - 1)d]$, for a

37 $s = \dfrac{ax^n - a}{x - 1}$, for a

38 $A = P + Prt$, for P

39 $\dfrac{1}{R} = \dfrac{1}{r} + \dfrac{1}{s}$, for R

40 $En = RI + \dfrac{nrI}{m}$, for I

In problems 41–50, find the solution set for each first-degree inequality and illustrate the solution on a number line.

41 $2x + 1 > 5$ **42** $-3x \le 12$

43 $3 - 2x \le 9$ **44** $3(x - 4) < 2(4 - x)$

45 $3x + 5 < 2x + 1$ **46** $\dfrac{2x + 3}{4} \ge 1 - x$

47 $\dfrac{1 - 3x}{3} \ge \dfrac{1 + 2x}{4}$ **48** $\dfrac{3x - 5}{19} > 1$

49 $5x + 2(3x - 1) > -1 - 5(x - 3)$ **50** $\dfrac{2x + 1}{5} - 2 \le \dfrac{3x + 1}{2}$

In problems 51–56, use the formula $d = |a - b|$ to find the distance between the numbers.

51 8 and -2 **52** 0 and -10

53 12 and 7   **54** -11 and 14

55 -4 and -11 **56** $7\frac{1}{2}$ and $3\frac{1}{4}$

In problems 57–66, find the solution set for each absolute value equation.

57 $|x| = 7$ **58** $|3x| = 15$

59 $|x + 2| = 5$ **60** $|4 - x| = -2$

61 $|3 - x| = 11$ **62** $|2(x - 3)| = 8$

63 $|3x + 4| = 12$ **64** $|5 - 4x| = 11$

65 $|3 - 2x| = 5$ **66** $|x - 1| = 2x$

In problems 67–76, find the solution set for each absolute value inequality and illustrate the solution on a number line.

67 $|x| < 7$ **68** $|x| \le 12$

69 $|x| \ge 9$ **70** $|3x| > 6$

71 $|x - 3| < 5$ **72** $|8 - x| \le 6$

73 $|4x + 7| > 9$ **74** $\left|\dfrac{3x}{4} - 1\right| > 2$

75 $|4 - 5x| \ge 14$ **76** $\left|\dfrac{1 - x}{5}\right| < 3$

77 In an election of the president of a faculty senate, there were two candidates. The successful candidate received 5 more votes than the defeated candidate. If 243 votes were cast all together, how many did each candidate receive?

78 How many pounds of water must be evaporated from 50 pounds of a 3 percent salt solution so that the remaining portion will be a 5 percent solution?

79 A student received grades of 80 and 92 on his first two tests. What grade must he obtain on his third test to average at least 90?

80 Two airplanes left airports which are 600 miles apart and flew toward each other. One plane flew 20 miles per hour faster than the other. If they passed each other at the end of 1 hour and 12 minutes, what were their speeds?

CHAPTER 5

Exponents, Radicals, and Complex Numbers

We are already familiar with the use of positive integers as exponents (see Section 2.1). If, however, we try to apply this meaning to cases such as a^{-2}, $b^{1/2}$, and c^0, we find it impossible to do so, since we cannot use a as a factor minus two times, b as a factor one-half times, or c as a factor zero times. Our objective in this chapter is to give meaning to a^n, $a \neq 0$, where n is an arbitrary integer or a rational number, so that the properties of exponents still hold. This will provide a clue to translating these properties into radical notation. The chapter also includes discussions of operations on radicals, radical equations, and complex numbers.

5.1 Zero and Negative Exponents

In Section 2.1 we verified and illustrated the properties of positive exponents. These properties were stated as follows:

If a and b are real numbers and m and n are positive integers, then:

1   $a^m a^n = a^{m+n}$

2   $(a^m)^n = a^{mn}$

3   $(ab)^n = a^n b^n$

4   $\left(\dfrac{a}{b}\right)^n = \dfrac{a^n}{b^n}, \quad b \neq 0$

5   $\dfrac{a^m}{a^n} = \begin{cases} a^{m-n} & \text{if } m > n \\ 1 & \text{if } m = n \\ \dfrac{1}{a^{n-m}} & \text{if } m < n \end{cases} \quad \text{for } a \neq 0$

Suppose that we wish to extend these properties to the nonpositive integers. Let a represent a nonzero real number and let $m = 0$; then Property 1 would give

$$a^0 a^n = a^{0+n} \qquad \text{or} \qquad a^0 a^n = a^n$$

This property is true if we assign the value 1 for a^0, since $1 \cdot a^n = a^n$. Properties 2 through 5 can also be verified for $a^0 = 1$. For example, if $m = n$ in Property 5, we have

$$\frac{a^m}{a^n} = \frac{a^n}{a^n} = a^{n-n} = a^0$$

Again, it follows that a^0 must be assigned the value 1, since $a^n/a^n = 1$. Thus, $3^0 = 1$, $(\frac{1}{2})^0 = 1$, and $(-5)^0 = 1$. Notice that 0^0 is not defined.

Now let us consider Property 1 for negative integral exponents; for instance, let $m = -n$, where n is a positive integer. Then $a^m a^n = a^{-n} a^n = a^0 = 1$. Thus, $a^{-n} a^n = 1$. After dividing both sides of this equation by a^n, we have $a^{-n} = 1/a^n$. For example, $3^2 \cdot 3^{-2} = 1$, but we know that $3^2 \cdot 1/3^2 = 1$. Therefore,

$$3^{-2} = \frac{1}{3^2} = \frac{1}{9}$$

The discussion above is formalized as follows:

DEFINITION 1   ZERO AND NEGATIVE EXPONENTS

If a is any real number, $a \neq 0$, and n is a positive integer, then:

a) $a^0 = 1$

b) $a^{-n} = \dfrac{1}{a^n}$ or $a^n = \dfrac{1}{a^{-n}}$

Two direct consequences of this definition are the following properties:

a) $\left(\dfrac{a}{b}\right)^{-n} = \left(\dfrac{b}{a}\right)^{n}$

b) $\dfrac{a^{-n}}{b^{-n}} = \dfrac{b^n}{a^n}$

The proof of these two properties is straightforward and is left as an exercise (see problem 66).

EXAMPLES

Write each of the following expressions with positive exponents and simplify.

1   6^{-2}

SOLUTION

$$6^{-2} = \frac{1}{6^2} = \frac{1}{36}$$

2 $(2x)^{-3}$

SOLUTION

$$(2x)^{-3} = \frac{1}{(2x)^3} = \frac{1}{2^3 x^3} = \frac{1}{8x^3}$$

3 $3^{-1} + 3^0$

SOLUTION

$$3^{-1} + 3^0 = \frac{1}{3} + 1 = \frac{1+3}{3} = \frac{4}{3}$$

4 $\left(\dfrac{18}{6x}\right)^0$

SOLUTION

$$\left(\frac{18}{6x}\right)^0 = 1$$

5 $(x^0 y)^{-3}$

SOLUTION

$$(x^0 y)^{-3} = \frac{1}{(x^0 \cdot y)^3} = \frac{1}{x^0 y^3} = \frac{1}{y^3}$$

6 $(5^{-1} + 5^0)^{-1}$

SOLUTION

$$(5^{-1} + 5^0)^{-1} = \left(\frac{1}{5} + 1\right)^{-1} = \left(\frac{6}{5}\right)^{-1} = \frac{5}{6}$$

7 $\dfrac{x^{-1} + y^{-1}}{x + y}$

SOLUTION

$$\frac{x^{-1} + y^{-1}}{x + y} = \frac{\dfrac{1}{x} + \dfrac{1}{y}}{x + y} = \frac{\dfrac{y + x}{xy}}{x + y}$$

$$= \frac{y + x}{xy} \cdot \frac{1}{x + y} = \frac{1}{xy}$$

Properties of Integral Exponents

The use of exponents can be extended to include the negative integers together with zero exponents so that Properties 1 through 5 (page 175) remain true. These properties are stated as follows: Let a and b be real numbers and p and q be integers; then:

1   $a^p a^q = a^{p+q}$

2   $(a^p)^q = a^{pq}$

3   $(ab)^p = a^p b^p$

4   $\left(\dfrac{a}{b}\right)^p = \dfrac{a^p}{b^p},\ b \neq 0$

5   $\dfrac{a^p}{a^q} = a^{p-q},\ a \neq 0$

Notice that Property 5 takes a different form than its corresponding property for positive exponents, because of the use of zero and negative exponents. For example,

$$\frac{x^3}{x^2} = x^{3-2} = x$$

$$\frac{x^2}{x^5} = x^{2-5} = x^{-3} = \frac{1}{x^3}$$

$$\frac{x^5}{x^5} = x^{5-5} = x^0 = 1$$

These properties will be illustrated further by the following examples.

EXAMPLES

Simplify the following expressions and express each answer in a form containing positive exponents.

8   $5^{-4} \cdot 5^2$

SOLUTION

$$5^{-4} \cdot 5^2 = 5^{-4+2} \qquad \text{(Property 1)}$$

$$= 5^{-2} = \frac{1}{5^2} = \frac{1}{25}$$

9   $x^6 \cdot x^{-2}$

SOLUTION

$$x^6 \cdot x^{-2} = x^{6-2} \qquad \text{(Property 1)}$$

$$= x^4$$

10 $(3^{-2})^3$

SOLUTION

$$(3^{-2})^3 = 3^{-2(3)} \qquad \text{(Property 2)}$$

$$= 3^{-6} = \frac{1}{3^6} = \frac{1}{729}$$

11 $(x^3)^{-2}$

SOLUTION

$$(x^3)^{-2} = x^{3(-2)} \qquad \text{(Property 2)}$$

$$= x^{-6} = \frac{1}{x^6}$$

12 $(5x)^{-3}$

SOLUTION

$$(5x)^{-3} = 5^{-3} \cdot x^{-3} \qquad \text{(Property 3)}$$

$$= \frac{1}{5^3} \cdot \frac{1}{x^3} = \frac{1}{125x^3}$$

13 $(6y^3)^{-2}$

SOLUTION

$$(6y^3)^{-2} = 6^{-2} \cdot (y^3)^{-2} \qquad \text{(Property 3)}$$

$$= \frac{1}{6^2} \cdot \frac{1}{(y^3)^2} = \frac{1}{36} \cdot \frac{1}{y^6} = \frac{1}{36y^6}$$

14 $\left(\dfrac{3}{5}\right)^{-4}$

SOLUTION

$$\left(\frac{3}{5}\right)^{-4} = \frac{3^{-4}}{5^{-4}} \qquad \text{(Property 4)}$$

$$= \frac{5^4}{3^4} \qquad \left(\text{Property } \frac{a^{-n}}{b^{-n}} = \frac{b^n}{a^n}\right)$$

$$= \frac{625}{81}$$

15    $\left(\dfrac{x^2}{y^3}\right)^{-5}$

SOLUTION

$$\left(\frac{x^2}{y^3}\right)^{-5} = \frac{(x^2)^{-5}}{(y^3)^{-5}} \qquad \text{(Property 4)}$$

$$= \frac{x^{-10}}{y^{-15}} \qquad \text{(Property 2)}$$

$$= \frac{y^{15}}{x^{10}} \qquad \left(\text{Property } \frac{a^{-n}}{b^{-n}} = \frac{b^n}{a^n}\right)$$

16    $\dfrac{3^{-7}}{3^{-5}}$

SOLUTION

$$\frac{3^{-7}}{3^{-5}} = 3^{-7-(-5)} \qquad \text{(Property 5)}$$

$$= 3^{-2} = \frac{1}{3^2} = \frac{1}{9}$$

17    $\dfrac{\left(\dfrac{3}{7}\right)^{13}}{\left(\dfrac{3}{7}\right)^{-5}}$

SOLUTION

$$\frac{\left(\dfrac{3}{7}\right)^{13}}{\left(\dfrac{3}{7}\right)^{-5}} = \left(\frac{3}{7}\right)^{13+5} = \left(\frac{3}{7}\right)^{18}$$

18    $\dfrac{3ab^{-2}}{c^3d^{-4}}$

SOLUTION

$$\frac{3ab^{-2}}{c^3d^{-4}} = \frac{3a}{c^3}\left(\frac{b^{-2}}{d^{-4}}\right)$$

$$= \frac{3a}{c^3}\left(\frac{d^4}{b^2}\right) \qquad \left(\text{Property: } \frac{a^{-n}}{b^{-n}} = \frac{b^n}{a^n}\right)$$

$$= \frac{3ad^4}{c^3b^2}$$

19 $\left[\dfrac{x^{-4}(-3x)}{(-5x)^{-2}}\right]^{-1}$

SOLUTION

$$\left[\frac{x^{-4}(-3x)}{(-5x)^{-2}}\right]^{-1} = \left[\frac{x^{-4}}{(-5x)^{-2}}(-3x)\right]^{-1}$$

$$= \left[\frac{(-5x)^2(-3x)}{x^4}\right]^{-1} \qquad \left(\text{Property: } \frac{a^{-n}}{b^{-n}} = \frac{b^n}{a^n}\right)$$

$$= \left[\frac{25x^2(-3x)}{x^4}\right]^{-1} = \left(\frac{-75x^3}{x^4}\right)^{-1}$$

$$= \left(\frac{-75}{x^{4-3}}\right)^{-1} = \left(-\frac{75}{x}\right)^{-1} = -\frac{x}{75}$$

Scientific Notation

Now that we have defined the exponential expression a^n for n any integer, we can apply this definition to examine a convenient way to represent very large or very small numbers. Such numbers can be expressed in an exponential form. For example, to write the following numbers in exponential notations, we have

$$38{,}765 = 3.8765 \times 10{,}000 = 3.8765 \times 10^4$$

$$1{,}234{,}000 = 1.234 \times 1{,}000{,}000 = 1.234 \times 10^6$$

$$0.0000803 = 8.03 \times \frac{1}{100{,}000} = 8.03 \times 10^{-5}$$

$$0.000000023 = 2.3 \times \frac{1}{100{,}000{,}000} = 2.3 \times 10^{-8}$$

We have expressed each number as the product of a number (greater than or equal to 1 but less than 10) and a power of 10. This form of a number is called *scientific form for real numbers* or *scientific notation*.

Notice that in each of the examples above, the power of 10 is the same as the number of places the decimal point was moved either to the right or to the left to obtain a number from 1 to 10. This provides us with a convenient way to express a number in scientific notation. We simply count the number of places necessary to move the decimal point. This number becomes the exponent on 10; it is positive if we move the decimal point to the left, and negative if we move it to the right. For instance, 27,658 can be expressed in scientific notation as

$$27{,}658 = 2.7658 \times 10^4 \qquad \text{(4 places to the left)}$$

EXAMPLE

20   Write each of the following numbers in scientific notation.
   a) 3,951   b) 21.33
   c) 0.0119   d) 0.000573

SOLUTION

   a) $3,951 = 3.951 \times 10^3$
   b) $21.33 = 2.133 \times 10^1$
   c) $0.0119 = 1.19 \times 10^{-2}$
   d) $0.000573 = 5.73 \times 10^{-4}$

To change a number from scientific notation to ordinary decimal notation, we reverse the process. That is, we move the decimal point the same number of places as the exponent on 10, to the right if the exponent is positive, and to the left if it is negative.

EXAMPLE

21   Write each of the following in ordinary decimal notation.
   a) 4.3×10^3 b) 3.7×10^6
   c) 5.13×10^{-4} d) 2.73×10^{-1}

SOLUTION

   a) $4.3 \times 10^3 = 4,300$
   b) $3.7 \times 10^6 = 3,700,000$
   c) $5.13 \times 10^{-4} = 0.000513$
   d) $2.73 \times 10^{-1} = 0.273$

PROBLEM SET 5.1

In problems 1–22, write each expression in a form with no negative exponents and simplify.

1	5^{-2}	**2**	3^{-4}
3	$(\frac{1}{2})^{-4}$	**4**	$(-10)^{-1}$
5	$(\frac{3}{4})^{-2}$	**6**	p^{-5}
7	x^{-100}	**8**	y^{-131}
9	$\dfrac{1}{7^{-1}}$	**10**	$\dfrac{1}{x^{-2}}$
11	$\dfrac{1}{x^{-5}}$	**12**	$\dfrac{1}{y^{-20}}$
13	$(-\frac{1}{3})^0$	**14**	$(-\frac{3}{5})^0$

15 $(-4x^2)^0$

16 $\left(\dfrac{x}{y}\right)^0$

17 $(x^{-10})^0$

18 $(7^{-1} + 3^{-2})^0$

19 $\dfrac{5^{-1} + 3^{-2}}{(45)^{-1}}$

20 $\dfrac{x^{-1} + y^{-1}}{(xy)^{-1}}$

21 $\dfrac{x^{-1}}{y^{-1}} + \left(\dfrac{x}{y}\right)^{-1}$

22 $\left(\dfrac{x}{y}\right)^{-1} + \left(\dfrac{y}{x}\right)^{-1}$

In problems 23–54, use the properties of integral exponents to simplify the given expression and express each answer in a form containing nonnegative exponents.

23 $7^{-1} \cdot 7^3$

24 $5^{-3} \cdot 5^{-2}$

25 $9^{-3} \cdot 9$

26 $(\tfrac{1}{2})^4 \cdot (\tfrac{1}{2})^{-2}$

27 $(\tfrac{2}{5})^{-3} \cdot (\tfrac{2}{5})^{-2}$

28 $(-\tfrac{2}{3})^4 \cdot (-\tfrac{2}{3})^{-6}$

29 $x^{-3} \cdot x^{-2}$

30 $x^{-7} \cdot x^{-2}$

31 $(2^{-3})^2$

32 $[(-5)^2]^{-1}$

33 $[(-4)^3]^{-2}$

34 $(x^{-2})^{-4}$

35 $(y^{-4})^3$

36 $(x^{-n})^4$; n a positive integer

37 $(3x)^{-4}$

38 $(3x^{-4})^2$

39 $(5^{-1}y^4)^{-2}$

40 $(6^{-2}y^{-3})^{-3}$

41 $(3^{-2}x)^3$

42 $(x^{-1}y^{-2})^5$

43 $(\tfrac{5}{3})^{-2}$

44 $\left(\dfrac{-2}{x^{-1}}\right)^{-1}$

45 $\left(\dfrac{-2}{x^2}\right)^{-3}$

46 $\left(\dfrac{4x^{-1}}{7y^2}\right)^{-2}$

47 $\left(\dfrac{x^2}{y^2}\right)^{-2}$

48 $\left(\dfrac{x}{x^6}\right)^{-1}$

49 $\dfrac{8^{-3}}{8^{-5}}$

50 $\dfrac{7^{-3}}{7^{-8}}$

51 $\dfrac{(\tfrac{2}{7})^{-4}}{(\tfrac{2}{7})^{-3}}$

52 $\dfrac{(x^{-2})^3}{x^{-4}}$

53 $\dfrac{xy^{-7}}{xy^{-4}}$

54 $\dfrac{(xy)^{-4}}{(xy)^{-7}}$

In problems 55–65, simplify the expressions. Express the answers without negative exponents.

55 $\left(\dfrac{x^2y^{-3}}{x^{-3}y^{-2}}\right)^{-3}$

56 $\left(\dfrac{3x^{-1}y^2}{5x^{-2}y^{-1}}\right)^{-2}$

57 $\left(\dfrac{x^{-1}y^{-2}}{x^{-2}y^3}\right)^2$

58 $\left(\dfrac{x^{-2}y^{-1}z^{-3}}{y^3z^{-1}}\right)^{-4}$

59 $\left[\dfrac{a^{-4}b^2c^{-6}}{(ab)^{-2}(bc)^{-3}}\right]^{-1}$

60 $\left[\dfrac{(ab)^{-2}(bc)^{-3}}{(ab)^3(bc)^{-1}}\right]^{-2}$

61 $\dfrac{(5a)^2(-c)^{-2}(2b)^{-1}}{20(abc)^{-1}}$

62 $\left[\dfrac{(x^{-2})^{-1}(y^{-2})^3}{(x^{-1})^2(y^{-1})^2}\right]^{-2}$

63 $\left[\left(\dfrac{x^{-1}}{y}\right)^{-1}\cdot\left(\dfrac{y^{-1}}{x}\right)^{-1}\right]^{-2}$

64 $\left[\dfrac{(3x)^{-1}(2x)^2y^{-2}}{(4x^{-2})(25y^{-3})}\right]^{-3}$

65 $\left[\dfrac{x^{-4}y^2z^{-3}}{x^3(yz)^{-2}}\right]^{-4}$

66 Use the properties of integral exponents to prove that

a) $\left(\dfrac{a}{b}\right)^{-n}=\left(\dfrac{b}{a}\right)^{n}$

b) $\dfrac{a^{-n}}{b^{-n}}=\dfrac{b^n}{a^n}$

In problems 67–76, write each number in scientific notation.

67 3,782

68 0.000132

69 38,173

70 375,000

71 137,100,000

72 0.0001321

73 0.000271312

74 681,000,000

75 0.0319

76 0.00127142281

In problems 77–86, write each number in ordinary decimal notation.

77 2.1×10^2

78 7.5×10^{-3}

79 3.14×10^{-5}

80 8.6×10^3

81 1.13×10^4

82 1.871×10^{-4}

83 5.41×10^{-6}

84 7.2×10^5

85 3.127×10^4

86 1.94×10^{-5}

5.2 Rational Exponents

Up to this point, we have considered the definition and the properties of integral exponents. In this section we consider exponents that are rational numbers. First, we begin by defining exponents that are positive rational numbers so that Properties 1 through 5 of Section 5.1 shall continue to hold. To do this, we consider rational exponents that are reciprocals of positive integers. If $a^{1/q}$ exists, where a is a real number, $a \neq 0$, and q is a positive integer, then it must satisfy the property $(a^m)^n = a^{mn}$. Since $(a^{1/q})^q = a^{q/q} = a$, then $a^{1/q}$ is defined in such a way that it represents one of q equal factors of a. The number $a^{1/q}$, if it exists, is called the qth *root of* a. Thus, $4^{1/2}$ is one of two equal factors of 4, because, $4^{1/2} \cdot 4^{1/2} = 4^{1/2+1/2} = 4$. Since squaring either 2 or -2 gives 4, we could define $4^{1/2}$ to be either 2 or -2. By convention, we agree that $4^{1/2}$ is 2 and call 2 the *principal square root* of 4. The symbol $-4^{1/2}$ means $-(4^{1/2}) = -2$. One should not confuse $-4^{1/2}$ with $(-4)^{1/2}$. Naturally, the latter one is meaningless in the set of real numbers. This is based on the fact that $(-4)^{1/2} \cdot (-4)^{1/2} = [(-4)^{1/2}]^2 = -4$, which contradicts the rule of multiplication of signed numbers (the square of any real number is either zero or positive). However, odd roots of negative numbers do exist and they are unique. For instance, $(-27)^{1/3} = -3$, since $(-27)^{1/3} = [(-3)^3]^{1/3} = (-3)^1 = -3$. Here -3 is called the *cube root* of -27. In general, we have the following rules for a positive integer q and when $a^{1/q}$ (the qth root of a) exists (of course, as a real number):

1 If a is a positive number, then $a^{1/q}$ is also a positive number (*principal qth root*).

2 If a is zero, then $a^{1/q}$ is also zero.

3 If a is a negative number and q is an *odd* positive integer, then $a^{1/q}$ is also negative.

Notice that if a is negative and q is an *even* positive integer, then $a^{1/q}$ does not exist, that is, $a^{1/q}$ is not a real number.

EXAMPLES

Compute the values of the following numbers.

1 $16^{1/2}$

SOLUTION. $16^{1/2} = 4$, since $16^{1/2} = (4^2)^{1/2} = 4$.

2 $(-64)^{1/3}$

SOLUTION. $(-64)^{1/3} = -4$, since $(-64)^{1/3} = [(-4)^3]^{1/3} = -4$.

3    $-16^{1/4}$

SOLUTION. $-16^{1/4} = -(16^{1/4}) = -2$, since $-(16^{1/4}) = -[(2^4)^{1/4}] = -(2)$
$= -2$.

4    $(-25)^{1/2}$

SOLUTION. $(-25)^{1/2}$ does not exist in the real number system, using the fact that $a^{1/q}$ is not a real number when a is negative and q is even.

The definition of $a^{1/q}$ can be extended to evaluate expressions whose exponents are arbitrary rational numbers such as $9^{3/2}$, $27^{4/3}$, and $32^{-3/5}$. This is formalized in the following definition of $a^{p/q}$.

DEFINITION 1   RATIONAL EXPONENT

Let a be a real number and r be a positive rational number of the form p/q, where p and q are positive integers with no common factors. We define a^r to be the number $(a^{1/q})^p$ provided that $a^{1/q}$ exists. a is called the *base* and r is called the *exponent*. If $a^{1/q}$ does not exist, we say that a^r is not defined in the real number system. For instance, using the definition, we have $9^{3/2} = (9^{1/2})^3$, so that $9^{3/2} = (9^{1/2})^3 = 3^3 = 27$. Also, $9^{3/2} = (9^3)^{1/2} = (729)^{1/2} = 27$. Therefore, $9^{3/2} = (9^{1/2})^3 = (9^3)^{1/2}$.

Generalizing the example above, we have: If a is a real number, q is a positive integer, and p is a positive integer, then $a^{p/q} = (a^{1/q})^p = (a^p)^{1/q}$, provided that $a^{1/q}$ exists. For example,

$$81^{3/4} = (81^{1/4})^3 = 3^3 = 27 \quad \text{or} \quad 81^{3/4} = (81^3)^{1/4} = (3^{12})^{1/4} = 3^3$$

If r is positive, we simply define $a^{-r} = 1/a^r$, $a \neq 0$, thereby ensuring that $a^{-r} \cdot a^r = a^0 = 1$. Using this definition, we point out that

$$a^{-p/q} = \frac{1}{a^{p/q}}, \, a \neq 0$$

For example,

$$8^{-1/3} = \frac{1}{8^{1/3}} = \frac{1}{(2^3)^{1/3}} = \frac{1}{2}$$

EXAMPLES

Compute the values of the following numbers.

5    $27^{4/3}$

SOLUTION

$$27^{4/3} = (27^{1/3})^4 = 3^4 = 81$$

6 $32^{-3/5}$

SOLUTION

$$32^{-3/5} = \frac{1}{32^{3/5}} = \frac{1}{(32^{1/5})^3} = \frac{1}{2^3} = \frac{1}{8}$$

7 $(0.064)^{-5/3}$

SOLUTION

$$(0.064)^{-5/3} = \left(\frac{64}{1,000}\right)^{-5/3}$$

$$= \left(\frac{1,000}{64}\right)^{5/3} = \frac{(1,000)^{5/3}}{(64)^{5/3}}$$

$$= \frac{(10^3)^{5/3}}{(4^3)^{5/3}} = \frac{10^5}{4^5} = \left(\frac{10}{4}\right)^5$$

$$= \left(\frac{5}{2}\right)^5 = \frac{3,125}{32}$$

Properties of Rational Exponents

We have defined rational exponents so that they satisfy the property $(a^m)^n = a^{mn}$. This definition ensures that they also satisfy the other properties. Thus, the properties of exponents that we originally developed for positive integral exponents, then extended to include all integers, are also true for rational exponents. Accordingly, we restate these properties as follows: Let a and b represent real numbers, and r and s represent reduced rational numbers, then (if the individual factors exist):

1 $a^r a^s = a^{r+s}$

2 $(a^r)^s = a^{rs}$

3 $(ab)^r = a^r b^r$

4 $\left(\dfrac{a}{b}\right)^r = \dfrac{a^r}{b^r}, b \neq 0$

5 $\dfrac{a^r}{a^s} = a^{r-s}, a \neq 0$

The properties of exponents are also true *for all real numbers* although the proof of this statement is beyond the scope of this book.

EXAMPLES

Use the properties of exponents to simplify the following expressions. Assume that all bases are positive.

8    $7^{-1/2} \cdot 7^{5/2}$

SOLUTION

$$7^{-1/2} \cdot 7^{5/2} = 7^{-1/2 + 5/2} \qquad \text{(Property 1)}$$

$$= 7^{4/2} = 7^2 = 49$$

9    $x^{-3/7} \cdot x^{9/14}$

SOLUTION

$$x^{-3/7} \cdot x^{9/14} = x^{-3/7 + 9/14} \qquad \text{(Property 1)}$$

$$= x^{3/14}$$

10    $(2^{-1/3})^{-21}$

SOLUTION

$$(2^{-1/3})^{-21} = 2^{(-1/3)(-21)} = 2^7 = 128 \qquad \text{(Property 2)}$$

11    $(x^{-3/4})^{-8/3}$

SOLUTION

$$(x^{-3/4})^{-8/3} = x^{(-3/4)(-8/3)} = x^2 \qquad \text{(Property 2)}$$

12    $(8x^3)^{2/3}$

SOLUTION

$$(8x^3)^{2/3} = 8^{2/3} \cdot (x^3)^{2/3} \qquad \text{(Property 3)}$$

$$= (2^3)^{2/3} \cdot (x^3)^{2/3}$$

$$= 2^2 \cdot x^2 = 4x^2$$

13    $(125x^{-18})^{-4/3}$

SOLUTION

$$(125x^{-18})^{-4/3} = (125)^{-4/3} \cdot (x^{-18})^{-4/3} \qquad \text{(Property 3)}$$

$$= (5^3)^{-4/3}(x^{-18})^{-4/3}$$

$$= 5^{-4}x^{24} = \frac{x^{24}}{5^4} = \frac{x^{24}}{625}$$

14 $\left(\dfrac{25}{4}\right)^{3/2}$

SOLUTION

$$\left(\frac{25}{4}\right)^{3/2} = \frac{25^{3/2}}{4^{3/2}} \qquad \text{(Property 4)}$$

$$= \frac{(5^2)^{3/2}}{(2^2)^{3/2}}$$

$$= \frac{5^3}{2^3} = \frac{125}{8}$$

15 $\left(\dfrac{32}{x^{-5}}\right)^{-2/5}$

SOLUTION

$$\left(\frac{32}{x^{-5}}\right)^{-2/5} = \frac{(32)^{-2/5}}{(x^{-5})^{-2/5}} \qquad \text{(Property 4)}$$

$$= \frac{(2^5)^{-2/5}}{(x^{-5})^{-2/5}} = \frac{2^{-2}}{x^2} = \frac{1}{2^2 x^2} = \frac{1}{4x^2}$$

16 $\dfrac{64^{2/3}}{(64)^{-1/2}}$

SOLUTION

$$\frac{64^{2/3}}{(64)^{-1/2}} = (64)^{2/3-(-1/2)} = (64)^{2/3+1/2} \qquad \text{(Property 5)}$$

$$= (64)^{7/6} = (2^6)^{7/6} = 2^7 = 128$$

17 $\dfrac{x^{2/3}}{x^{-4/5}}$

SOLUTION

$$\frac{x^{2/3}}{x^{-4/5}} = x^{2/3-(-4/5)} \qquad \text{(Property 5)}$$

$$= x^{10/15-(-12/15)} = x^{22/15}$$

18 $\left(\dfrac{8x^6}{y^3}\right)^{2/3}$

SOLUTION

$$\left(\frac{8x^6}{y^3}\right)^{2/3} = \frac{8^{2/3}(x^6)^{2/3}}{(y^3)^{2/3}} = \frac{(8^{1/3})^2 x^4}{y^2} = \frac{2^2 x^4}{y^2} = \frac{4x^4}{y^2}$$

19    $\left(\dfrac{x^2y^{-3}}{z^4}\right)^{-1/2}$

SOLUTION

$$\left(\dfrac{x^2y^{-3}}{z^4}\right)^{-1/2} = \dfrac{(x^2)^{-1/2}(y^{-3})^{-1/2}}{(z^4)^{-1/2}} = \dfrac{x^{-1}y^{3/2}}{z^{-2}} = \dfrac{z^2y^{3/2}}{x}$$

20    $\dfrac{x^{-3}y^2z^{1/2}}{x^2y^{-3}z^{-1}}$

SOLUTION

$$\dfrac{x^{-3}y^2z^{1/2}}{x^2y^{-3}z^{-1}} = x^{-3-2} \cdot y^{2-(-3)} \cdot z^{(1/2)-(-1)}$$

$$= x^{-5}y^5z^{3/2} = \dfrac{y^5z^{3/2}}{x^5}$$

PROBLEM SET 5.2

In problems 1–16, compute the value of each expression.

1	$64^{1/2}$	2	$(144)^{-1/2}$
3	$(-8)^{-4/3}$	4	$(-32)^{3/5}$
5	$(-8)^{5/3}$	6	$\left(\dfrac{121}{100}\right)^{3/6}$
7	$9^{-3/2}$	8	$(27)^{-2/3}$
9	$(81)^{-3/4}$	10	$(32)^{-8/10}$
11	$(128)^{-5/7}$	12	$(16)^{-5/4}$
13	$(0.04)^{15/10}$	14	$(0.216)^{-4/6}$
15	$(0.125)^{-3/9}$	16	$(0.000027)^{4/3}$

In problems 17–42, use the properties of exponents to simplify each expression. Assume that all bases represent positive real numbers.

17	$2^{1/3} \cdot 2^{2/3}$	18	$5^{1/2} \cdot 5^{-3/2}$
19	$8^{1/3} \cdot 8^{-2/3}$	20	$x^{-2/3} \cdot x^{5/3}$
21	$x^{1/2} \cdot x^{4/5}$	22	$y^{2/15} \cdot y^{-7/60}$
23	$(5^{1/7})^7$	24	$(8^{-2/3})^3$
25	$(x^{-7/9})^{18/7}$	26	$(y^{-2})^{-15/2}$
27	$(y^{-7})^{-2/21}$	28	$(x^{-3/8})^{-8/3}$

29 $(8x^9)^{4/3}$

30 $(32x^{-5})^{-3/5}$

31 $(2^{-1/3}x^{-1/7})^{-21}$

32 $(81x^{12})^{-3/4}$

33 $\left(\dfrac{125}{x^3}\right)^{-1/3}$

34 $\left(\dfrac{1{,}024}{x^6}\right)^{-1/10}$

35 $\left(\dfrac{x^{-5}}{32}\right)^{4/5}$

36 $\left(\dfrac{x^{11}}{y^{11}}\right)^{-5/11}$

37 $\dfrac{5^{2/3}}{5^{-1/7}}$

38 $\dfrac{3^{1/2}}{3^{-2/5}}$

39 $\dfrac{x^{1/3}}{x^{-1/6}}$

40 $\dfrac{7^{0.5}}{7^{-0.2}}$

41 $\dfrac{(x^{-3/4})^{-2}}{(x^2)^{-5/8}}$

42 $\dfrac{(x^{-3})^{2/3}}{(x^6)^{-4/3}}$

In problems 43–53, use the properties of exponents to simplify each expression. Express the answer in a form containing nonnegative exponents.

43 $\dfrac{5^{2/3}\cdot 3^{1/2}}{5^{-1}}$

44 $\dfrac{3^{5/2}\cdot 2^{-1/5}}{3^{-2}}$

45 $\left(\dfrac{5^2\cdot 4^6}{5^{-4}\cdot 4^7}\right)^{-1/2}$

46 $\left(\dfrac{x^{-1}y^{-2/3}}{z^{-2}}\right)^{-3}$

47 $\left[\dfrac{-5x^{-3}}{(16y^2)^{-1/2}}\right]^{-1}$

48 $\dfrac{x^{-1/4}y^{-5/2}}{x^3y^{-3}}$

49 $\left(\dfrac{81x^{-12}}{y^{16}}\right)^{-1/4}$

50 $(x^{5/7}y^{3/2})\cdot(x^{2/3}y^{5/2})$

51 $\left[\dfrac{(xy)^{-r/s}}{x^{1/r}y^{1/s}}\right]^{-1}$

52 $\dfrac{a^{1/s}\cdot b^{1/r}}{a^r b^s}$

53 $\left(\dfrac{x^{n/2}y^{n/4}z^{6n/5}}{z^{n/5}}\right)^{1/n}$; n a positive integer

54 Criticize the following proof:
$(-1)^{1/3} = (-1)^{2/6} = [(-1)^{1/6}]^2 = [(-1)^2]^{1/6} = 1^{1/6} = 1$
However, $(-1)^{1/3} = -1$. Therefore, $1 = -1$.

5.3 Radicals

In Section 5.2 we discussed rational exponents and their properties. There are times when radical notation is more appropriate than

exponential notation; so this section is devoted to the radicals ar
their basic properties.

The symbol $\sqrt{a}$ is another way of writing $a^{1/2}$, the *principal squa*
root of a. If n is greater than 2, the symbol $\sqrt[n]{a}$ is written as $a^{1/n}$, the n
root of a. In $\sqrt[n]{a}$, the symbol $\sqrt[n]{}$ is called a *radical*, n is called the *ind*
(note that for the square root of a the index 2 is omitted), and a
called the *radicand*. Thus, $\sqrt{4} = 4^{1/2} = 2$, $\sqrt[3]{8} = 8^{1/3} = 2$, and $\sqrt[5]{\cdot}$
$= (32)^{1/5} = 2$. We indicated in Section 5.2 that $a^{p/q} = (a^p)^{1/q}$. Replacin
p and q by m and n, we have $a^{m/n} = (a^m)^{1/n}$, which is written in radic
form as $\sqrt[n]{a^m}$, so that

$$a^{m/n} = \sqrt[n]{a^m}$$

In fact, the rules listed for $a^{1/q}$ on page 185 can be expressed
equivalent radical form as follows:

1   If $a > 0$ and n is any positive integer, then $\sqrt[n]{a}$ is positive.

2   If $a < 0$ and n is any positive odd integer, then $\sqrt[n]{a}$ is negativ

3   If $a < 0$ and n is any positive even integer, then $\sqrt[n]{a}$ is not define
in the real numbers system.

EXAMPLES

Express each of the following expressions in exponential form ar
compute the value of each expression.

1   $\sqrt{25}$

SOLUTION
$$\sqrt{25} = 25^{1/2} = (5^2)^{1/2} = 5$$

2   $\sqrt[3]{-64}$

SOLUTION
$$\sqrt[3]{-64} = (-64)^{1/3} = [(-4)^3]^{1/3} = -4$$

3   $\sqrt[4]{16}$

SOLUTION
$$\sqrt[4]{16} = 16^{1/4} = (2^4)^{1/4} = 2$$

4   $\sqrt[5]{y^5}$

SOLUTION
$$\sqrt[5]{y^5} = (y^5)^{1/5} = y$$

5 $\sqrt[7]{x^{21}}$

SOLUTION

$$\sqrt[7]{x^{21}} = (x^{21})^{1/7} = x^3$$

EXAMPLES

Use the fact that $a^{m/n} = \sqrt[n]{a^m}$ to write each of the following expressions in radical form. Assume that all variables represent positive real numbers.

6 $7^{3/5}$

SOLUTION

$$7^{3/5} = \sqrt[5]{7^3}$$

7 $x^{3/4}$

SOLUTION

$$x^{3/4} = \sqrt[4]{x^3}$$

8 $(-y)^{5/7}$

SOLUTION

$$(-y)^{5/7} = \sqrt[7]{(-y)^5} = \sqrt[7]{-y^5}$$

A negative rational power of a number can be expressed in radical form by considering the denominator of the exponent to be positive. For example, $4^{-3/2} = \sqrt{4^{-3}} = \sqrt{\frac{1}{64}} = \frac{1}{8}$ and $8^{-2/3} = \sqrt[3]{8^{-2}} = \sqrt[3]{\frac{1}{64}} = \frac{1}{4}$.

If a is a real number, $\sqrt{a^2} = a$ if and only if a is positive or zero. For example, $\sqrt{2^2} = 2$, since $\sqrt{2^2} = \sqrt{4} = 2$. However, $\sqrt{(-2)^2} \neq -2$, since $\sqrt{(-2)^2} = \sqrt{4} = 2$ (remember our discussion on principal roots). In this case, $\sqrt{(-2)^2} = -(-2)$. In general,

$$\sqrt{a^2} = \begin{cases} a & \text{if } a \text{ is positive or zero} \\ -a & \text{if } a \text{ is negative} \end{cases}$$

Recall the definition of the absolute value of a real number. The above two statements are equivalent to the following:

$$\sqrt{a^2} = |a|$$

Properties of Radicals

Since radicals provide alternative symbols for expressions of the form $a^{1/n}$ all properties of exponents also apply to radicals. In particular, we have the following: Let m and n represent positive integers an assume that each of the roots exist; then:

1   $\sqrt[n]{a}\,\sqrt[n]{b} = \sqrt[n]{ab}$

2   $\dfrac{\sqrt[n]{a}}{\sqrt[n]{b}} = \sqrt[n]{\dfrac{a}{b}},\ b \neq 0$

3   $\sqrt[n]{a^m} = (\sqrt[n]{a})^m$

4   $\sqrt[m]{\sqrt[n]{a}} = \sqrt[n]{\sqrt[m]{a}} = \sqrt[mn]{a}$

5   $\sqrt[cn]{a^{cm}} = \sqrt[n]{a^m}$, where c is a positive integer

These properties can be verified using the equivalent exponential notations. For instance, to verify Property 1, we have

$$\sqrt[n]{a}\,\sqrt[n]{b} = a^{1/n} \cdot b^{1/n} = (ab)^{1/n} = \sqrt[n]{ab}$$

Radical expressions are expressed in various ways. To "simplify a radical expression we use the properties above and write the ex pression in a form satisfying the following conditions:

1   The power of any factor under the radical is less than the inde of the radical.

2   The exponents on factors under the radicals and the index c the radical have no common factors.

3   The radicand contains no fractions.

EXAMPLES

Use the properties of radicals to perform the following operation and simplify the resulting expressions.

9   $\sqrt{3} \cdot \sqrt{12}$

SOLUTION
$$\sqrt{3} \cdot \sqrt{12} = \sqrt{3 \times 12} \qquad \text{(Property 1)}$$
$$= \sqrt{36} = 6$$

10   $\sqrt{72}$

SOLUTION
$$\sqrt{72} = \sqrt{36 \times 2} = \sqrt{36} \cdot \sqrt{2} = 6\sqrt{2}$$

11 $\sqrt[3]{x^5}$

SOLUTION

$$\sqrt[3]{x^5} = \sqrt[3]{x^3 \cdot x^2} = \sqrt[3]{x^3} \cdot \sqrt[3]{x^2}$$
$$= x\sqrt[3]{x^2}$$

12 $\sqrt{\dfrac{25}{81}}$

SOLUTION

$$\sqrt{\frac{25}{81}} = \frac{\sqrt{25}}{\sqrt{81}} \qquad \text{(Property 2)}$$
$$= \frac{5}{9}$$

13 $\sqrt[3]{\dfrac{27}{64}}$

SOLUTION

$$\sqrt[3]{\frac{27}{64}} = \frac{\sqrt[3]{27}}{\sqrt[3]{64}} = \frac{3}{4}$$

14 $\dfrac{\sqrt[5]{32x^4}}{\sqrt[5]{x^9}}$

SOLUTION

$$\frac{\sqrt[5]{32x^4}}{\sqrt[5]{x^9}} = \sqrt[5]{\frac{32x^4}{x^9}} = \sqrt[5]{\frac{32}{x^5}} = \frac{2}{x}$$

15 $\sqrt{25^3}$

SOLUTION

$$\sqrt{25^3} = (\sqrt{25})^3 \qquad \text{(Property 3)}$$
$$= 5^3 = 125$$

16 $\sqrt[3]{(-8)^2}$

SOLUTION

$$\sqrt[3]{(-8)^2} = (\sqrt[3]{-8})^2 = (-2)^2 = 4$$

17 $\sqrt[7]{x^{35}}$

SOLUTION

$$\sqrt[7]{x^{35}} = \sqrt[7]{(x^7)^5} = (\sqrt[7]{x^7})^5 = x^5$$

18    $\sqrt[3]{\sqrt{64}}$

SOLUTION

$$\sqrt[3]{\sqrt{64}} = \sqrt[6]{64} = 2 \qquad \text{(Property 4)}$$

This expression can also be simplified as follows:

$$\sqrt[3]{\sqrt{64}} = [(64)^{1/2}]^{1/3} = 8^{1/3} = 2$$

19    $\sqrt[3]{\sqrt[5]{-x^{15}}}$

SOLUTION

$$\sqrt[3]{\sqrt[5]{-x^{15}}} = \sqrt[15]{-x^{15}} = \sqrt[15]{(-x)^{15}} = -x$$

Another way of simplifying this expression is

$$\sqrt[3]{\sqrt[5]{-x^{15}}} = [(-x^{15})^{1/5}]^{1/3} = (-x^3)^{1/3} = -x$$

20    $\sqrt[10]{32}$

SOLUTION

$$\sqrt[10]{32} = \sqrt[10]{2^5}$$
$$= \sqrt{2} \qquad \text{(Property 5)}$$

21    $\sqrt[14]{x^{21}}$

SOLUTION

$$\sqrt[14]{x^{21}} = \sqrt{x^3} = x\sqrt{x}$$

22    $\sqrt[3]{-125x^8y^{10}}$

SOLUTION

$$\sqrt[3]{-125x^8y^{10}} = \sqrt[3]{-125x^6y^9x^2y}$$
$$= \sqrt[3]{-125x^6y^9} \cdot \sqrt[3]{x^2y}$$
$$= -5x^2y^3\sqrt[3]{x^2y}$$

Property 5 may be used to express a given radical as anoth
radical whose index is an integral multiple of the given index. T
following examples will illustrate.

EXAMPLES

23 Write the radical $\sqrt[3]{x^2 y}$ as a radical having an index 6.

SOLUTION

$$\sqrt[3]{x^2 y} = \sqrt[3 \times 2]{x^{2 \times 2} y^{2 \times 1}} \qquad \text{(Property 5)}$$

$$= \sqrt[6]{x^4 y^2}$$

24 Express the radicals $\sqrt[4]{x^3}$ and $\sqrt[3]{x^2}$ as radicals having a common index 12; then find $\sqrt[4]{x^3} \cdot \sqrt[3]{x^2}$.

SOLUTION

$$\sqrt[4]{x^3} \cdot \sqrt[3]{x^2} = \sqrt[4 \times 3]{x^{3 \times 3}} \cdot \sqrt[3 \times 4]{x^{2 \times 4}} = \sqrt[12]{x^9} \cdot \sqrt[12]{x^8}$$

$$= \sqrt[12]{x^9 \cdot x^8} = \sqrt[12]{x^{17}} = x \sqrt[12]{x^5}$$

PROBLEM SET 5.3

In problems 1–16, write each expression in exponential form, and find the value of each expression. Assume that all variables represent positive real numbers.

1 $\sqrt{121}$

2 $\sqrt{0.01}$

3 $\sqrt[3]{-125}$

4 $\sqrt[3]{-216}$

5 $\sqrt[4]{81}$

6 $\sqrt[4]{625}$

7 $\sqrt[5]{-32}$

8 $\sqrt[5]{243}$

9 $\sqrt[3]{-0.125}$

10 $\sqrt[3]{512}$

11 $\sqrt{x^4}$

12 $\sqrt[3]{x^9}$

→13 $\sqrt[5]{-x^{10}}$

14 $\sqrt[7]{-x^{14}}$

15 $\sqrt[7]{(128)^2}$

16 $\sqrt[3]{(8)^{-3}}$

In problems 17–28, use the fact that $a^{m/n} = \sqrt[n]{a^m}$ to write each expression in radical form. Assume that all variables represent positive real numbers.

17 $3^{5/6}$

18 $11^{2/3}$

19 $26^{3/4}$

20 $25^{3/7}$

21 $85^{2/3}$

22 $17^{3/8}$

→23 $(32)^{-3/5}$

24 $(64)^{-5/8}$

25 $x^{5/11}$ **26** $x^{3/4}$

27 $x^{3/22}$ **28** $y^{8/11}$

In problems 29–74, use the properties of radicals to simplify each expression. Assume that all variables represent positive real numbers.

29 $\sqrt{2} \cdot \sqrt{32}$ **30** $\sqrt[3]{2} \cdot \sqrt[3]{4}$

31 $\sqrt[4]{3} \cdot \sqrt[4]{27}$ **32** $\sqrt[5]{-2} \cdot \sqrt[5]{-16}$

33 $\sqrt{x} \cdot \sqrt{x^3}$ **34** $\sqrt[3]{x^2} \cdot \sqrt[3]{x^3}$

35 $\sqrt[3]{-9x^5}$ **36** $\sqrt[4]{x^3} \cdot \sqrt[4]{x}$

37 $\sqrt[3]{8x^4}$ **38** $\sqrt[5]{64x^5}$

39 $\sqrt[3]{-243x^{10}}$ **40** $\sqrt[7]{-128x^{15}}$

41 $\sqrt{\dfrac{49}{9}}$ **42** $\sqrt[3]{\dfrac{-27}{64}}$

43 $\dfrac{\sqrt{1,000}}{\sqrt{40}}$ **44** $\dfrac{\sqrt{147}}{\sqrt{3}}$

45 $\dfrac{\sqrt[5]{-64}}{\sqrt[5]{2}}$ **46** $\dfrac{\sqrt[7]{256}}{\sqrt[7]{2}}$

47 $\dfrac{\sqrt[3]{54}}{\sqrt[3]{18}}$ **48** $\dfrac{\sqrt[3]{x^4}}{\sqrt[3]{8x}}$

49 $\dfrac{\sqrt[4]{32x^5}}{\sqrt[4]{2x}}$ **50** $\sqrt[3]{\dfrac{x}{y^3}}$

51 $\dfrac{\sqrt[5]{x^6}}{\sqrt[5]{-x}}$ **52** $\dfrac{\sqrt[7]{x^{10}}}{\sqrt[7]{x^3}}$

53 $\sqrt{3^2}$ **54** $\sqrt[3]{(-27)^2}$

55 $\sqrt[3]{(-64)^2}$ **56** $\sqrt[5]{(32)^3}$

57 $\sqrt[4]{x^{12}}$ **58** $\sqrt[5]{x^{14}}$

59 $\sqrt[3]{(-x^3)^2}$ **60** $\sqrt[7]{-x^{21}}$

61 $\sqrt[4]{\sqrt{32}}$ **62** $\sqrt{\sqrt[5]{1,024}}$

63 $\sqrt[3]{\sqrt[4]{x^{12}}}$ **64** $\sqrt[5]{\sqrt[3]{x^{15}}}$

65 $\sqrt{\sqrt[4]{x^{16}}}$ **66** $\sqrt[3]{\sqrt[10]{x^{90}}}$

67 $\sqrt[6]{16}$ **68** $\sqrt[4]{64}$

69 $\sqrt[8]{x^{12}} \cdot \sqrt[8]{x^5 y^{-8}} \cdot \sqrt[8]{x^2 y^9}$

70 $\dfrac{\sqrt[3]{a^{-1}b^3} \cdot \sqrt[3]{125a^{-2}b^{-1}}}{\sqrt[3]{8a^3 b^{-4}}}$

71 $\dfrac{\sqrt[3]{x^{11}} \cdot \sqrt[3]{x^5}}{\sqrt[3]{x}}$

72 $\dfrac{\sqrt{x^2 y} \cdot \sqrt{xy^4}}{\sqrt{xy^3}}$

$\Rightarrow$ **73** $\sqrt[4]{\dfrac{x^4 y^3}{x^3 y}} \cdot \dfrac{\sqrt[4]{x^5 y}}{\sqrt[4]{xy^{-1}}}$

74 $\dfrac{\sqrt{324 x^5 y} \cdot \sqrt{9x^3}}{\sqrt{25 x^2 y}}$

In problems 75–80, write each radical as a radical with the indicated index. Assume that all variables represent positive real numbers.

75 $\sqrt[3]{3xy^2}$; index 6

76 $\sqrt{5x}$; index 6

77 $\sqrt{7x}$; index 4

78 $\sqrt[3]{25 x^2 y^3}$; index 9

79 $\sqrt[5]{3x^4 y^3}$; index 10

80 $\sqrt[7]{5x^4 y^2}$; index 14

In problems 81–84, express each expression as a common radical, then simplify the result. Assume that all variables represent positive real numbers. (See Example 24, page 197.)

81 $\sqrt[3]{xy^2} \cdot \sqrt[4]{x^2 y}$

82 $\sqrt[4]{x^3 y^5} \cdot \sqrt[6]{xy^4}$

83 $\sqrt[5]{xy^2} \cdot \sqrt[3]{x^2 y^5}$

84 $\sqrt[5]{x^2 y^4} \cdot \sqrt[3]{x^4 y} \cdot \sqrt[3]{x^3 y^4}$

5.4 Operations of Radical Expressions

Expressions such as $\sqrt{2} + \sqrt{5}$, $\sqrt[3]{16} - \sqrt[3]{2}$, $\sqrt{2}(\sqrt{5} + \sqrt{3})$, and $(\sqrt{3} - 2)/\sqrt{2}$ are called *radical expressions*. In this section we shall perform the operations of addition, subtraction, multiplication, and division of radical expressions.

Addition and Subtraction of Radical Expressions

Recall from Section 2.3 the basic principles that only "like" or "similar" terms can be added or subtracted. These ideas apply to terms involving radicals. *Similar radicals* by definition are those whose radicands and indices are alike. Thus, $\sqrt{7}$ and $-3\sqrt{7}$ are similar radicals, while $\sqrt{5}$ and $\sqrt{3}$ are not. The distributive property provides the means for adding or subtracting similar radicals. Thus, $3\sqrt{2} + 4\sqrt{2} = (3+4)\sqrt{2} = 7\sqrt{2}$ and $5\sqrt{x} - 2\sqrt{x} = (5-2)\sqrt{x} = 3\sqrt{x}$. Of course, there are times when we have to simplify radical expressions to combine similar radicals. For instance, to perform the addition

$\sqrt{72} + \sqrt{50}$, we first express each term in a simpler form, so that $\sqrt{72} = \sqrt{36 \cdot 2} = \sqrt{36} \cdot \sqrt{2} = 6\sqrt{2}$ and $\sqrt{50} = \sqrt{25 \cdot 2} = \sqrt{25} \cdot \sqrt{2} = 5\sqrt{2}$. Thus, $\sqrt{72} + \sqrt{50} = 6\sqrt{2} + 5\sqrt{2} = (6 + 5)\sqrt{2} = 11\sqrt{2}$.

EXAMPLES

Simplify the following radical expressions by combining similar terms.

1   $2\sqrt{3} + 5\sqrt{3}$

SOLUTION

$$2\sqrt{3} + 5\sqrt{3} = (2 + 5)\sqrt{3}$$
$$= 7\sqrt{3}$$

2   $\sqrt{8} + \sqrt{32}$

SOLUTION

$$\sqrt{8} + \sqrt{32} = \sqrt{4 \cdot 2} + \sqrt{16 \cdot 2}$$
$$= \sqrt{4} \cdot \sqrt{2} + \sqrt{16} \cdot \sqrt{2}$$
$$= 2\sqrt{2} + 4\sqrt{2} = (2 + 4)\sqrt{2}$$
$$= 6\sqrt{2}$$

3   $3\sqrt{99} - 2\sqrt{44}$

SOLUTION

$$3\sqrt{99} - 2\sqrt{44} = 3\sqrt{9 \cdot 11} - 2\sqrt{4 \cdot 11}$$
$$= 3\sqrt{9} \cdot \sqrt{11} - 2\sqrt{4} \cdot \sqrt{11}$$
$$= 3 \cdot 3\sqrt{11} - 2 \cdot 2\sqrt{11}$$
$$= 9\sqrt{11} - 4\sqrt{11} = 5\sqrt{11}$$

4   $\sqrt{50} + \sqrt[4]{64}$

SOLUTION

$$\sqrt{50} + \sqrt[4]{64} = \sqrt{2 \cdot 25} + \sqrt[4]{16 \cdot 2^2} = \sqrt{25} \cdot \sqrt{2} + \sqrt[4]{16} \cdot \sqrt[4]{2^2}$$
$$= 5\sqrt{2} + 2\sqrt[4]{4}$$
$$= 5\sqrt{2} + 2\sqrt{2} \quad \text{[since } 2\sqrt[4]{4} = 2(2^2)^{1/4}$$
$$= 7\sqrt{2} \qquad\qquad\qquad = 2(2^{1/2}) = 2\sqrt{2}]$$

5   $4\sqrt{12} + 5\sqrt{8} - \sqrt{50}$

SOLUTION

$$4\sqrt{12} + 5\sqrt{8} - \sqrt{50} = 4\sqrt{4 \cdot 3} + 5\sqrt{4 \cdot 2} - \sqrt{25 \cdot 2}$$

$$= 4\sqrt{4} \cdot \sqrt{3} + 5\sqrt{4} \cdot \sqrt{2} - \sqrt{25} \cdot \sqrt{2}$$

$$= 4 \cdot 2\sqrt{3} + 5 \cdot 2\sqrt{2} - 5\sqrt{2}$$

$$= 8\sqrt{3} + 10\sqrt{2} - 5\sqrt{2}$$

$$= 8\sqrt{3} + (10 - 5)\sqrt{2}$$

$$= 8\sqrt{3} + 5\sqrt{2}$$

6   $7\sqrt{4x} - 5\sqrt{9x}$; $x > 0$

SOLUTION

$$7\sqrt{4x} - 5\sqrt{9x} = 7\sqrt{4} \cdot \sqrt{x} - 5\sqrt{9} \cdot \sqrt{x}$$

$$= 7(2\sqrt{x}) - 5(3\sqrt{x})$$

$$= 14\sqrt{x} - 15\sqrt{x}$$

$$= -\sqrt{x}$$

7   $8\sqrt[3]{xy^3} - 2y\sqrt[3]{x}$

SOLUTION

$$8\sqrt[3]{xy^3} - 2y\sqrt[3]{x} = 8\sqrt[3]{x} \cdot \sqrt[3]{y^3} - 2y\sqrt[3]{x}$$

$$= 8y\sqrt[3]{x} - 2y\sqrt[3]{x}$$

$$= (8y - 2y)\sqrt[3]{x}$$

$$= 6y\sqrt[3]{x}$$

Multiplication of Radical Expressions

Expressions involving radicals are multiplied in the same way that polynomial expressions are multiplied. The property $\sqrt[n]{a} \cdot \sqrt[n]{b} = \sqrt[n]{ab}$ and the distributive property provide the means for simplifying the results. The following examples will be used as a guide to illustrate this procedure.

EXAMPLES

Find the products and express the results in simplest form. Assume that all variables represent positive real numbers.

8   $\sqrt{5}(\sqrt{15} + \sqrt{25})$

SOLUTION

$$\sqrt{5}(\sqrt{15} + \sqrt{25}) = \sqrt{5} \cdot \sqrt{15} + \sqrt{5} \cdot \sqrt{25}$$
$$= \sqrt{75} + 5\sqrt{5}$$
$$= \sqrt{25 \cdot 3} + 5\sqrt{5} = \sqrt{25} \cdot \sqrt{3} + 5\sqrt{5}$$
$$= 5\sqrt{3} + 5\sqrt{5}$$

9    $(\sqrt{3} - \sqrt{2})(2\sqrt{3} + \sqrt{2})$

SOLUTION

$$(\sqrt{3} - \sqrt{2})(2\sqrt{3} + \sqrt{2}) = 2\sqrt{3} \cdot \sqrt{3} + \sqrt{3} \cdot \sqrt{2}$$
$$- 2\sqrt{3} \cdot \sqrt{2} - \sqrt{2} \cdot \sqrt{2}$$
$$= 2(3) + \sqrt{6} - 2\sqrt{6} - 2$$
$$= 6 - \sqrt{6} - 2 = 4 - \sqrt{6}$$

10    $(\sqrt{x} + 2\sqrt{y})^2$

SOLUTION.   Using the special product $(a + b)^2 = a^2 + 2ab + b^2$, we have

$$(\sqrt{x} + 2\sqrt{y})^2 = (\sqrt{x})^2 + 2\sqrt{x}(2\sqrt{y}) + (2\sqrt{y})^2$$
$$= x + 4\sqrt{x} \cdot \sqrt{y} + 4y$$
$$= x + 4\sqrt{xy} + 4y$$

11    $(\sqrt{10} + \sqrt{2})(\sqrt{10} - \sqrt{2})$

SOLUTION.   Using the special product $(a + b)(a - b) = a^2 - b^2$, we have

$$(\sqrt{10} + \sqrt{2})(\sqrt{10} - \sqrt{2}) = (\sqrt{10})^2 - (\sqrt{2})^2$$
$$= 10 - 2 = 8$$

12    $(3\sqrt{x} - 5\sqrt{y})(3\sqrt{x} + 5\sqrt{y})$

SOLUTION

$$(3\sqrt{x} - 5\sqrt{y})(3\sqrt{x} + 5\sqrt{y}) = (3\sqrt{x})^2 - (5\sqrt{y})^2$$
$$= 9x - 25y$$

Division of Radical Expressions — Rationalizing Denominators

The rules for dividing radical expressions follow from the property $\sqrt[n]{a}/\sqrt[n]{b} = \sqrt[n]{a/b}$, where $b \neq 0$. Thus,

$$\sqrt{6} \div \sqrt{2} = \frac{\sqrt{6}}{\sqrt{2}} = \sqrt{\frac{6}{2}} = \sqrt{3}$$

and

$$\sqrt{3} \div \sqrt{6} = \frac{\sqrt{3}}{\sqrt{6}} = \sqrt{\frac{3}{6}} = \sqrt{\frac{1}{2}} = \frac{1}{\sqrt{2}}$$

In the latter expression the denominator $\sqrt{2}$ is an irrational number. The expression $1/\sqrt{2}$ can be transformed into an equivalent form in which the denominator is a rational number by multiplying the numerator and the denominator of the expression by $\sqrt{2}$, so that

$$\frac{1}{\sqrt{2}} = \frac{(1)(\sqrt{2})}{(\sqrt{2})(\sqrt{2})} = \frac{\sqrt{2}}{2}$$

This method is called *rationalizing the denominator*. Often in computation we encounter radical fractions whose denominators are binomial expressions containing a radical in either one or both terms. Examples of such fractions are $2/(\sqrt{3} - \sqrt{2})$, $7/(\sqrt{5} - 1)$, and $11/(2\sqrt{7} + 3\sqrt{5})$. To rationalize the denominator of such fractions, we recall the special product $(a + b)(a - b) = a^2 - b^2$. For instance, to rationalize the denominator of $(2 + \sqrt{3})/(\sqrt{5} - \sqrt{3})$, we note that

$$(\sqrt{5} - \sqrt{3})(\sqrt{5} + \sqrt{3}) = (\sqrt{5})^2 - (\sqrt{3})^2 = 5 - 3 = 2$$

Thus,

$$\frac{2 + \sqrt{3}}{\sqrt{5} - \sqrt{3}} = \frac{(2 + \sqrt{3})(\sqrt{5} + \sqrt{3})}{(\sqrt{5} - \sqrt{3})(\sqrt{5} + \sqrt{3})} = \frac{(2 + \sqrt{3})(\sqrt{5} + \sqrt{3})}{(\sqrt{5})^2 - (\sqrt{3})^2}$$

$$= \frac{2\sqrt{5} + 2\sqrt{3} + \sqrt{3}\sqrt{5} + 3}{2}$$

$$= \frac{2\sqrt{5} + 2\sqrt{3} + \sqrt{15} + 3}{2}$$

EXAMPLES

Rationalize the denominators of the following radical expressions. Assume that all variables represent positive real numbers.

13    $\dfrac{7}{\sqrt{3}}$

SOLUTION

$$\frac{7}{\sqrt{3}} = \frac{7\sqrt{3}}{\sqrt{3} \cdot \sqrt{3}} = \frac{7\sqrt{3}}{3}$$

14    $\dfrac{3}{5\sqrt{2x}}$

SOLUTION

$$\frac{3}{5\sqrt{2x}} = \frac{3\sqrt{2x}}{5\sqrt{2x}\cdot\sqrt{2x}} = \frac{3\sqrt{2x}}{5(2x)} = \frac{3\sqrt{2x}}{10x}$$

15    $\dfrac{2}{\sqrt[3]{5}}$

SOLUTION

$$\frac{2}{\sqrt[3]{5}} = \frac{2\sqrt[3]{5}\cdot\sqrt[3]{5}}{\sqrt[3]{5}\cdot\sqrt[3]{5}\cdot\sqrt[3]{5}} = \frac{2\sqrt[3]{5(5)}}{5} = \frac{2\sqrt[3]{25}}{5}$$

16    $\dfrac{5}{\sqrt{3}-1}$

SOLUTION

$$\frac{5}{\sqrt{3}-1} = \frac{5(\sqrt{3}+1)}{(\sqrt{3}-1)(\sqrt{3}+1)}$$

$$= \frac{5\sqrt{3}+5}{(\sqrt{3})^2-1^2} = \frac{5\sqrt{3}+5}{3-1}$$

$$= \frac{5\sqrt{3}+5}{2}$$

17    $\dfrac{3\sqrt{5}+7\sqrt{2}}{6\sqrt{5}-3\sqrt{2}}$

SOLUTION

$$\frac{3\sqrt{5}+7\sqrt{2}}{6\sqrt{5}-3\sqrt{2}} = \frac{(3\sqrt{5}+7\sqrt{2})}{(6\sqrt{5}-3\sqrt{2})}\cdot\frac{(6\sqrt{5}+3\sqrt{2})}{(6\sqrt{5}+3\sqrt{2})}$$

$$= \frac{18(\sqrt{5})^2+51(\sqrt{5}\cdot\sqrt{2})+21(\sqrt{2})^2}{(6\sqrt{5})^2-(3\sqrt{2})^2}$$

$$= \frac{90+51\sqrt{10}+42}{180-18}$$

$$= \frac{132+51\sqrt{10}}{162} = \frac{44+17\sqrt{10}}{54}$$

18    $\dfrac{3}{\sqrt{x}-1}$

SOLUTION

$$\frac{3}{\sqrt{x}-1} = \frac{3(\sqrt{x}+1)}{(\sqrt{x}-1)(\sqrt{x}+1)}$$

$$= \frac{3\sqrt{x}+3}{(\sqrt{x})^2-(1)^2}$$

$$= \frac{3\sqrt{x}+3}{x-1}$$

19 $\dfrac{\sqrt{x}+\sqrt{y}}{\sqrt{x}-\sqrt{y}}$

SOLUTION

$$\frac{\sqrt{x}+\sqrt{y}}{\sqrt{x}-\sqrt{y}} = \frac{(\sqrt{x}+\sqrt{y})(\sqrt{x}+\sqrt{y})}{(\sqrt{x}-\sqrt{y})(\sqrt{x}+\sqrt{y})}$$

$$= \frac{(\sqrt{x})^2+2\sqrt{x}\cdot\sqrt{y}+(\sqrt{y})^2}{(\sqrt{x})^2-(\sqrt{y})^2}$$

$$= \frac{x+2\sqrt{xy}+y}{x-y}$$

PROBLEM SET 5.4

In problems 1–20, simplify each radical expression by combining similar terms. Assume that all variables represent positive real numbers.

1 $5\sqrt{7}+3\sqrt{7}$ 2 $8\sqrt{3}-2\sqrt{3}$

3 $\sqrt{8}+\sqrt{18}$ 4 $\sqrt{18}+\sqrt{2}$

5 $\sqrt{72}-\sqrt{2}$ 6 $\sqrt{50}+\sqrt{128}$

7 $\sqrt{50}+\sqrt{32}$ 8 $\sqrt{75}+\sqrt{108}$

9 $5\sqrt{125}+\sqrt{320}$ 10 $\sqrt{18}-\sqrt{8}+\sqrt{32}$

11 $\sqrt{20}+\sqrt{45}-\sqrt{80}$ 12 $\sqrt{294}+\sqrt{486}-\sqrt{24}$

13 $\sqrt{75x}-\sqrt{3x}-\sqrt{12x}$ 14 $2\sqrt{108x}-\sqrt{27x}+\sqrt{363x}$

15 $\sqrt{x^3}+\sqrt{25x^3}+\sqrt{9x}$ 16 $\sqrt{x^3}-2x\sqrt{x^5}+3x\sqrt{x^7}$

17 $2\sqrt[3]{3}+\sqrt[3]{24}+3\sqrt[3]{192}$ 18 $\sqrt[3]{8x^3}-\sqrt[3]{125x^2}-\sqrt[3]{x^2}$

19 $\sqrt[3]{2xy}-8\sqrt[3]{16xy}-\sqrt[3]{2x^4y}$ 20 $\sqrt{x^3}-2x\sqrt[4]{x^2}+3x\sqrt[6]{x^3}$

In problems 21–40, find each product and express the result in simplest form. Assume that all variables represent positive real numbers.

21 $\sqrt{3}(\sqrt{18} - \sqrt{2})$

22 $\sqrt{5}(4\sqrt{5} - 3\sqrt{2})$

23 $\sqrt{x}(\sqrt{x} + 3)$

24 $\sqrt{x}(2\sqrt{x} - \sqrt{y})$

25 $(\sqrt{3} + 1)(\sqrt{3} - 2)$

26 $(2\sqrt{3} + 5)(\sqrt{3} - 2)$

27 $(\sqrt{x} + 7)(2\sqrt{x} + 1)$

28 $(\sqrt{x} - 1)(3\sqrt{x} + 5)$

29 $(2\sqrt{2} + 5)^2$

30 $(\sqrt{3} + 5\sqrt{2})^2$

31 $(2\sqrt{x} - \sqrt{y})^2$

32 $(x + 3\sqrt{y})^2$

33 $(2\sqrt{3} - 1)(2\sqrt{3} + 1)$

34 $(5\sqrt{7} + \sqrt{2})(5\sqrt{7} - \sqrt{2})$

35 $(3\sqrt{5} - \sqrt{3})(3\sqrt{5} + \sqrt{3})$

36 $(2\sqrt{x} + 7)(2\sqrt{x} - 7)$

37 $(3\sqrt{x} - 11)(3\sqrt{x} + 11)$

38 $(4\sqrt{x} - \sqrt{y})(4\sqrt{x} + \sqrt{y})$

39 $(\sqrt{x} + \sqrt{3})(\sqrt{x} - \sqrt{3})$

40 $(\sqrt{7x} + \sqrt{2x})(\sqrt{7x} - \sqrt{2x})$

In problems 41–60, rationalize the denominator of each expression and simplify the result. Assume that all variables represent positive real numbers.

41 $\dfrac{2}{\sqrt{3}}$

42 $\dfrac{9}{\sqrt{21}}$

43 $\dfrac{8}{7\sqrt{11x}}$

44 $\dfrac{10x}{3\sqrt{5x}}$

45 $\dfrac{5}{\sqrt[3]{7}}$

46 $\dfrac{8}{\sqrt[3]{36}}$

47 $\dfrac{10}{\sqrt{5} - 1}$

48 $\dfrac{36}{\sqrt{3} + 1}$

49 $\dfrac{\sqrt{2}}{1 + \sqrt{2}}$

50 $\dfrac{\sqrt{2} + 1}{\sqrt{3} + \sqrt{2}}$

51 $\dfrac{2 - 2\sqrt{3}}{\sqrt{7} - \sqrt{5}}$

52 $\dfrac{8}{6\sqrt{5} - 5\sqrt{3}}$

53 $\dfrac{\sqrt{y}}{3\sqrt{x} - 2\sqrt{y}}$

54 $\dfrac{3\sqrt{2} - \sqrt{3}}{2\sqrt{3} - 7\sqrt{2}}$

55 $\dfrac{x}{x + \sqrt{y}}$

56 $\dfrac{5}{y - \sqrt{y}}$

57 $\dfrac{4\sqrt{2} + 3\sqrt{5}}{7\sqrt{5} - 3\sqrt{2}}$

58 $\dfrac{1}{\sqrt{5} + \sqrt{3} + \sqrt{2}}$

59 $\dfrac{1}{\sqrt{7} - \sqrt{5} + 2}$

60 $\dfrac{1}{\sqrt[3]{5} - \sqrt[3]{2}}$ [*Hint:* Use the special product $(a-b)(a^2 + ab + b^2) = a^3 - b^3$.]

5.5 Equations Involving Radicals

In previous sections we discussed rational exponents and radicals. In this section we shall consider the solutions of equations which involve radicals. Examples of such equations are $\sqrt{x} = 3$, $\sqrt{5x+1} = 4$, $\sqrt{x+2} = 2$, and $\sqrt[3]{5x+2} = 3$. To find the solution of radical equations such as $\sqrt{x} = 3$, we square both sides of the equation, so that $(\sqrt{x})^2 = 3^2$ or $x = 9$. Therefore, the solution set of the equation is $\{9\}$, for it is true that $\sqrt{9} = 3$. In general, to find the solution set of an equation that involves square roots, we square both sides of the equation. It is important to notice that the process of squaring both sides of an equation to remove the radicals can introduce an "apparent solution" which does not satisfy the original equation. Therefore, it is necessary to check all solutions that result from this process to determine which of the proposed solutions are to be accepted or rejected. For example, to solve the equation $\sqrt{x+1} = -3$, we square both sides of the equation, so that $(\sqrt{x+1})^2 = (-3)^2$ or $x + 1 = 9$ or $x = 8$. To check the solution in the original equation we substitute 8 for x in the equation, so that $\sqrt{8+1} = \sqrt{9} \ne -3$. (*Note:* $\sqrt{9} = 3$, not ± 3, because of the agreement on principal square roots.) Therefore, $x = 8$ is not a solution to the original equation. The solution set of this equation is $\varnothing$, the empty set. The solution $x = 8$ of the above equation is called an *extraneous solution*.

EXAMPLES

Find the solution set of each of the following equations.

1 $\sqrt{2x+5} = 3$

SOLUTION. We first eliminate the radical by squaring both sides to get $(\sqrt{2x+5})^2 = 3^2$, so that $2x + 5 = 9$. Solving the latter equation, we have

$$2x = 4 \quad \text{or} \quad x = 2$$

Check: We are to determine if $x = 2$ is a solution of the given equation $\sqrt{2x + 5} = 3$. Substituting 2 for x, we have

$$\sqrt{2(2) + 5} = \sqrt{4 + 5} = \sqrt{9} = 3$$

Hence, $\{3\}$ is the solution set.

2    $\sqrt{4x^2 - 3} = 2x + 1$

SOLUTION.   Squaring both sides of the equation, we have $(\sqrt{4x^2 - 3})^2 = (2x + 1)^2$, so that

$$4x^2 - 3 = 4x^2 + 4x + 1 \qquad \text{or} \qquad -3 = 4x + 1$$

Therefore, $4x = -4$, or $x = -1$.

Check: We are to determine if $x = -1$ is a solution of the given equation $\sqrt{4x^2 - 3} = 2x + 1$. Substituting -1 for x, we have

$$\sqrt{4(-1)^2 - 3} \stackrel{?}{=} 2(-1) + 1 \qquad \text{or} \qquad 1 \neq -1$$

Hence, the solution set is $\varnothing$.

This procedure can be generalized to solve equations involving cube roots, fourth roots, and so forth. This can be shown as follows:

EXAMPLES

Find the solution sets of the following equations.

3    $\sqrt[3]{3x - 1} = 2$

SOLUTION.   Cubing both sides of the equation, we have $(\sqrt[3]{3x - 1})^3 = 2^3$, so that

$$3x - 1 = 8 \qquad \text{or} \qquad 3x = 9 \qquad \text{or} \qquad x = 3$$

Check: We are to determine if $x = 3$ is a solution of the given equation $\sqrt[3]{3x - 1} = 2$. Substituting 3 for x, we have

$$\sqrt[3]{3(3) - 1} = \sqrt[3]{9 - 1} = \sqrt[3]{8} = 2$$

Hence, $\{3\}$ is the solution set.

4    $\sqrt[4]{x^2 - 5x + 6} = \sqrt{x - 2}$

SOLUTION. Raising both sides of the equation to the fourth power, we have

$$(\sqrt[4]{x^2 - 5x + 6})^4 = (\sqrt{x - 2})^4$$
$$x^2 - 5x + 6 = x^2 - 4x + 4$$

so that $-5x + 4x = 4 - 6$ or $-x = -2$, which is equivalent to $x = 2$.

Check: For $x = 2$, we have $\sqrt[4]{(2)^2 - 5(2) + 6} \overset{?}{=} \sqrt{2 - 2}$, so that

$$\sqrt[4]{4 - 10 + 6} \overset{?}{=} \sqrt{0}$$

or

$$\sqrt[4]{0} = \sqrt{0}$$
$$0 = 0$$

Therefore, the solution set is $\{2\}$.

PROBLEM SET 5.5

In problems 1–30, find the solution set of each equation.

1 $\sqrt{2x + 5} = 4$

2 $\sqrt{6x - 3} = 27$

3 $5 + \sqrt{x - 5} = 0$

4 $\sqrt{x^2 + 3} + x = 3$

5 $\sqrt{x - 3} - 6 = 0$

6 $5 + \sqrt{x^2 + 7} = x$

7 $8 - \sqrt{x - 1} = 6$

8 $2 + \sqrt{7x - 5} = 6$

9 $\sqrt{11 - x} = \sqrt{x + 6}$

10 $\sqrt{9x^2 - 7} - 3x = 2$

11 $\sqrt{3x + 1} = \sqrt{x + 3}$

12 $\sqrt{x^2 - 3x + 3} = x + 1$

13 $\sqrt{3x + 4} = 8$

14 $\sqrt{4x^2 - 11} = 2x + 3$

15 $\sqrt{x - 14} = \sqrt{5x}$

16 $\sqrt{x^2 + 5x + 2} = x$

17 $\sqrt{x^2 + 3x} = x + 1$

18 $\sqrt{2x + 1} + 1 = 0$

19 $\sqrt[3]{x - 3} = 3$

20 $\sqrt[3]{x + 2} = -2$

21 $\sqrt[3]{3x - 4} = 2$

22 $\sqrt[3]{x^2 + 2x - 6} = \sqrt[3]{x^2}$

23 $\sqrt[3]{x^2 + 7x - 21} = \sqrt[3]{x^2}$

24 $\sqrt[4]{x - 1} = 2$

25 $\sqrt[4]{3x - 6} = 3$

26 $\sqrt{x + 4} = \sqrt[4]{x^2 - 5x + 6}$

27 $\sqrt[4]{x^2 - 7x + 1} = \sqrt{x - 5}$

28 $\sqrt[4]{x^2 + 1} = \sqrt{x + 1}$

29 $\sqrt[4]{x - 8} = \sqrt[4]{2x}$

30 $\sqrt[3]{2x - 1} + 3 = 0$

5.6   Complex Numbers

In Section 5.3 it was pointed out that the square root of a negative number is not a real number. For example, $\sqrt{-4}$ and $\sqrt{-5}$ are not real numbers. In order to have a set of numbers that contain the square roots of all real numbers (positive, zero, and negative), a new set, called the set of *complex numbers,* will be considered.

We define $\sqrt{-x}$, $x > 0$, to be a number such that $\sqrt{-x}\,\sqrt{-x} = -x$. In particular, for $x = 1$, we have $\sqrt{-1}\,\sqrt{-1} = -1$. The number $\sqrt{-1}$ is usually designated by the symbol i. Thus, $i = \sqrt{-1}$, or $i^2 = -1$. Using the symbol i, we can represent the square root of negative numbers such as $\sqrt{-4}$ by $\sqrt{-4} = \sqrt{4(-1)} = \sqrt{4}\,\sqrt{-1} = 2i$. Numbers of the form $z = a + bi$, where a and b are real numbers, are called *complex numbers.* Thus, $2i$ and $4 + 5i$ are complex numbers. If $b = 0$, then $a + bi = a + 0i$, which is denoted by a. Hence, the real numbers are elements of the set of complex numbers.

It should be noticed that positive integral powers of complex numbers have the same meaning in terms of repeated multiplication as real numbers. Thus, we can extend the definition of positive integral exponents to include complex numbers. In particular, we have

$$i^1 = i$$
$$i^2 = i \cdot i = -1$$
$$i^3 = i \cdot i \cdot i = -1 \cdot i = -i$$
$$i^4 = i \cdot i \cdot i \cdot i = (-1)(-1) = 1$$

Also, using properties of exponents, we have

$$i^5 = i^4 \cdot i = 1 \cdot i = i$$
$$i^6 = i^4 \cdot i^2 = 1 \cdot (-1) = -1$$
$$i^7 = i^4 \cdot i^3 = 1 \cdot (-i) = -i$$
$$i^8 = (i^4)^2 = 1^2 = 1$$
$$.$$
$$.$$
$$.$$

Notice the repetition of the numbers $i, -1, -i, 1$ in the values of these powers of i. Using the fact that $i^4 = 1$ and the properties of exponents, we can write any positive integral power of i as $i, -1, -i,$ or 1.

EXAMPLES

Express each of the following powers of i as $i, -1, -i,$ or 1.

1. i^{27}

SOLUTION

$$i^{27} = i^{24+3} = i^{24} \cdot i^3 = (i^4)^6 i^3 = 1^6(-i) = -i$$

2 i^{33}

SOLUTION

$$i^{33} = i^{32} \cdot i = (i^4)^8 i = 1^8 \cdot i = i$$

Operations of Complex Numbers

The complex numbers are closed under the operations of addition, subtraction, multiplication, and division. Also, complex numbers possess Properties 1 through 12 of the real numbers from Chapter 1 (page 17).

The real number a of the complex number $z = a + bi$ is called the *real part of z*, and the real number b is called the *imaginary part of z*. It would perhaps be better to identify these numbers as the "non-i part" and the "i part" of the complex numbers, since both kinds of numbers exist, but the choice of words "real part" and "imaginary part" is always accepted for historical reasons. We define equality of complex numbers, and the operations of addition, subtraction, and multiplication for complex numbers as follows:

DEFINITION 1 COMPLEX NUMBERS

Let a, b, c, and d be real numbers; then

a) *Equality:* $a + bi = c + di$ if and only if $a = c$ and $b = d$

b) *Addition:* $(a + bi) + (c + di) = (a + c) + (b + d)i$

c) *Subtraction:* $(a + bi) - (c + di) = (a - c) + (b - d)i$

d) *Multiplication:* $(a + bi)(c + di) = (ac - bd) + (ad + bc)i$

As an immediate consequence of this definition of equality, we conclude that, if $z = a + bi = 0$, then $a = 0$ and $b = 0$. (Note that $z = 0$ means that $z = 0 + 0i$.)

EXAMPLES

In Examples 3–5, indicate the real and imaginary parts of the following complex numbers.

3 $3 + 5i$

SOLUTION. 3 is the real part and 5 is the imaginary part of the complex number.

4   4

SOLUTION.   Since $4 = 4 + 0i$, 4 is the real part and 0 is the imagina
part of the complex number.

5     $-6i$

SOLUTION.   Since $-6i = 0 + (-6)i$, 0 is the real part and -6 is t
imaginary part of the complex number.

In Examples 6–9, find (a) the sum, (b) the difference, and (c) th
product of each of the following pairs of complex numbers.

6   i and i

SOLUTION

a) $i + i = 2i$
b) $i - i = 0$
c) $i \cdot i = i^2 = -1$

7     $5 + 6i$ and $9 + 3i$

SOLUTION

a) $(5 + 6i) + (9 + 3i) = (5 + 9) + (6 + 3)i = 14 + 9i$
b) $(5 + 6i) - (9 + 3i) = (5 - 9) + (6 - 3)i = -4 + 3i$
c) $(5 + 6i)(9 + 3i) = [(5)(9) - (6)(3)] + [(5)(3) + (6)(9)]i$
         $= (45 - 18) + (15 + 54)i = 27 + 69i$

8     $4 + 2i^3$ and $-3 + i^5$

SOLUTION. $4 + 2i^3 = 4 - 2i$ and $-3 + i^5 = -3 + i$, so that

a) $(4 - 2i) + (-3 + i) = 1 - i$
b) $(4 - 2i) - (-3 + i) = 7 - 3i$
c) $(4 - 2i)(-3 + i) = (-12 + 2) + (6 + 4)i = -10 + 10i$

9     $-3 + i^5$ and $4 + 3i^9$

SOLUTION. $-3 + i^5 = -3 + i$ and $4 + 3i^9 = 4 + 3i$, so that

a) $(-3 + i) + (4 + 3i) = 1 + 4i$
b) $(-3 + i) - (4 + 3i) = -7 - 2i$
c) $(-3 + i)(4 + 3i) = (-12 - 3) + (-9 + 4)i = -15 - 5i$

10 Find the values of x and y so that the statement $2x + yi = 2 + 3i$ is true.

SOLUTION. By the definition of equality of complex numbers, we have $2x = 2$ and $y = 3$. Hence, $x = 1$ and $y = 3$.

A method for dividing complex numbers is obtained from the fact that for each complex number $a + bi$, there is another complex number $a - bi$ such that their product is a real number. For example, if we want to find the product of $1 + 2i$ and $1 - 2i$, we have

$$(1 + 2i)(1 - 2i) = [1 \cdot 1 - 2(-2)] + [1(-2) + 2 \cdot 1]i$$
$$= 5 + 0i = 5$$

which is a real number.

In general, if $z = a + bi$ is a complex number, we define the *conjugate* of z, denoted by $\bar{z}$, to be $\bar{z} = a - bi$. Thus, for any complex number $z = a + bi$, we have

$$z\bar{z} = (a + bi)(a - bi) = a^2 + b^2$$

which is a real number.

To divide two complex numbers, we multiply the numerator and denominator by the conjugate of the complex number in the denominator. For instance, to divide $1 + 2i$ by $2 + 3i$, we have

$$\frac{1 + 2i}{2 + 3i} = \frac{(1 + 2i)(2 - 3i)}{(2 + 3i)(2 - 3i)} = \frac{[2 - (-6)] + (4 - 3)i}{2^2 + 3^2}$$

$$= \frac{8 + i}{13} = \frac{8}{13} + \frac{1}{13}i$$

Guided by the example above, we establish a general procedure for dividing two complex numbers. If w and z are complex numbers, with $z \neq 0$, then $w/z = (w \cdot \bar{z})/(z \cdot \bar{z})$, where $\bar{z}$ is the conjugate of z. That is, if $w = c + di$ and $z = a + bi$, then $w/z = (c + di)/(a + bi)$, so that

$$\frac{w\bar{z}}{z\bar{z}} = \frac{(c + di)(a - bi)}{(a + bi)(a - bi)} = \frac{[ac - (-b)d] + [ad + (-b)c]i}{a^2 + b^2}$$

Therefore,

$$\frac{w}{z} = \frac{c + di}{a + bi} = \frac{ac + bd}{a^2 + b^2} + \frac{ad - bc}{a^2 + b^2}i$$

EXAMPLES

11   Identify the conjugate in each of the following.

   a) $3 + 2i$ b) -4

   c) $5i$ d) $-1 - i$

   e) $6i - 5$

SOLUTION

   a) $\overline{3 + 2i} = 3 - 2i$
   b) $\overline{-4} = \overline{-4 + 0i} = -4 - 0i = -4$
   c) $\overline{5i} = -5i$
   d) $\overline{-1 - i} = -1 + i$
   e) $\overline{6i - 5} = -6i - 5$

In Examples 12–15, write each of the following expressions in the form of $a + bi$.

12   $\dfrac{1}{3 - 2i}$

SOLUTION

$$\frac{1}{3 - 2i} = \frac{1}{3 - 2i} \cdot \frac{3 + 2i}{3 + 2i} = \frac{3 + 2i}{9 + 4} = \frac{3}{13} + \frac{2}{13}i$$

13   $\dfrac{1 + i}{1 - i}$

SOLUTION

$$\frac{1 + i}{1 - i} = \frac{1 + i}{1 - i} \cdot \frac{1 + i}{1 + i} = \frac{(1 + i)^2}{1 + 1} = \frac{2i}{2} = i$$

14   $\dfrac{3 + i^9}{1 - i^7}$

SOLUTION. $3 + i^9 = 3 + i$ and $1 - i^7 = 1 + i$, so that

$$\frac{3 + i^9}{1 - i^7} = \frac{3 + i}{1 + i} \cdot \frac{1 - i}{1 - i} = \frac{[3 - (-1)] + (-3 + 1)i}{1 + 1}$$

$$= \frac{4 - 2i}{2} = 2 - i$$

15   $\dfrac{1}{z}$, $z = a + bi$, $z \neq 0$

SOLUTION

$$\frac{1}{z} = \frac{1}{a+bi} \cdot \frac{a-bi}{a-bi} = \frac{a-bi}{a^2+b^2} = \frac{a}{a^2+b^2} - \frac{b}{a^2+b^2}i$$

PROBLEM SET 5.6

In problems 1–12, replace each expression by an equivalent expression in terms of $i, -i, 1,$ or -1.

1	i^{29}	**2**	i^{37}
3	i^{54}	**4**	i^{49}
5	i^{65}	**6**	i^{74}
7	i^{108}	**8**	i^{119}
9	i^{213}	**10**	i^{403}
11	$(i^{205})^3$	**12**	$(i^{409})^2$

In problems 13–38, write each expression in the form $a + bi$, a and b are real numbers.

13	$3\sqrt{-25}$	**14**	$-5\sqrt{-8}$
15	$5 + \sqrt{-81}$	**16**	$-8 + \sqrt{-72}$
17	$-\sqrt{-28x^2},\ x > 0$	**18**	$6 - \sqrt{-16^2},\ x > 0$
19	$(-3 + 6i) + (2 + 3i)$	**20**	$(-2 + 5i) + (-5 + i)$
21	$(-5 + 3i) + (5i)$	**22**	$(10 - 24i) + (3 + 7i)$
23	$(6 - 8i^3) + (6 + 8i^3)$	**24**	$(3 + 2i^5) + (-7 + 2i^5)$
25	$(-2 - i) - (-3 - 2i)$	**26**	$(7 + 24i) - (-3 - 4i)$
27	$(10 - 8i) - (10 + 8i)$	**28**	$(5 - 7i) - (5 - 13i)$
29	$(6 - 8i^7) - (5 - 3i^{11})$	**30**	$(4 - \sqrt{-25}) - (2 + \sqrt{-36})$
31	$(2 + 3i)(1 - 3i)$	**32**	$(1 - 6i)(2 + 5i)$
33	$(3 - 7i)(2 + 3i)$	**34**	$(-2 + 7i)(-2 - 7i)$
35	$(5 - 4i)(5 + 4i)$	**36**	$(3 - i^3)(5 + i^7)$
37	$(2 + i^{11})(3 - i^{13})$	**38**	$(2 + \sqrt{3}i^3)(2 - \sqrt{3}i^3)$

In problems 39–42, determine the real values of x and y for which each equation is satisfied.

39 $x + iy = 3 - 4i$ **40** $5x - 3yi = 4 + yi$

41 $-2x + 5yi = 6 + 15i$ **42** $(3x + 7) + 5i = 4 + 15yi$

In problems 43–46, find $\bar{z}$ (the conjugate of z) for each given

43 $z = 2 + 4i$ **44** $z = 3 - 4i^3$

45 $z = 3i^7$ **46** $z = 2i^{100}$

In problems 47–58, write each expression in the form of $a + b$

47 $\dfrac{2 + i}{2 + 2i}$ **48** $\dfrac{1 - i}{-1 + i}$

49 $\dfrac{-3 + 4i}{-2 - i}$ **50** $\dfrac{2}{3 + 2i}$

51 $\dfrac{i}{-1 - i}$ **52** $\dfrac{1 + i}{2 + \sqrt{3}i}$

53 $\dfrac{5 - 2i^3}{3 + 2i^7}$ **54** $\dfrac{(3 - 2i)^2}{5 - 2i^3}$

55 $\dfrac{3 + 2i^{11}}{5 - 2i^5}$ **56** $\dfrac{3 + 2i - 4i^2}{5 - 2i + i^3}$

57 $\dfrac{1}{1 + 2i^{77}}$ **58** $\dfrac{2 + \sqrt{-3}}{1 + \sqrt{-2}}$

REVIEW PROBLEM SET

In problems 1–16, write each expression with nonnegative exponents and simplify.

1 3^{-2} **2** 7^{-2}

3 10^{-3} **4** $\dfrac{1}{5^{-2}}$

5 $\dfrac{1}{2^{-6}}$ **6** $(0.01)^{-3}$

7 $\dfrac{1}{8^{-2}}$ **8** x^{-4}

9 $\dfrac{x^{-2}}{y^{-2}}$ **10** $a^{-2}b^{-1}$

11 $\dfrac{a^{-3}}{b^2}$

12 $\dfrac{m^{-2}}{m^{-4}}$

13 3^0

14 $(-32x)^0, \ x \neq 0$

15 $a^0, \ a \neq 0$

16 $\left(\dfrac{7}{5}\right)^0$

In problems 17–38, simplify each expression and write each answer in a form containing nonnegative exponents.

17 $3^{-21} \cdot 3^{17}$

18 $19^{-3} \cdot 19^5$

19 $x^{-10} \cdot x^{31}$

20 $x^{31} \cdot x^{-45}$

21 $(-2^4)^{-5}$

22 $(3^{-2})^{-1}$

23 $(x^{-1})^{-17}$

24 $(y^{-2})^4$

25 $(5x)^{-3}$

26 $(ab^2)^{-2}$

27 $(a^2 b^{-4})^{-1}$

28 $(x^{-3} y^2)^{-4}$

29 $\left(\dfrac{2}{3}\right)^{-4}$

30 $\left(\dfrac{7}{3}\right)^{-1}$

31 $\left(\dfrac{-3}{x}\right)^{-2}$

32 $\left(\dfrac{a^{-2}}{b^3}\right)^{-2}$

33 $\dfrac{5^{-4}}{5^{-6}}$

34 $\dfrac{11^{-3}}{11^{-2}}$

35 $\dfrac{(xy)^2}{(xy)^{-3}}$

36 $\dfrac{-(xy^{-1}z)^{-2}}{-(xy^{-1}z)^{-4}}$

37 $(3x^{-2}y^3)^{-3}$

38 $(x^{-2}y^{-3}z^{-1})^{-3}$

In problems 39–42, express each number in scientific notation.

39 46,800,000

40 432,000,000

41 0.0000012

42 0.000000326

In problems 43–46, express each number in ordinary decimal notation.

43 5.6×10^3

44 3.21×10^4

45 1.92×10^{-4}

46 5.21×10^{-6}

In problems 47–58, compute the value of each expression.

47 $(-32)^{3/5}$

48 $(-27)^{-4/3}$

49 $16^{-3/4}$ **50** $-(-8)^{1/3}$

51 $-4^{3/2}$ **52** $-25^{-1/2}$

53 $125^{-4/3}$ **54** $216^{-2/3}$

55 $(0.125)^{-2/3}$ **56** $(0.008)^{4/3}$

57 $4^{-7/2}$ **58** $(-128)^{-3/7}$

In problems 59–84, use the properties of exponents to simplify each expression. Assume that all bases represent positive real numbers.

59 $7^{-2/3} \cdot 7^{5/3}$ **60** $9^{-4/7} \cdot 9^{11/7}$

61 $x^{3/2} \cdot x^{-5/4}$ **62** $y^{5/11} \cdot y^{-16/11}$

63 $(2^{-1/3})^9$ **64** $(8^2)^{-5/12}$

65 $(x^{-11/3})^{3/22}$ **66** $(y^{-51})^{3/17}$

67 $(-8x^6)^{1/3}$ **68** $(16x^4)^{-1/2}$

69 $(-32x^5)^{2/5}$ **70** $(4x^2y^8)^{3/2}$

71 $(-27x^{-27})^{-2/3}$ **72** $(16x^{-16}y^{12})^{3/4}$

73 $\left(\dfrac{8}{x^9}\right)^{-2/3}$ **74** $\left(\dfrac{128}{x^7}\right)^{-2/7}$

75 $\left(\dfrac{x^{-5/2}}{y^{-2/5}}\right)^{-30}$ **76** $\left(\dfrac{16x^4}{y^8}\right)^{-3/4}$

77 $\left(\dfrac{-x^5y^{10}}{32z^{15}}\right)^{-3/5}$ **78** $\left(\dfrac{8x^{-6}y^{-30}}{27y^{-12}}\right)^{-1/3}$

79 $\dfrac{15^{-3/2}}{15^{-3/4}}$ **80** $\dfrac{7^{0.5}}{7^{-0.3}}$

81 $\dfrac{x^{-2/3}}{x^{-5/7}}$ **82** $\dfrac{(y^2)^{-3/2}}{y^{7/2}}$

83 $\dfrac{(x^{-1}y^{-2/3})^{-3}}{(x^{-1}y^{-2/3})^5}$ **84** $\dfrac{(x^{-7/3})^{2/7}}{x^{-3}}$

In problems 85–94, write each expression in exponential form. Assume that all variables represent positive real numbers.

85 $\sqrt[6]{8}$ **86** $\sqrt[3]{17}$

87 $(\sqrt{13})^{-2}$ **88** $(\sqrt[5]{7})^2$

89 $\sqrt[3]{x}$ **90** $\sqrt{ab}$

91 $\sqrt[6]{x^{11}}$

92 $(\sqrt[7]{x})^2$

93 $(\sqrt[5]{x})^4$

94 $\sqrt[11]{x^{12}}$

In problems 95–106, use the fact that $a^{m/n} = \sqrt[n]{a^m}$ to write each expression in radical form. Assume that all variables represent positive real numbers.

95 $10^{2/5}$

96 $25^{1/7}$

97 $5^{-3/2}$

98 $7^{-2/3}$

99 $11^{2/9}$

100 $17^{5/6}$

101 $y^{-2/3}$

102 $a^{-3/4}$

103 $b^{-3/4}$

104 $x^{-2/7}$

105 $(5xy^{-2})^{-1/2}$

106 $2^{-1/2}x^{-1/2}$

In problems 107–142, use the properties of radicals to simplify each expression. Assume that all variables represent positive real numbers.

107 $\sqrt[3]{5} \cdot \sqrt[3]{25}$

108 $\sqrt[7]{2} \cdot \sqrt[7]{64}$

109 $\sqrt[5]{x^4} \cdot \sqrt[4]{x}$

110 $\sqrt[100]{x^{98}} \cdot \sqrt[100]{x^2}$

111 $\sqrt{125}$

112 $\sqrt[5]{-64}$

113 $\sqrt[4]{16x^8}$

114 $\sqrt[8]{256x^{16}}$

115 $\sqrt{4x^2y^4}$

116 $\sqrt[n]{x^{2n}}$

117 $\sqrt{\dfrac{1}{625}}$

118 $\sqrt[3]{\dfrac{2}{125}}$

119 $\sqrt[4]{\dfrac{1}{16}}$

120 $\sqrt{\dfrac{x^4}{y^8}}$

121 $\sqrt[3]{-64}$

122 $\sqrt[5]{5^{10}}$

123 $\sqrt[7]{(-x^2)^7}$

124 $\sqrt[13]{(x^3)^{-39}}$

125 $\sqrt{\sqrt{16}}$

126 $\sqrt[3]{\sqrt[3]{x^9}}$

127 $\sqrt{\sqrt[4]{x^{24}}}$

128 $\sqrt[5]{\sqrt{x^{-50}}}$

129 $\sqrt[4]{x^2}$

130 $\sqrt[10]{x^5}$

131 $\sqrt[4]{32x^5y^{10}}$

132 $\sqrt[3]{-125x^5}$

133 $\sqrt[3]{54x^4y^5z^7}$

134 $\sqrt[4]{32x^5y^{10}z^{15}}$

135 $\sqrt{\dfrac{125xy^3}{4x^2y^4}}$

136 $\sqrt[3]{\dfrac{16x^4y^7}{27x^3y^6}}$

137 $\sqrt[4]{\dfrac{405x^7}{16y^8}}$

138 $\sqrt[n]{\dfrac{x^{2n+1}}{x^{-n+2}}}$

139 $\sqrt[3]{\dfrac{x^3\sqrt{x^5}}{x^4}}$

140 $\sqrt[5]{x^{-3}\sqrt[3]{x^4\sqrt{x^{-5}}}}$

141 $\sqrt[6]{x\sqrt[3]{y\sqrt{z}}}$

142 $\sqrt{64x^{12}y^{18}z^{16}}$

In problems 143–146, write each expression as a radical with the indicated index. Assume that all variables represent positive real numbers.

143 $\sqrt{17x}$; index 4

144 $\sqrt[3]{5ab^2}$; index 6

145 $\sqrt[5]{3x^2y}$; index 10

146 $\sqrt[4]{xy^2}$; index 14

In problems 147–174, perform the indicated operations and simplify
the results. Assume that all variables represent positive real numbers.

147 $7\sqrt{2} - 3\sqrt{2} + 4\sqrt{2}$

148 $\sqrt{5} - 6\sqrt{5} + 2\sqrt{5}$

149 $4\sqrt{x} + 5\sqrt{x} - 5\sqrt{x}$

150 $3\sqrt[3]{5} - 2\sqrt[3]{5} + 5\sqrt[3]{5}$

151 $\sqrt{128} - \sqrt{8}$

152 $\sqrt{48} - \sqrt{12}$

153 $\sqrt{108} - \sqrt{27}$

154 $\sqrt{45} + \sqrt{80}$

155 $\sqrt{32x} + \sqrt{72x}$

156 $\sqrt{128x} + \sqrt{8x}$

157 $\sqrt{63} + 2\sqrt{112} - \sqrt{252}$

158 $\sqrt[3]{16} - \sqrt[3]{54} + \sqrt[3]{250}$

159 $\sqrt{3}(\sqrt{2} + \sqrt{5})$

160 $\sqrt{5}(\sqrt{7} - \sqrt{2})$

161 $\sqrt{3}(\sqrt{6} - \sqrt{8})$

162 $\sqrt{5}(\sqrt{15} + \sqrt{10})$

163 $(\sqrt{3} + \sqrt{6})(2\sqrt{3} - \sqrt{6})$

164 $(\sqrt{6} + 1)(\sqrt{8} + \sqrt{18})$

165 $(\sqrt{x} - \sqrt{y})(2\sqrt{x} + \sqrt{y})$

166 $(3\sqrt{x} - y)(5\sqrt{x} + 2y)$

167 $(\sqrt{10} + \sqrt{2})^2$

168 $(2\sqrt{6} - 3\sqrt{2})^2$

169 $(\sqrt{8} - \sqrt{2})(\sqrt{8} + \sqrt{2})$

170 $(\sqrt{a} - \sqrt{b})(\sqrt{a} + \sqrt{b})$

171 $(\sqrt{3x} + \sqrt{2})(\sqrt{3x} - \sqrt{2})$

172 $(\sqrt{ab} + \sqrt{c})(\sqrt{ab} - \sqrt{c})$

173 $(\sqrt{xyz} - \sqrt{x})(\sqrt{xyz} - \sqrt{x})$

174 $(\sqrt{10x} - \sqrt{2y})(\sqrt{10x} + \sqrt{2y})$

In problems 175–188, rationalize the denominator of each expression. Assume that all variables represent positive real numbers.

175 $\dfrac{4}{\sqrt{3}}$

176 $\dfrac{7}{\sqrt{14}}$

177 $\dfrac{5}{\sqrt{32}}$

178 $\dfrac{8}{\sqrt{18x}}$

179 $\dfrac{6x}{\sqrt{27xy}}$

180 $\dfrac{2}{\sqrt{7}-1}$

181 $\dfrac{5}{\sqrt{11}+2}$

182 $\dfrac{6}{\sqrt{2}+3\sqrt{5}}$

183 $\dfrac{10}{5-\sqrt{15}}$

184 $\dfrac{\sqrt{8}}{3+\sqrt{2}}$

185 $\dfrac{\sqrt{7}-\sqrt{6}}{\sqrt{7}+\sqrt{6}}$

186 $\dfrac{2\sqrt{13}-\sqrt{2}}{\sqrt{13}+\sqrt{2}}$

187 $\dfrac{x-\sqrt{y}}{x-2\sqrt{y}}$

188 $\dfrac{2}{1+\sqrt{2}-\sqrt{3}}$

In problems 189–202, find the solution set of each equation.

189 $\sqrt{x}=5$

190 $\sqrt{x-1}=-5$

191 $\sqrt{x-1}=4$

192 $\sqrt{2x+5}-7=-4$

193 $\sqrt{x+17}+1=5$

194 $1+\sqrt{x-3}=4$

195 $2\sqrt{2x-3}+4=1$

196 $\sqrt{x}-7=2\sqrt{x}$

197 $\sqrt[3]{3x-2}=4$

198 $1-\sqrt[3]{x-3}=4$

199 $\sqrt[3]{5x-9}-4=0$

200 $\sqrt[4]{6x}=\sqrt[4]{2x}$

201 $\sqrt[4]{5x+7}=\sqrt[4]{x}$

202 $\sqrt[4]{4x-3}=\sqrt[4]{5x-7}$

In problems 203–208, replace each expression by i, $-i$, 1, or -1.

203 i^{601}

204 i^{117}

205 i^{-21}

206 i^{-32}

207 i^{-51}

208 i^{-204}

In problems 209–224, write each expression in the form $a+bi$.

209 $(-4-i)+(3+17i)$

210 $(3-42i)+(17+4i)$

211 $(-7+6i)+(3-2i)$

212 $(12+6i)+(12-6i)$

213 $(-24+i)-(3-19i)$

214 $(65-i)-(21-3i)$

215 $(8\sqrt{6} + 6\sqrt{5}i) - (5\sqrt{6} - \sqrt{5}i)$ **216** $(-3 - 11i) - (7 - 2i)$

217 $(6 - 2i)(1 + 3i)$ **218** $(2 + 3i)(1 - 17i)$

219 $(3 - 8i)(2 - 11i)$ **220** $(3 - 2\sqrt{7}i)(3 + 2\sqrt{7}i)$

221 $\dfrac{3 + 2i}{4 + i}$ **222** $\dfrac{1 - i}{2 - 3i}$

223 $\dfrac{7 + 3i}{3 - 2i}$ **224** $\dfrac{2 + 7i}{2 + 2i}$

CHAPTER 6

Quadratic Equations and Inequalities

In this chapter we shall be concerned with solving second-degree equations or *quadratic equations* such as $x^2 + 1 = 0$, $3x^2 - 2x + 4 = 0$, and $x^2 + 4x - 5 = 0$. That is, we shall find values of x that make each equation a true statement. Each such value is called a *solution* or *root* of the equation. To solve quadratic equations, we shall use the factoring method, completing-the-square method, or the quadratic formula. In order to employ these methods, it is desirable to express the quadratic equation in the general form $ax^2 + bx + c = 0$, where a, b, and c are real numbers and $a \neq 0$. Although we are restricting the coefficients a, b, and c of the quadratic equations to the set of real numbers, we shall see that the solution sets of such equations may contain complex numbers. The chapter also includes solutions of equations that are quadratic in form and quadratic inequalities.

6.1 Solving Quadratic Equations by Factoring

In Section 2.6 we discussed the procedure of factoring some second-degree polynomials. Here we employ factoring to solve those quadratic equations $ax^2 + bx + c = 0$, $a \neq 0$, whose left member can be expressed in factored form. For instance, to solve the quadratic equation $x^2 - 6x + 5 = 0$, we write $x^2 - 6x + 5 = 0$ in factored form as $(x-1)(x-5) = 0$. To solve the latter equation, we note (from Section 1.3, Property 12) that if a and b are real numbers, then

$$ab = 0 \quad \text{if and only if} \quad a = 0 \quad \text{or} \quad b = 0 \quad \text{or} \quad \text{both}$$

Thus, the equation $(x-1)(x-5) = 0$ holds, whenever $x - 1 = 0$ or $x - 5 = 0$, so that $x = 1$ or $x = 5$. It follows that the solution set of $x^2 - 6x + 5 = 0$ is $\{1, 5\}$.

In general, if the left member of $ax^2 + bx + c = 0$, $a \neq 0$, can be expressed in the factored form $(dx + e)(fx + g) = 0$, where d, e, f, and g are integers, then its solution set is $\{-e/d, -g/f\}$.

EXAMPLES

Find the solution set of each of the following equations. In Examples 1–7, use the factoring method.

1   $x(x - 2) = 0$

SOLUTION.   Setting each factor equal to zero, we have $x = 0$ or $x - 2 = 0$, so that $x = 0$ or $x = 2$. The solution set is $\{0,2\}$.

2   $(2x - 3)(x + 2) = 0$

SOLUTION.   Setting each factor equal to zero, we have $2x - 3 = 0$ or $x + 2 = 0$. Then $x = \frac{3}{2}$ or $x = -2$. The solution set is $\{-2, \frac{3}{2}\}$.

3   $x^2 - 5x + 6 = 0$

SOLUTION.   Factoring the left member of the equation $x^2 - 5x + 6 = 0$, we have $(x - 2)(x - 3) = 0$. Setting each factor equal to zero, we have $x - 2 = 0$ or $x - 3 = 0$, so that $x = 2$ or $x = 3$. The solution set is $\{2,3\}$.

4   $6x^2 + 5x - 4 = 0$

SOLUTION.   Factoring the left member of the equation $6x^2 + 5x - 4 = 0$, we have $(2x - 1)(3x + 4) = 0$, so that $2x - 1 = 0$ or $3x + 4 = 0$. Therefore, $x = \frac{1}{2}$ or $x = -\frac{4}{3}$. The solution set is $\{\frac{1}{2}, -\frac{4}{3}\}$.

5   $x^2 + 3x = 0$

SOLUTION.   Factoring the left member of the equation $x^2 + 3x = 0$, we have $x(x + 3) = 0$, so that $x = 0$ or $x = -3$. The solution set is $\{0, -3\}$. Notice that both 0 and -3 satisfy the equation $x^2 + 3x = 0$, but if both sides of the equation had been divided by x, the solution $x = 0$ would have been lost. (Why?)

6   $x^2 - 4x + 4 = 0$

SOLUTION.   First, factor the left member of the equation $x^2 - 4x + 4 = 0$, so that $(x - 2)(x - 2) = 0$. Setting each factor equal to zero, we have $x - 2 = 0$ or $x - 2 = 0$. Therefore, $x = 2$ or $x = 2$. The solution set is $\{2\}$.

7   $\dfrac{5}{x + 4} - \dfrac{3}{x - 2} = 4$

SOLUTION. We first clear the equation from fractions by multiplying each side by the L.C.D. $(x+4)(x-2)$. Thus,

$$(x+4)(x-2)\left(\frac{5}{x+4} - \frac{3}{x-2}\right) = 4(x+4)(x-2)$$

$$5(x-2) - 3(x+4) = 4(x^2 + 2x - 8)$$

$$5x - 10 - 3x - 12 = 4x^2 + 8x - 32$$

$$2x - 22 = 4x^2 + 8x - 32$$

so that

$$4x^2 + 6x - 10 = 0 \qquad \text{or} \qquad 2x^2 + 3x - 5 = 0$$

Factoring the left member of this equation, we have $(2x+5)(x-1) = 0$, so that $2x+5 = 0$ or $x-1 = 0$. Therefore, $x = -\frac{5}{2}$ or $x = 1$. The solution set is $\{-\frac{5}{2}, 1\}$.

Forming Quadratic Equations with Given Solutions

If the solutions or the roots of a quadratic equation are given, we can construct the equation by reversing the procedure above. For example, if the solution set of a quadratic equation is $\{-2, 5\}$, then we can form the quadratic equation as follows: $x = -2$ or $x = 5$, so that $x + 2 = 0$ or $x - 5 = 0$. These two equations are equivalent to the equation

$$(x+2)(x-5) = 0 \qquad \text{or} \qquad x^2 - 3x - 10 = 0$$

In general, if the solution set of a quadratic equation is $\{r_1, r_2\}$, then the quadratic equation is formed as follows: $x = r_1$ or $x = r_2$, so that $x - r_1 = 0$ or $x - r_2 = 0$, which is equivalent to the equation

$$(x - r_1)(x - r_2) = 0 \qquad \text{or} \qquad x^2 - (r_1 + r_2)x + r_1 r_2 = 0$$

EXAMPLES

Find a quadratic equation for each of the following solution sets.

8 $\{1, 3\}$

SOLUTION. $x = 1$ or $x = 3$. Then $x - 1 = 0$ or $x - 3 = 0$, so that

$$(x-1)(x-3) = 0 \qquad \text{or} \qquad x^2 - 4x + 3 = 0$$

9    $\{-\frac{1}{2}, \frac{2}{3}\}$

SOLUTION.   Using the general equation $x^2 - (r_1 + r_2)x + r_1 r_2 = 0$, with $r_1 = -\frac{1}{2}$ and $r_2 = \frac{2}{3}$, we have $x^2 - (-\frac{1}{2} + \frac{2}{3})x + (-\frac{1}{2})(\frac{2}{3}) = 0$, so that $x^2 - \frac{1}{6}x - \frac{2}{6} = 0$, or, equivalently, $6x^2 - x - 2 = 0$.

10    $\{2 + \sqrt{3}i, 2 - \sqrt{3}i\}$

SOLUTION.   Using the general equation $x^2 - (r_1 + r_2)x + r_1 r_2 = 0$, with $r_1 = 2 + \sqrt{3}i$ and $r_2 = 2 - \sqrt{3}i$, we have

$$r_1 + r_2 = (2 + \sqrt{3}i) + (2 - \sqrt{3}i) = 4$$

and

$$r_1 r_2 = (2 + \sqrt{3}i)(2 - \sqrt{3}i) = 4 + 3 = 7$$

Therefore, the equation is $x^2 - 4x + 7 = 0$.

PROBLEM SET 6.1

In problems 1–16, find the solution set of each equation.

1	$(x - 1)(x - 2) = 0$	**2**	$(2x + 1)(x - 1) = 0$
3	$(2x + 1)(2x - 1) = 0$	**4**	$x(x - 5) = 0$
5	$(3x - 2)(x + \frac{2}{3}) = 0$	**6**	$(5x - 3)(-2x + 1) = 0$
7	$(5x - 2)(5x + 2) = 0$	**8**	$(3x - 7)(x + 2) = 0$
9	$(42x - 7)(6x + 1) = 0$	**10**	$(3x - 2)(-2x - 1) = 0$
11	$(x + \sqrt{3})(x - \sqrt{3}) = 0$	**12**	$(7x - 1)(-3x - 2) = 0$
13	$x(x + 2\sqrt{2}) = 0$	**14**	$(9x - 5)(-5x + 1) = 0$
15	$(3x + 2)(3x + 7) = 0$	**16**	$(8x - 5)(5x + 3) = 0$

In problems 17–52, use the factoring method to find the solution set of each equation.

17	$x^2 - 6x + 8 = 0$	**18**	$x^2 + 6x - 16 = 0$
19	$x^2 + 2x - 3 = 0$	**20**	$x^2 + 5x = 66$
21	$x^2 - x = 20$	**22**	$x^2 + 3x = 10$
23	$x^2 + 4x = 21$	**24**	$x^2 + 8x = 0$
25	$x^2 - 7x = 0$	**26**	$3x^2 - 7x = 0$

27 $4x^2 - 16 = 0$

28 $6x^2 = 5(x + 25)$

29 $3x^2 - 2x = 5$

30 $15x^2 - 19x + 6 = 0$

31 $10x^2 + x - 2 = 0$

32 $4x^2 - (x + 1)^2 = 0$

33 $2x^2 - x - 3 = 0$

34 $6x^2 + 17x - 3 = 0$

35 $9x^2 + 6x = 8$

36 $5x^2 + 34x = 7$

37 $4x^2 - 21x + 20 = 0$

38 $12x^2 + 5x = 2$

39 $6x^2 + 7x = 20$

40 $4x^2 = 27x + 7$

41 $10x^2 - 31x - 14 = 0$

42 $18x^2 + 61x - 7 = 0$

43 $4x^2 + 25x + 6 = 0$

44 $(x + 3)(2x + 5) = 11(x + 3)$

45 $2x^2 - 19x - 33 = 0$

46 $x(x - 2) = 9 - 2x$

47 $x^2 - 2ax - 15a^2 = 0$;
$a = $ constant

48 $12x^2 - 10mx = 12m^2$;
$m = $ constant

49 $\dfrac{12}{x} - 7 = \dfrac{12}{1 - x}$

50 $\dfrac{5}{4(x + 4)} - 1 = \dfrac{3}{4(x - 2)}$

51 $\dfrac{15}{(x - 2)^2} + \dfrac{2}{x - 2} = 1$

52 $\dfrac{3a - 2x}{4x} - 4 = \dfrac{3a}{4x - 3a}$;
$a = $ constant

In problems 53–60, find a quadratic equation whose solution set is given.

53 $\{5,6\}$

54 $\{-1,7\}$

55 $\left(-\frac{2}{3},\frac{1}{3}\right)$

56 $\{-\frac{7}{5},\frac{2}{3}\}$

57 $\{-2,-1\}$

58 $\{\sqrt{2},-\sqrt{3}\}$

59 $\{2i,-2i\}$

60 $\{1 - 3i, 1 + 3i\}$

6.2 Solving Quadratic Equations by Completing the Square

In Section 6.1, we solved quadratic equations by factoring. However, not every quadratic expression in a second-degree equation is factorable in the sense discussed in Section 2.6. $4x^2 - 7 = 0$, $x^2 + 5x + 9 = 0$, and $x^2 - x - 3 = 0$ are examples of such quadratic equations. In this section we shall develop a method called *completing the square*, which enables us to solve such equations. In preparation for presenting the method of solution by completing the square, let us consider the solutions of equations such as $x^2 = p$ and $(x+a)^2 = p$, where a and p are real numbers.

Extraction of Roots

To solve equations such as $x^2 = 9$, we use the fact that if $x^2 = p$, then $x = \sqrt{p}$ or $x = -\sqrt{p}$. Following this argument, the solution of the equation $x^2 = 9$ is $x = -\sqrt{9}$ or $x = \sqrt{9}$, so that $x = -3$ or $x = 3$. The solution set of the equation is $\{-3, 3\}$.

The above method can also be used to solve equations of the type $(x+a)^2 = p$. For example, if $(x-2)^2 = 25$, then

$$x - 2 = -\sqrt{25} \qquad\qquad x - 2 = \sqrt{25}$$
$$\text{or}$$
$$x - 2 = -5 \qquad\qquad x - 2 = 5$$

so that $x = -3$ or $x = 7$. The solution set is $\{-3, 7\}$.

The procedure of solving quadratic equations as outlined in the preceding two examples is called the *extraction-of-roots method.*

EXAMPLES

Find the solution sets of the following quadratic equations by the extraction-of-roots method.

1    $x^2 - 4 = 0$

SOLUTION. $x^2 - 4 = 0$, so that $x^2 = 4$. Therefore, $x = -2$ or $x = 2$, and the solution set is $\{-2, 2\}$.

2    $x^2 + 25 = 0$

SOLUTION. $x^2 + 25 = 0$, so that $x^2 = -25$. Therefore, $x = -\sqrt{-25}$ or $x = \sqrt{-25}$. That is, $x = -5i$ or $x = 5i$. Hence, the solution set is $\{-5i, 5i\}$.

3    $(x - 1)^2 = 9$

SOLUTION. $(x - 1)^2 = 9$, so that $x - 1 = -3$ or $x - 1 = 3$. That is, $x = -2$ or $x = 4$. Hence, the solution set is $\{-2, 4\}$.

Completing-the-Square Method

The equation $(x - 1)^2 = 9$ in Example 3 of Section 2.1 can also be written as $x^2 - 2x - 8 = 0$. (Why?) Thus, another method for solving equations of the form $ax^2 + bx + c = 0$, $a, b \neq 0$, is suggested by the above equivalence. For example, to solve the equation $x^2 - 2x - 8 = 0$, first change it to the equivalent equation $(x - 1)^2 = 9$, then solve this equation by the method of extraction of roots in the previous section. Thus, the equation $x^2 - 2x - 8 = 0$ is written in the equivalent form, $x^2 - 2x = 8$. Now observe that $(x - 1)^2 = x^2 - 2x + 1$; so if we add 1 to both sides of the equation, we have $x^2 - 2x + 1 = 9$, so that $(x - 1)^2 = 9$. By extraction of roots, we have $x - 1 = 3$ or $x - 1 = -3$, so that $x = 1 + 3 = 4$ or $x = 1 - 3 = -2$. Hence, the solution set is $\{4, -2\}$.

This technique is called the *completing-the-square method*. Its use involves finding the constant necessary to form a perfect quadratic square. From the identity $(x \pm d)^2 = x^2 \pm 2dx + d^2$, it appears that in order to complete the square when the two terms $x^2 \pm 2dx$ are given, we must add the number d^2. But $d^2 = [\frac{1}{2}(2d)]^2$; therefore, the term to be added is the square of one half the coefficient of x. Hence, the quadratic expression $x^2 - 6x$ can be written as a complete quadratic square by adding $[\frac{1}{2}(6)]^2$ or $(3)^2 = 9$. Thus, $x^2 - 6x + 9 = (x - 3)^2$ is a perfect quadratic square.

EXAMPLES

Find the solution sets of the following quadratic equations using the completing-the-square method.

4   $x^2 + 4x + 2 = 0$

SOLUTION. $x^2 + 4x + 2 = 0$ is equivalent to the equation $x^2 + 4x = -2$. Now, if the square of one half the coefficient of the first-degree term, that is, $[\frac{1}{2}(4)]^2$, is added to each side of the equation, we have

$$x^2 + 4x + [\tfrac{1}{2}(4)]^2 = -2 + [\tfrac{1}{2}(4)]^2$$

so that

$$x^2 + 4x + 4 = -2 + 4 \qquad \text{or} \qquad x^2 + 4x + 4 = 2$$

The left side of this equation is written as $(x + 2)^2$, so that $(x + 2)^2 = 2$. Then $x + 2 = -\sqrt{2}$ or $x + 2 = \sqrt{2}$, so that $x = -2 + \sqrt{2}$ or $x = -2 - \sqrt{2}$. Hence, the solution set is $\{-2 - \sqrt{2}, -2 + \sqrt{2}\}$.

5   $x^2 + 2x - 4 = 0$

SOLUTION. $x^2 + 2x - 4 = 0$ is written in the equivalent form $x^2 + 2x = 4$. Thus,

$$x^2 + 2x + [\tfrac{1}{2}(2)]^2 = 4 + [\tfrac{1}{2}(2)]^2 \qquad \text{(completing the square)}$$

or $x^2 + 2x + 1 = 4 + 1$, so that $(x + 1)^2 = 5$. Therefore, $x + 1 = -\sqrt{5}$ or $x + 1 = \sqrt{5}$, so that $x = -1 - \sqrt{5}$ or $x = -1 + \sqrt{5}$. Hence, the solution set is $\{-1 - \sqrt{5}, -1 + \sqrt{5}\}$.

In solving a quadratic equation such as $5x^2 - 20x - 13 = 0$, we express the equation in the equivalent form $5x^2 - 20x = 13$, then divide both sides of the equation by 5 to obtain $x^2 - 4x = \frac{13}{5}$. From there on, we proceed as before. In general, to solve a quadratic equation $ax^2 + bx + c = 0$, $a, b \neq 0$, first write it in the equivalent form $ax^2 + bx = -c$, then divide through by a to obtain $x^2 + (b/a)x = -c/a$. Add $(b/2a)^2$ to both sides of the equation, then use the extraction of roots to obtain the solution. The following examples illustrate the procedure.

EXAMPLES

Find the solution sets of the following quadratic equations by the completing-the-square method.

6    $3x^2 - 10x - 2 = 0$

SOLUTION.   Write $3x^2 - 10x - 2 = 0$ in the equivalent form $3x^2 - 10x = 2$. Divide both sides of the equation by 3, so that $x^2 - \frac{10}{3}x = \frac{2}{3}$ Completing the square, we get

$$x^2 - \frac{10}{3}x + \left(\frac{5}{3}\right)^2 = \frac{2}{3} + \left(\frac{5}{3}\right)^2$$

so that $\left(x - \frac{5}{3}\right)^2 = \frac{31}{9}$.

Therefore, $x - \dfrac{5}{3} = -\dfrac{\sqrt{31}}{3}$ or $x - \dfrac{5}{3} = \dfrac{\sqrt{31}}{3}$,

so that $x = \dfrac{5}{3} - \dfrac{\sqrt{31}}{3}$ or $x = \dfrac{5}{3} + \dfrac{\sqrt{31}}{3}$,

or, equivalently, $x = \dfrac{5 - \sqrt{31}}{3}$ or $x = \dfrac{5 + \sqrt{31}}{3}$.

The solution set is $\left\{\dfrac{5 - \sqrt{31}}{3}, \dfrac{5 + \sqrt{31}}{3}\right\}$.

7    $2x^2 + 3x + 2 = 0$

SOLUTION.   Write $2x^2 + 3x + 2 = 0$ in the equivalent form $2x^2 + 3x = -2$. Divide both sides of the equation by 2, so that $x^2 + \frac{3}{2}x = -1$.

Adding $[\frac{1}{2}(\frac{3}{2})]^2$ to both sides of the equation, we have

$$x^2 + \frac{3}{2}x + \left[\frac{1}{2}\left(\frac{3}{2}\right)\right]^2 = -1 + \left[\frac{1}{2}\left(\frac{3}{2}\right)\right]^2$$

or

$$x^2 + \frac{3}{2}x + \frac{9}{16} = -1 + \frac{9}{16}$$

That is, $x^2 + \frac{3}{2}x + \frac{9}{16} = -\frac{7}{16}$ or $(x + \frac{3}{4})^2 = -\frac{7}{16}$, which implies that

$$x + \frac{3}{4} = -\sqrt{-\frac{7}{16}} \qquad\qquad x + \frac{3}{4} = \sqrt{-\frac{7}{16}}$$

or

$$x + \frac{3}{4} = -\frac{\sqrt{7}}{4}i \qquad\qquad x + \frac{3}{4} = \frac{\sqrt{7}}{4}i$$

so that

$$x = -\frac{3}{4} - \frac{\sqrt{7}}{4}i \qquad \text{or} \qquad x = -\frac{3}{4} + \frac{\sqrt{7}}{4}i$$

or, equivalently,

$$x = \frac{-3 - \sqrt{7}i}{4} \qquad \text{or} \qquad x = \frac{-3 + \sqrt{7}i}{4}$$

The solution set is $\left\{\dfrac{-3 - \sqrt{7}i}{4}, \dfrac{-3 + \sqrt{7}i}{4}\right\}$.

PROBLEM SET 6.2

In problems 1–16, find the solution set of each equation by using the extraction-of-roots method.

1. $x^2 = 64$

2. $9x^2 - 4 = 0$

3. $x^2 + 81 = 0$

4. $5x^2 + 49 = 0$

5. $(x - 1)^2 = 5$

6. $(2x + 1)^2 - 25 = 0$

7. $(x - 6)^2 + 25 = 0$

8. $(2x + 5)^2 + 11 = 0$

9. $(6x - 5)^2 - 4 = 0$

10. $(x - \frac{3}{8})^2 = 0$

11 $(4x - 3)^2 = 16$ **12** $(3x - 2)^2 + 49 = 0$

13 $x^2 = 63 - 6x^2$ **14** $5x^2 + 3 = 3x^2 - 2$

15 $12x^2 - 125 = 7x^2$ **16** $2x(x + 2) = 4x + 15$

In problems 17–26, find the number that makes each expression a perfect quadratic square.

17 $x^2 + 4x +$ _____   **18** $t^2 + 8t +$ _____

19 $x^2 + 3x +$ _____   **20** $y^2 - \frac{4}{3}y +$ _____

21 $x^2 - 2x +$ _____   **22** $x^2 + 5x +$ _____

23 $x^2 + 20x +$ _____   **24** $x^2 - 9x +$ _____

25 $3x^2 - 6x +$ _____   **26** $5x^2 + 30x +$ _____

In problems 27–50, find the solution set of each equation by using the completing-the-square method.

27 $x^2 + 2x - 3 = 0$ **28** $x^2 + 21x + 10 = 0$

29 $x^2 - 12x + 35 = 0$ **30** $x^2 - 3x + 13 = 0$

31 $x^2 - 6x + 11 = 0$ **32** $x^2 - 13x + 30 = 0$

33 $5x^2 - 8x + 17 = 0$ **34** $x^2 - 4x + 7 = 0$

35 $4x^2 - 8x + 3 = 0$ **36** $9x^2 - 27x + 14 = 0$

37 $3x^2 - 7x - 13 = 0$ **38** $4x^2 + 16x + 15 = 0$

39 $4x^2 + 7x + 5 = 0$ **40** $2x^2 + 7x + 3 = 0$

41 $2x^2 + 12x + 5 = 0$ **42** $5x^2 + 2x + 7 = 0$

43 $5x^2 + 10x - 3 = 0$ **44** $25y^2 - 25y - 14 = 0$

45 $3t^2 - 8t + 4 = 0$ **46** $7p^2 + 5p + 1 = 0$

47 $9x^2 - 6x - 1 = 0$ **48** $25x^2 - 50x + 21 = 0$

49 $ax^2 + bx + 3b = 0$;   **50** $a^2x^2 - 3bx + c^2 = 0$;
    a and $b =$ constants, $a \neq 0$ a, b, and $c =$ constants, $a \neq 0$

6.3   Solving Quadratic Equations by the Quadratic Formula

In Section 6.2 we solved quadratic equations by the completing-the-square method. This method will now be generalized to develop a formula that enables us to solve any quadratic equation. Later in the section we investigate the nature of solutions by examining the quantity that occurs under the radical in the formula.

The Quadratic Formula

Consider the quadratic equation $ax^2 + bx + c = 0$, with a, b, and c real numbers and $a \neq 0$. By applying the completing-the-square method, we can obtain its solution set in terms of a, b, and c as follows:

$$ax^2 + bx + c = 0 \quad \text{is equivalent to} \quad ax^2 + bx = -c$$

Dividing both sides of the equation by a, we have $x^2 + (b/a)x = -(c/a)$. Adding $[\frac{1}{2}(b/a)]^2 = b^2/4a^2$ to both sides of the equation, we obtain

$$x^2 + \frac{b}{a}x + \frac{b^2}{4a^2} = \frac{-c}{a} + \frac{b^2}{4a^2} = \frac{b^2}{4a^2} - \frac{c}{a} = \frac{b^2 - 4ac}{4a^2}$$

The left side of the equation is the square of $x + (b/2a)$, so that

$$\left(x + \frac{b}{2a}\right)^2 = \frac{b^2 - 4ac}{4a^2}$$

which implies that

$$x + \frac{b}{2a} = \pm\sqrt{\frac{b^2 - 4ac}{4a^2}} \quad \text{or} \quad x + \frac{b}{2a} = \frac{\pm\sqrt{b^2 - 4ac}}{2a}$$

so that

$$x = \frac{-b}{2a} \pm \frac{\sqrt{b^2 - 4ac}}{2a} = \frac{-b \pm \sqrt{b^2 - 4ac}}{2a}$$

or, equivalently,

$$x = \frac{-b + \sqrt{b^2 - 4ac}}{2a} \quad \text{or} \quad x = \frac{-b - \sqrt{b^2 - 4ac}}{2a}$$

The solution set is $\left\{ \dfrac{-b + \sqrt{b^2 - 4ac}}{2a}, \dfrac{-b - \sqrt{b^2 - 4ac}}{2a} \right\}$.

This result, which provides a formula for solving any quadratic equation, is called the *quadratic formula*. The solutions of $ax^2 + bx + c = 0$, $a \neq 0$, are

$$x = \frac{-b \pm \sqrt{b^2 - 4ac}}{2a}$$

which is an abbreviated way of writing the two possible solutions.

EXAMPLES

Find the solution sets of the following equations by using the quadratic formula.

1    $3x^2 - 4x + 1 = 0$

SOLUTION.   The equation $ax^2 + bx + c = 0$ becomes $3x^2 - 4x + 1 = 0$ if $a = 3$, $b = -4$, and $c = 1$. Substituting into the quadratic formula,

$$x = \frac{-b \pm \sqrt{b^2 - 4ac}}{2a},$$

gives $x = \dfrac{-(-4) \pm \sqrt{(-4)^2 - 4(3)(1)}}{2(3)}$ or $x = \dfrac{4 \pm \sqrt{4}}{6}$

Thus, $x = \dfrac{4 + 2}{6} = 1$ or $x = \dfrac{4 - 2}{6} = \dfrac{1}{3}$

The solution set is $\{\frac{1}{3}, 1\}$.

2    $2x^2 + 3x - 1 = 0$

SOLUTION.   The equation $ax^2 + bx + c = 0$ becomes $2x^2 + 3x - 1 = 0$ if $a = 2$, $b = 3$, and $c = -1$, so that

$$x = \frac{-3 \pm \sqrt{(3)^2 - 4(2)(-1)}}{2(2)} \qquad \text{or} \qquad x = \frac{-3 \pm \sqrt{17}}{4}$$

The solution set is $\left\{ \dfrac{-3 - \sqrt{17}}{4}, \dfrac{-3 + \sqrt{17}}{4} \right\}$.

3    $x^2 + 2x + 2 = 0$

SOLUTION.   Comparing $x^2 + 2x + 2 = 0$ with the general form $ax^2 + bx + c = 0$, we have $a = 1$, $b = 2$, and $c = 2$, so that

$$x = \frac{-2 \pm \sqrt{(2)^2 - 4(1)(2)}}{2(1)} \qquad \text{or} \qquad x = \frac{-2 \pm \sqrt{-4}}{2}$$

or, equivalently, $x = (-2 \pm 2i)/2$, that is, $x = -1 \pm i$. The solution set is $\{-1 - i, -1 + i\}$.

Sum and Product of Roots of Quadratic Equations

Consider the quadratic equation $5x^2 + 9x - 2 = 0$. Using the quadratic formula, with $a = 5$, $b = 9$, and $c = -2$, we have

$$x = \frac{-9 \pm \sqrt{9^2 - 4(5)(-2)}}{2(5)} = \frac{-9 \pm \sqrt{121}}{10} = \frac{-9 \pm 11}{10}$$

so that $x = -2$ or $x = \frac{1}{5}$.

If we add these two roots, we have $-2 + \frac{1}{5} = -\frac{9}{5} = -b/a$. Similarly, if we multiply the two roots, we have $(-2)\frac{1}{5} = -\frac{2}{5} = c/a$. Guided by this example, we have the following general result.

THEOREM 1   SUM AND PRODUCT OF ROOTS

If

$$r_1 = \frac{-b - \sqrt{b^2 - 4ac}}{2a} \quad \text{and} \quad r_2 = \frac{-b + \sqrt{b^2 - 4ac}}{2a}$$

are the roots of a quadratic equation $ax^2 + bx + c = 0$, $a \neq 0$, then

$$r_1 + r_2 = -\frac{b}{a} \quad \text{and} \quad r_1 r_2 = \frac{c}{a}$$

The proof will be left as an exercise (see problem 44).

EXAMPLES

Find the sum and the product of the roots in each of the following equations.

4    $6x^2 + 11x - 35 = 0$

SOLUTION. $a = 6$, $b = 11$, and $c = -35$. Using Theorem 1 above, we have

$$r_1 + r_2 = -\frac{b}{a} = -\frac{11}{6} \quad \text{and} \quad r_1 r_2 = \frac{c}{a} = -\frac{35}{6}$$

5    $2x^2 - 2x + 1 = 0$

SOLUTION. $a = 2$, $b = -2$, and $c = 1$. Using Theorem 1 above, we have

$$r_1 + r_2 = -\frac{b}{a} = -\frac{-2}{2} = 1 \quad \text{and} \quad r_1 r_2 = \frac{c}{a} = \frac{1}{2}$$

Quadratic Discriminant

The solutions of any quadratic equation of the form $ax^2 + bx + c = 0$, where a, b, and c are real numbers, $a \neq 0$, are given by the formulas:

$$x = \frac{-b + \sqrt{b^2 - 4ac}}{2a} \quad \text{or} \quad x = \frac{-b - \sqrt{b^2 - 4ac}}{2a}$$

The kind of solutions obtained for a quadratic equation, whether they are real or complex, can be determined by the expression $b^2 - 4ac$ under the radical sign. This expression is called the *discriminant* of the quadratic equation. If the discriminant $b^2 - 4ac$ is nonnegative (positive or zero), the solutions are real numbers. However, if the discriminant $b^2 - 4ac$ is negative, the solutions are complex numbers.

For example, given the equation $x^2 + 5x + 6 = 0$ with $a = 1$, $b = 5$, and $c = 6$, the discriminant $b^2 - 4ac = 5^2 - 4(1)(6) = 1$, so that the solutions are real numbers. Also, the solutions of the equation $x^2 - 6x + 9 = 0$ are real numbers, since $b^2 - 4ac = (-6)^2 - 4(1)(9) = 0$. The solutions of the equation $2x^2 - x + 2 = 0$ are complex numbers, since the discriminant $b^2 - 4ac = (-1)^2 - 4(2)(2) = -15$. These results are summarized as follows:

Discriminant	*Kind of solutions*
$b^2 - 4ac \geq 0$	Real numbers
$b^2 - 4ac < 0$	Complex numbers

EXAMPLES

Determine the kind of solutions of the following equations by using the discriminant of the quadratic equation.

6 $2x^2 - 4x + 1 = 0$

SOLUTION. Since $a = 2$, $b = -4$, and $c = 1$,
$$b^2 - 4ac = (-4)^2 - 4(2)(1) = 8.$$
Therefore, the solutions are real numbers.

7 $x^2 - 4x + 4 = 0$

SOLUTION. Since $a = 1$, $b = -4$, and $c = 4$,
$$b^2 - 4ac = (-4)^2 - 4(1)(4) = 0.$$
Therefore, the solutions are real numbers.

8 $2x^2 + 3x + 2 = 0$

SOLUTION. Since $a = 2$, $b = 3$, and $c = 2$,
$$b^2 - 4ac = (3)^2 - 4(2)(2) = -7.$$
Therefore, the solutions are complex numbers.

It should be noticed that in Example 6, $b^2 - 4ac > 0$, so the solutions are unequal real numbers, whereas in Example 7, $b^2 - 4ac = 0$, so that the solutions are equal real numbers.

PROBLEM SET 6.3

In problems 1–28, find the solution set of each equation by using the quadratic formula.

1 $x^2 - 5x + 4 = 0$ **2** $x^2 - x + 1 = 0$

3 $x^2 - 6x + 8 = 0$ **4** $x^2 - 12x + 20 = 0$

5 $x^2 + 3x + 10 = 0$ **6** $x^2 + 3x + 13 = 0$

7 $x^2 - 2x - 35 = 0$ **8** $x^2 - \frac{1}{2}x - \frac{5}{6} = 0$

9 $2x^2 - 5x + 1 = 0$ **10** $6x^2 + 3x + 5 = 0$

11 $6x^2 - 7x - 5 = 0$ **12** $6x^2 + 5x + 2 = 0$

13 $5x^2 - 2x + 7 = 0$ **14** $9x^2 - 2x + 11 = 0$

15 $2x^2 - x + 4 = 0$ **16** $12x^2 - 4x + 3 = 0$

17 $7x^2 - 8x + 3 = 0$ **18** $4x^2 + 11x - 3 = 0$

19 $15x^2 + 2x + 8 = 0$ **20** $6x^2 + 17x - 14 = 0$

21 $4x(x - 1) = 19$ **22** $x(x + 8) + 8(x - 8) = 0$

23 $(t - 2)(t - 3) = 5$ **24** $(3y - 1)(2y + 5) = 3$

25 $(y - 1)(y - 5) = 9$ **26** $(t + 1)(2t - 1) = 4$

27 $Lx^2 + Rx + \dfrac{1}{C} = 0$; $L, R,$ and $C = $ constant, $C \neq 0$ **28** $Mx^2 + 2Rx + K = 0$; $M, R,$ and $K = $ constants, $M \neq 0$

In problems 29–43, find the sum and the product of the roots of each equation without solving.

29 $6x^2 + 5x - 6 = 0$ **30** $12x^2 = x + 20$

31 $x^2 - 100x = 0$ **32** $2x^2 - x + 17 = 0$

33 $x^2 - 7x + 13 = 0$ **34** $6x^2 + x = 2$

35 $8 = 3x^2 - 2x$ **36** $1 + x - 6x^2 = 0$

37 $x^2 + 2x + 3 = 0$ **38** $2x^2 - 5x + 1 = 0$

39 $3x^2 + 1 = 6x$ **40** $5x^2 + 7x - 2 = 0$

41 $x(7x - 12) = 14$ **42** $x(6 - 3x) = 5$

43 $5x^2 + 3x + 2 = 0$

44   Show that if r_1 and r_2 are the roots of $ax^2 + bx + c = 0$, $a \neq 0$, then $r_1 + r_2 = -b/a$ and $r_1 r_2 = c/a$.

In problems 45–60, determine the kind of roots of each equation by using the discriminant.

45 $x^2 + 6x - 7 = 0$ **46** $6x^2 - x + 1 = 0$

47 $4x^2 + 12x + 9 = 0$ **48** $16x^2 - 40x + 25 = 0$

49 $x^2 + 3x + 5 = 0$ **50** $x^2 + x + 5 = 0$

51 $x(x - 4) = x - 2$ **52** $x^2 - 6x + 6 = 0$

53 $3x^2 - 7x + 11 = 0$ **54** $6x^2 + x + 2 = 0$

55 $3x^2 + x - 14 = 0$ **56** $4x^2 + 3x + 2 = 0$

57 $8x^2 - 10x - 3 = 0$ **58** $18x^2 + 9x - 20 = 0$

59 $2x^2 - 5x + 13 = 0$ **60** $20x^2 - 11x + 3 = 0$

6.4   Equations in Quadratic Form

Some equations that are not quadratic equations can still be solved by the methods of this chapter. These equations have a form that is equivalent to $au^2 + bu + c = 0$, where u is an expression involving another variable. For example, $3x^4 + 2x^2 - 1 = 0$, $x - \sqrt{x} + 2 = 0$, and $(x^2 - 1)^2 + 2(x^2 - 1) - 3 = 0$ are equations that are quadratic in form. We can solve equations of this type by first making a substitution to put them in the form $au^2 + bu + c = 0$ and then solving for u. This technique is illustrated in the following examples.

EXAMPLES

Find the solution sets of the following equations.

1    $x^4 - 5x^2 + 6 = 0$

SOLUTION.   We first write $x^4 - 5x^2 + 6 = 0$ as $(x^2)^2 - 5(x^2) + 6 = 0$. Letting $u = x^2$, we have the quadratic in the form $u^2 - 5u + 6 = 0$, so that $(u - 2)(u - 3) = 0$, $u - 2 = 0$, or $u - 3 = 0$. That is, $u = 2$ or $u = 3$. Now, since $u = x^2$ and $u = 2$ or $u = 3$, we have $x^2 = 2$ or $x^2 = 3$. Thus, $x = \pm\sqrt{2}$ or $x = \pm\sqrt{3}$. By substituting these values in the original equation, we can verify that they satisfy the equation. The solution set is $\{-\sqrt{2}, \sqrt{2}, -\sqrt{3}, \sqrt{3}\}$.

2   $x^{-2} + x^{-1} - 6 = 0$

SOLUTION.   Since $x^{-2} + x^{-1} - 6 = (x^{-1})^2 + x^{-1} - 6 = 0$, the equation is quadratic in form. Letting $u = x^{-1}$, we have $u^2 + u - 6 = 0$, so that $(u + 3)(u - 2) = 0$, $u + 3 = 0$ or $u - 2 = 0$. That is, $u = -3$ or $u = 2$. Replacing u by x^{-1}, we have $x^{-1} = -3$ or $x^{-1} = 2$. That is, $1/x = -3$ or $1/x = 2$, so that $-3x = 1$ or $2x = 1$. Therefore, $x = -\frac{1}{3}$ or $x = \frac{1}{2}$. The solution set is $\{-\frac{1}{3}, \frac{1}{2}\}$.

3   $\left(x - \dfrac{8}{x}\right)^2 + \left(x - \dfrac{8}{x}\right) = 42$

SOLUTION.   Let $y = x - (8/x)$ in the equation $[x - (8/x)]^2 + [x - (8/x)] = 42$. We have $y^2 + y = 42$ or $y^2 + y - 42 = 0$, so that $(y + 7)(y - 6) = 0$. That is, $y + 7 = 0$ or $y - 6 = 0$, which implies $y = -7$ or $y = 6$, so that $x - (8/x) = -7$ or $x - (8/x) = 6$. Therefore, $x^2 + 7x - 8 = 0$ or $x^2 - 6x - 8 = 0$, so that $x = -8$ or $x = 1$ are solutions to the equation $x^2 + 7x - 8 = 0$, and $x = 3 - \sqrt{17}$ or $x = 3 + \sqrt{17}$ are solutions of $x^2 - 6x - 8 = 0$. Hence, the solution set is $\{-8, 1, 3 - \sqrt{17}, 3 + \sqrt{17}\}$.

4   $x + 2 + \sqrt{x + 2} - 2 = 0$

SOLUTION. $x + 2 + \sqrt{x + 2} - 2 = 0$ or $(\sqrt{x + 2})^2 + \sqrt{x + 2} - 2 = 0$. Letting $u = \sqrt{x + 2}$ in the equation, we have $u^2 + u - 2 = 0$, so that $(u - 1)(u + 2) = 0$. That is, $u - 1 = 0$ or $u + 2 = 0$. Therefore, $u = 1$ or $u = -2$. Replacing u by $\sqrt{x + 2}$, we get $\sqrt{x + 2} = 1$ or $\sqrt{x + 2} = -2$. These two radical equations can be solved by squaring both sides of each equation, so that

$$x + 2 = 1 \qquad\qquad x + 2 = 4$$

$$\text{or}$$

$$x = -1 \qquad\qquad x = 2$$

To ensure that we reject the extraneous roots (if any exist), we check both solutions in the original equation.

Check: For $x = -1$, we have

$$[(-1) + 2] + \sqrt{(-1) + 2} - 2 = 1 + \sqrt{1} - 2 = 2 - 2 = 0$$

Therefore, $x = -1$ is a solution.

For $x = 2$, we have $[(2) + 2] + \sqrt{2 + 2} - 2 = 4 + 2 - 2 \neq 0$

Therefore, $x = 2$ is an extraneous solution and must be rejected. Hence, the solution set is $\{-1\}$.

When solving an equation involving radicals such as $\sqrt{1 - 5x} + \sqrt{1 - x} = 2$, write the equation so that the term involving the most complicated radical belongs to one side of the equation and all other terms belong to the other side. Then square both sides of the equation to reduce it to an equivalent equation that does not involve radicals. It is sometimes necessary to square both sides of one equation twice before it is reduced to an equation that is quadratic in form. Again, care must be taken to check the validity of the solution obtained to reject the extraneous solutions. This is illustrated in the following example.

EXAMPLE

5   Find the solution set of the radical equation $\sqrt{1 - 5x} + \sqrt{1 - x} = 2$.

SOLUTION.   Before squaring, we add $-\sqrt{1 - x}$ to both sides of the equation in order to simplify the subsequent steps, so that

$$\sqrt{1 - 5x} = 2 - \sqrt{1 - x}$$

Squaring both sides, we have $(\sqrt{1 - 5x})^2 = (2 - \sqrt{1 - x})^2$, so that

$$1 - 5x = 4 - 4\sqrt{1 - x} + 1 - x$$
$$-4 - 4x = -4\sqrt{1 - x}$$

Dividing both sides of the equation by -4, we have $1 + x = \sqrt{1 - x}$. Again, squaring both sides, we have

$$(1 + x)^2 = (\sqrt{1 - x})^2 \quad \text{or} \quad 1 + 2x + x^2 = 1 - x$$

That is, $x^2 + 3x = 0$ or $x(x + 3) = 0$. Solving for x, we have $x + 3 = 0$ or $x = 0$, that is, $x = -3$ or $x = 0$.

Check: For $x = 0$, we have

$$\sqrt{1 - 5(0)} + \sqrt{1 - 0} \overset{?}{=} 2$$
$$\sqrt{1} + \sqrt{1} \overset{?}{=} 2$$
$$1 + 1 = 2$$

Therefore, $x = 0$ is a solution.

For $x = -3$, we have

$$\sqrt{1 - 5(-3)} + \sqrt{1 - (-3)} = \sqrt{16} + \sqrt{4} = 4 + 2 = 6 \neq 2$$

Therefore, $x = -3$ is an extraneous root, so that the solution set is $\{0\}$.

PROBLEM SET 6.4

In problems 1–22, find the solution set of each equation by reducing each to a quadratic equation using the suggested substitution for u.

1 $x^4 - 13x^2 + 36 = 0; \; u = x^2$ **2** $x^4 - 17x^2 + 16 = 0; \; u = x^2$

3 $x^4 - 3x^2 - 4 = 0; \; u = x^2$ **4** $x^4 - 10x^2 + 9 = 0; \; u = x^2$

5 $x^{-2} - 2x^{-1} - 8 = 0; \; u = x^{-1}$ **6** $6x^{-2} + 5x^{-1} - 4 = 0; \; u = x^{-1}$

7 $x^{-4} - 9x^{-2} + 20 = 0; \; u = x^{-2}$ **8** $x^{-4} + 63x^{-2} - 64 = 0; \; u = x^{-2}$

9 $x^3 - 9x^{3/2} + 8 = 0; \; u = x^{3/2}$ **10** $x^5 - 33x^{5/2} + 32 = 0; \; u = x^{5/2}$

11 $x^{1/3} - 1 - 12x^{-1/3} = 0; \; u = x^{1/3}$ **12** $x^{1/4} + 2 - 8x^{-1/4} = 0; \; u = x^{1/4}$

13 $(x^2 + 1)^2 - 3(x^2 + 1) + 2 = 0; \; u = x^2 + 1$

14 $(2x^2 + 7x)^2 - 3(2x^2 + 7x) = 10; \; u = 2x^2 + 7x$

15 $(x^2 + 2x)^2 - 2(x^2 + 2x) = 3; \; u = x^2 + 2x$

16 $(x^2 - x)^2 + 12 = 8(x^2 - x); \; u = x^2 - x$

17 $\left(3x - \dfrac{2}{x}\right)^2 + 6\left(3x - \dfrac{2}{x}\right) + 5 = 0; \; u = 3x - \dfrac{2}{x}$

18 $\left(x - \dfrac{5}{x}\right)^2 - 2x + \dfrac{10}{x} = 8; \; u = x - \dfrac{5}{x}$

19 $x^2 + x + \dfrac{36}{x^2 + x} = 15; \; u = x^2 + x$

20 $\dfrac{x^2 + 1}{x} + \dfrac{4x}{x^2 + 1} - 4 = 0; \; u = \dfrac{x^2 + 1}{x}$

21 $\dfrac{x + 1}{x} + 2 = \dfrac{3x}{x + 1}; \; u = \dfrac{x}{x + 1}$

22 $\dfrac{x^2}{x + 1} + \dfrac{2(x + 1)}{x^2} = 3; \; u = \dfrac{x^2}{x + 1}$

In problems 23–28, find the solution set of each equation by reducing it to a quadratic equation using the suggested substitution for u. Check the solutions to reject any extraneous roots.

23 $x + 7 - \sqrt{x + 7} - 2 = 0;\ u = \sqrt{x + 7}$

24 $x^2 + 3x + \sqrt{x^2 + 3x - 2} = 22;\ u = \sqrt{x^2 + 3x - 2}$

25 $\sqrt{x + 20} - 4\sqrt[4]{x + 20} + 3 = 0;\ u = \sqrt[4]{x + 20}$

26 $2\sqrt[3]{1 - x} + 3\sqrt[6]{1 - x} = 2;\ u = \sqrt[6]{1 - x}$

27 $2x^2 + x - 4\sqrt{2x^2 + x + 4} = 1;\ u = \sqrt{2x^2 + x + 4}$

28 $x^2 + 6x - 6\sqrt{x^2 + 6x - 2} + 3 = 0;\ u = \sqrt{x^2 + 6x - 2}$

In problems 29–44, find the solution set of each equation by squaring both sides of each equation once or twice to reduce it to an equivalent equation that does not involve radicals. Also, check the solutions to reject extraneous roots.

29 $\sqrt{x + 12} = 2 + \sqrt{x}$

30 $\sqrt{t} = \sqrt{t + 16} - 2$

31 $\sqrt{x^2 + 6x} = x + \sqrt{2x}$

32 $\sqrt{2x^2 + 4} + 2 = 2x$

33 $\sqrt{y} + 1 = \sqrt{y + 5}$

34 $\sqrt{x + 7} = 5 + \sqrt{x - 2}$

35 $\sqrt{5t + 1} = 1 + \sqrt{3t}$

36 $\sqrt{3 - x} - \sqrt{2 + x} = 3$

37 $\sqrt{2x - 5} = \sqrt{x - 2} + 2$

38 $\sqrt{3y + 1} - 1 = \sqrt{3y - 8}$

39 $\sqrt{t + 2} + \sqrt{t - 3} = 5$

40 $\sqrt{6x + 7} = \sqrt{3x + 3} + 1$

41 $\sqrt{x + 4} + 1 = \sqrt{x + 11}$

42 $\sqrt{7x - 6} = \sqrt{7x + 22} - 2$

43 $\sqrt{2x + \sqrt{3 + x}} = 2$

44 $\sqrt{2x} + \sqrt{2x + 12} = 2\sqrt{4x + 1}$

6.5 Quadratic Inequalities

In Chapter 4 we used the properties of inequalities to solve first-degree inequalities in one variable. These properties can also be applied to solving inequalities such as $x^2 - 5x + 6 < 0$, $2x^2 - x + 1 > 0$, $3x^2 - x - 5 \le 0$, and $x^2 - x - 2 \ge 0$. Such inequalities are called *quadratic inequalities*. The procedure of finding the solution set of quadratic inequalities is illustrated by the following examples.

Consider the inequality $x^2 - 5x + 6 < 0$. By factoring the quadratic expression, we can rewrite the inequality as $(x - 2)(x - 3) < 0$. The solution set will be those values of x for which the product of $x - 2$ and $x - 3$ is a negative number. Recalling the rules for multiplying signed numbers, either $x - 2$ is a positive number and $x - 3$ is a negative number, or $x - 2$ is a negative number and $x - 3$ is a positive number. That is, either

1 $x - 2 > 0$ and simultaneously $x - 3 < 0$, or

2 $x - 2 < 0$ and simultaneously $x - 3 > 0$.

Let us consider these two cases:

  Case 1: $x - 2 > 0$ *and* $x - 3 < 0$. Solving these first-degree inequalities, we have $x > 2$ and $x < 3$, or

$$\{x \mid x > 2\} \cap \{x \mid x < 3\} = \{x \mid 2 < x < 3\}$$

  Case 2: $x - 2 < 0$ *and* $x - 3 > 0$. Solving these first-degree inequalities, we have $x < 2$ and $x > 3$, or, equivalently, $\{x \mid x < 2\} \cap \{x \mid x > 3\}$. But x cannot be both less than 2 and at the same time greater than 3. That is,

$$\{x \mid x < 2\} \cap \{x \mid x > 3\} = \varnothing$$

Therefore, the solution set of the inequality $x^2 - 5x + 6 < 0$ is $\{x \mid 2 < x < 3\}$. Figure 6.1 illustrates these two cases where the set of values of x for which the two expressions $x - 2$ and $x - 3$ are positive or negative, and the values of x for which the product $(x - 2)(x - 3)$ is negative, become obvious.

Figure 6.1

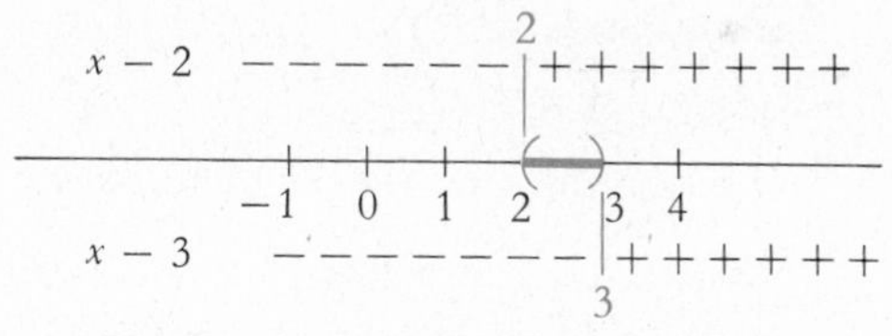

To find the solution set of the inequality $x^2 + x - 2 > 0$, we may start by factoring the left side, so that $(x + 2)(x - 1) > 0$. Since the product of these two expressions is positive, we have either:

1 $x + 2 > 0$ and $x - 1 > 0$, or

2 $x + 2 < 0$ and $x - 1 < 0$.

To find the solution sets of these inequalities, we consider the following cases:

Case 1: $x + 2 > 0$ and $x - 1 > 0$, so that $x > -2$ and $x > 1$, or, equivalently,

$$\{x \,|\, x > -2\} \cap \{x \,|\, x > 1\} = \{x \,|\, x > 1\}$$

Case 2: $x + 2 < 0$ and $x - 1 < 0$, so that $x < -2$ and $x < 1$, or, equivalently,

$$\{x \,|\, x < -2\} \cap \{x \,|\, x < 1\} = \{x \,|\, x < -2\}$$

Thus, the solution set for $x^2 + x - 2 > 0$ is $\{x \,|\, x > 1\} \cup \{x \,|\, x < -2\}$ (Figure 6.2).

Figure 6.2

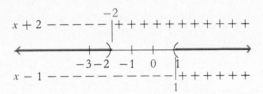

The techniques for solving quadratic inequalities of the form $x^2 - (p + q)x + pq < 0$ or $x^2 - (p + q)x + pq > 0$ [such inequalities could also be of the form $x^2 - (p + q)x + pq \leq 0$ or $x^2 - (p + q)x + pq \geq 0$] can be simplified by adopting the following steps:

1   Change the given inequality into an equation by replacing the sign of inequality with an equal sign. In each case, the resulting equation is $x^2 - (p + q)x + pq = 0$.

2   Find the two real solutions of the quadratic equation in step 1. These solutions are $x = p$ and $x = q$.

3   Locate the two solutions $x = p$ and $x = q$ of the quadratic equation $x^2 - (p + q)x + pq = 0$ on the number line. These solutions will separate the number line into three parts, which we denote by A, B, and C, each corresponding to a set of real numbers (Figure 6.3).

Figure 6.3

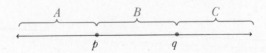

4   Select a number from parts A, B, and C in step 3 as a trial number and substitute it for x in the given inequality. If the trial number

satisfies the given inequality, then all other numbers in the set it represents satisfy it.

To illustrate this technique, we give the following examples.

EXAMPLES

Use steps 1–4 above to find the solution sets of the following inequalities.

1 $x^2 - 5x + 6 < 0$

SOLUTION

Step 1 Set $x^2 - 5x + 6 = (x - 2)(x - 3) = 0$.

Step 2 Solve the equation in step 1, so that $x - 2 = 0$ or $x - 3 = 0$, so that $x = 2$ or $x = 3$.

Step 3 The numbers 2 and 3 from step 2 separate the number line into three parts, denoted by A, B, and C (Figure 6.4).

Figure 6.4

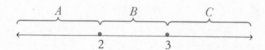

Step 4 Select a trial number, $x = 0$, from part A; then $(0 - 3)(0 - 2) = 6 \not< 0$. Therefore, the numbers in part A do not belong to the solution set. Select a trial number $x = \frac{5}{2}$ from part B; then $(\frac{5}{2} - 3)(\frac{5}{2} - 2) = -\frac{1}{4} < 0$. Hence, the numbers in part B belong to the solution set. Select a trial number, $x = 4$, from part C; then $(4 - 3)(4 - 2) = 2 \not< 0$, and the numbers in part C do not belong to the solution set. From these results, we see that the solution set includes all numbers in part B. This can be written $\{x \mid 2 < x < 3\}$ (Figure 6.5).

Figure 6.5

2 $x^2 + x - 2 > 0$

SOLUTION

Step 1 Set $x^2 + x - 2 = (x + 2)(x - 1) = 0$.

Step 2 Solve the equation $(x + 2)(x - 1) = 0$ in step 1 to obtain $x = -2$ and $x = 1$.

Step 3   The numbers -2 and 1 divide the number line into three parts, which we denote by A, B, and C (Figure 6.6).

Figure 6.6

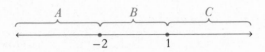

Step 4   Select numbers from each part and check them as follows:

A: If $x = -3$, then $(-3 + 2)(-3 - 1) = 4 > 0$.

B: If $x = 0$, then $(0 + 2)(0 - 1) = -2 \not> 0$.

C: If $x = 2$, then $(2 + 2)(2 - 1) = 4 > 0$.

Thus, the solution set includes the numbers in both parts A and C. This can be written as $\{x \mid x < -2 \text{ or } x > 1\}$ (Figure 6.7).

Figure 6.7

3    $2x^2 + 13x - 7 \geq 0$

SOLUTION

Step 1   Set $2x^2 + 13x - 7 = (2x - 1)(x + 7) = 0$.

Step 2   Solve the equation $(2x - 1)(x + 7) = 0$ in step 1 to obtain $x = -7$ and $x = \frac{1}{2}$.

Step 3   The numbers -7 and $\frac{1}{2}$ divide the number line into three parts, indicated by A, B, and C (Figure 6.8).

Figure 6.8

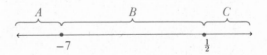

Step 4   We select numbers from each part and check them as follows:

A: If $x = -8$, then $(-16 - 1)(-8 + 7) = 17 > 0$.

B: If $x = 0$, then $(0 - 1)(0 + 7) = -7 \not> 0$.

C: If $x = 1$, then $(2 - 1)(1 + 7) = 8 > 0$.

Therefore, the solution set includes the numbers in both parts A and C (it also includes -7 and $\frac{1}{2}$). This can be written as $\{x \mid x \leq -7 \text{ or } x \geq \frac{1}{2}\}$ (Figure 6.9).

Figure 6.9

$$-7 \qquad\qquad\qquad \tfrac{1}{2}$$

To solve the inequality $(x - 1)/(x + 2) > 0$, we might be tempted to multiply both of its sides by $x + 2$ without considering that $x + 2$ can be either a positive or a negative number (for $x \neq -2$). Since $x + 2$ is either positive or negative, we cannot be sure whether or not to reverse the sign of the inequality. However, since $(x + 2)^2$ is always positive, we can multiply both sides of the inequality by $(x + 2)^2$, so that

$$(x + 2)^2 \cdot \left(\frac{x - 1}{x + 2}\right) > 0 \cdot (x + 2)^2 \qquad \text{or} \qquad (x + 2)(x - 1) > 0$$

The latter inequality is a quadratic inequality that happens to be the same as the one in Example 2. Its solution is written as

$$\{x \mid x < -2 \text{ or } x > 1\} \qquad \text{(Figure 6.10)}$$

Figure 6.10

$$-2 \qquad\qquad 1$$

EXAMPLE

4   Find the solution set of the inequality $(3x - 12)/(x + 2) \leq 0$.

SOLUTION. We multiply both sides of the inequality by $(x + 2)^2$. Since $(x + 2)^2$ is always positive for $x \neq -2$, we have

$$(x + 2)^2 \left(\frac{3x - 12}{x + 2}\right) \leq (x + 2)^2 \cdot 0$$

so that

$$(x + 2)(3x - 12) \leq 0$$

To find the solution set for this quadratic inequality, we follow the steps, as before.

Step 1   Set $(x + 2)(3x - 12) = 0$.

Step 2 $x + 2 = 0$ or $3x - 12 = 0$, so that

$$x = -2 \quad \text{or} \quad x = 4$$

Step 3   The numbers -2 and 4 divide the number line into three parts, denoted by A, B, and C (Figure 6.11).

Figure 6.11

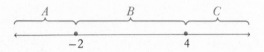

Step 4   Select numbers from each part and check them as follows:

A: If $x = -3$, then $(-3 + 2)(-9 - 12) = 21 \not< 0$.

B: If $x = 0$, then $(0 + 2)(0 - 12) = -24 < 0$.

C: If $x = 5$, then $(5 + 2)(15 - 12) = 21 \not< 0$.

Therefore, the solution set includes the numbers in part B (it also includes 4). This can be written $\{x \mid -2 < x \le 4\}$ (Figure 6.12).

Figure 6.12

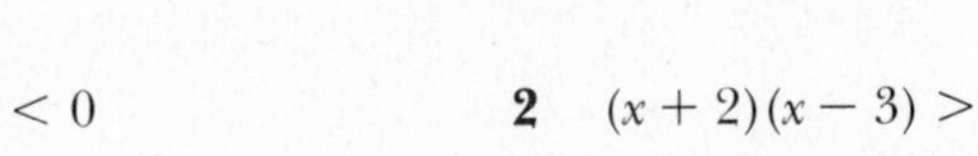

PROBLEM SET 6.5

In problems 1–18, find the solution set of each quadratic inequality and illustrate the solution on the number line.

1    $(x - 1)(x + 2) < 0$

2    $(x + 2)(x - 3) > 0$

3    $2x^2 + x - 1 \le 0$

4    $2x^2 + 9x - 5 \ge 0$

5    $x^2 \ge x + 2$

6    $x^2 + 5x \le -6$

7    $x^2 \ge 4x + 12$

8    $x^2 + 2x < 3$

9    $x^2 - 10x + 25 < 0$

10    $16x^2 + 8x + 1 > 0$

11    $3x^2 \le 2 - 5x$

12    $4x^2 + 11x - 3 \ge 0$

13    $5x^2 - 6 \ge 7x$

14    $6x^2 + 5x - 14 \le 0$

15 $x^2 + 2x - 3 > 0$

16 $40 - 3x - x^2 < 0$

17 $x^2 - x - 6 \leq 0$

18 $x^2 - 6x + 8 \geq 0$

In problems 19–30, find the solution set of each quotient inequality and illustrate the solution on the number line.

19 $\dfrac{x-1}{x-4} < 0$

20 $\dfrac{x+3}{x-3} \leq 0$

21 $\dfrac{x+1}{x-2} \geq 0$

22 $\dfrac{2x+3}{4x+8} > 0$

23 $\dfrac{x-4}{x+2} \leq 0$

24 $\dfrac{x-2}{3x-1} \geq 0$

25 $\dfrac{5x-1}{x-2} > 0$

26 $\dfrac{2x-1}{x+2} < 0$

27 $\dfrac{1-x}{3-x} \geq 0$

28 $\dfrac{1}{(1-x)(1-3x)} > 0$

29 $\dfrac{1-x}{x} < 1$

30 $\dfrac{1}{x+2} \leq 3$

6.6 Applications of Quadratic Equations

Quadratic equations may be used in some mathematical and physical applications. Again, in solving such problems we suggest (as we did in Chapter 4) the following steps:

1 Read the problem carefully, determining what facts are given and precisely what is to be found.

2 Draw diagrams, if possible, to help interpret the given information and the nature of the solution.

3 Express the given information and the solution in mathematical symbols and/or numbers.

4 Express the problem in the form of an equation.

5 Solve the equation.

6 Check the solution to see if it satisfies the equation and to see if it makes sense with regard to the story problem.

If the equation that represents the word problem happens to be quadratic, we may have two different solutions. Often, only one of the solutions to the quadratic equation is actually a solution to the word problem. For example, if the problem is to find two consecutive positive integers whose product is 12, the equation that represents the problem is $x(x + 1) = 12$, where x represents the first integer and $x + 1$ represents the second. Solving the equation, we have

$$x(x + 1) = 12 \quad \text{or} \quad x^2 + x = 12$$

so that

$$x^2 + x - 12 = 0 \quad \text{or} \quad (x + 4)(x - 3) = 0$$

Therefore,

$$x + 4 = 0 \quad \text{or} \quad x - 3 = 0$$

or, equivalently,

$$x = -4 \quad \text{or} \quad x = 3$$

The solution set of $x(x + 1) = 12$ is $\{-4, 3\}$, but only the value 3 satisfies the word problem, since the condition of positiveness was placed on the consecutive integers. Hence, the two positive integers are 3 and 4.

EXAMPLES

1   Find two numbers whose sum is 7 and whose product is 12.

SOLUTION.   If x represents one number, then $7 - x$ represents the other number. (Why?) The product of the two numbers is 12; therefore, $x(7 - x) = 12$, so that

$$7x - x^2 = 12 \quad \text{or} \quad x^2 - 7x + 12 = 0$$

Then, $(x - 3)(x - 4) = 0$ or, equivalently,

$$x - 3 = 0 \quad \text{or} \quad x - 4 = 0$$

so that

$$x = 3 \quad \text{or} \quad x = 4$$

If $x = 3$, then $7 - x = 7 - 3 = 4$, and $3(4) = 12$. Also, if $x = 4$, then $7 - x = 7 - 4 = 3$. Hence, the two numbers are 3 and 4.

2   The area of a rectangle is 120 square feet. Its length is 7 feet more than its width. Find its dimensions.

SOLUTION.   Let x feet be the length of the rectangle; then the width is $x - 7$ feet (Figure 6.13). The area of a rectangle is given by

$$\text{area} = \text{length} \times \text{width}$$

so that $120 = (x - 7)x$. Solving this equation, we get

$$(x - 7)x = 120 \qquad \text{or} \qquad x^2 - 7x = 120$$

so that $x^2 - 7x - 120 = 0$. Then, $(x - 15)(x + 8) = 0$, or, equivalently, $x - 15 = 0$ or $x + 8 = 0$, so that $x = 15$ or $x = -8$. Since the dimensions of the rectangle are always positive numbers, we only accept the solution $x = 15$, so that $x - 7 = 15 - 7 = 8$. Therefore, the length and the width of the rectangle are 15 feet and 8 feet, respectively.

Figure 6.13

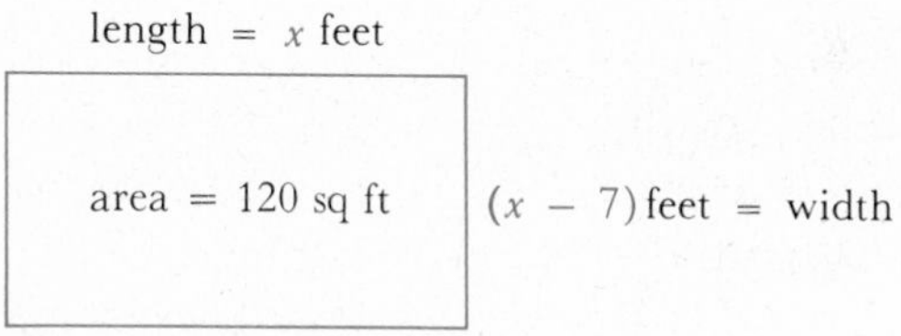

length = x feet

area = 120 sq ft | $(x - 7)$ feet = width

3   A rectangular garden 40 feet long and 26 feet wide has a walk of uniform width around it. If the area of the walk is 432 square feet, find its width.

SOLUTION.   Let x feet represent the width of the walk (Figure 6.14). Then the area of the walk will be the difference of the area of the two rectangles; thus, $(40 + 2x)(26 + 2x) - (40)(26) = 432$ or

Figure 6.14

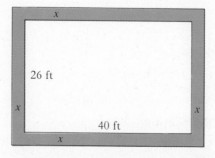

$4x^2 + 132x - 432 = 0$, so that $x^2 + 33x - 108 = 0$. Then, $(x-3)(x+36) = 0$, or, equivalently, $x = 3$ or $x = -36$. Since the width is a positive number, we only accept the solution $x = 3$. Therefore, the width of the walk is 3 feet.

4   A rocket is fired upward from the ground at an initial velocity of 560 feet per second, and its distance s feet above the ground t seconds later is given by the equation $s = 560t - 16t^2$. In how many seconds will it reach a height of 3,136 feet?

SOLUTION.   To determine t when $s = 3,136$, we have $s = 560t - 16t^2$, so that $3,136 = 560t - 16t^2$. Solving for t, we have $16t^2 - 560t + 3,136 = 0$ or $t^2 - 35t + 196 = 0$. Then, $(t-28)(t-7) = 0$, so that $t - 28 = 0$ or $t - 7 = 0$, or, equivalently, $t = 28$ or $t = 7$. Both solutions make sense, since at $t = 7$ seconds, the rocket is $560(7) - 16(7^2) = 3,136$ feet high. It goes on up in its path, reaches its peak, and 28 seconds after the start it is back at a height of 3,136 feet, which can be shown by finding s when $t = 28$.

5   A merchant bought a certain number of mirrors for a total cost of \$36. After breaking one mirror, he sold the remainder at a profit of 50 cents each, for a total profit of \$2.50. How many mirrors did the merchant buy?

SOLUTION.   Let x be the number of mirrors, so that \$36/x is the cost per mirror. $x - 1$ is the number of mirrors left after breaking one of them. Thus, $(36/x) + 0.50$ is the price of each mirror with 50 cents over cost. Hence,

$$(x-1)\left(\frac{36}{x} + 0.50\right) = 36 + 2.50 = 38.50$$

or

$$36x + 0.50x^2 - 36 - 0.50x = 38.5x$$

so that $x^2 - 6x - 72 = 0$ or $(x-12)(x+6) = 0$. Then, $x - 12 = 0$ or $x + 6 = 0$, or, equivalently, $x = 12$ or $x = -6$. Therefore, the number of mirrors he bought is 12.

PROBLEM SET 6.6

1   Find two consecutive integers whose product is 30.

2   Find two consecutive negative integers whose product is 56.

3   Find two positive integers whose sum is 14 and whose product is 45.

4 Find two consecutive positive integers such that the sum of their squares is 13.

5 Find the number such that the sum of the number and its reciprocal is $\frac{10}{3}$.

6 Find a number that is 1 less than its square.

7 Find a number that exceeds its reciprocal by $1\frac{1}{5}$.

8 The units' digit of a two-digit whole number is 1 more than its tens' digit. If the number is multiplied by the number with its digits reversed, the result is 736. Find the number.

9 Find the dimensions of a rectangle if its area is 60 square feet and its perimeter is 32 feet.

10 A box without a top is to be constructed from a square piece of tin by removing a 3-inch square from each corner and bending up the sides. If the box is to have a volume of 300 cubic inches, what is the length of the side of the original tin square?

11 A certain rectangle is twice as long as it is wide. If each dimension were increased by 2 inches, the new area would be 144 square inches. Find the original dimensions.

12 A rectangular garden whose length is 6 feet more than its width is surrounded by a strip 4 feet wide. If the strip is removed, the area of the garden is increased by 272 square feet. Find the dimensions of the original garden.

13 A skating rink is 100 feet long and 70 feet wide. To increase its area to 13,000 square feet by adding rectangular strips of equal width to one side and one end while maintaining its rectangular shape, what should the width of the strips be?

14 If increasing the sides of a square by 8 feet results in a square whose area is 25 times the area of the original square, what was the length of the side of the original square?

15 The area of a rectangular garden is 1,320 square feet. If the perimeter is 148 feet, what are the dimensions of the garden?

16 A lawn 30 feet long and 20 feet wide has a flower border of uniform width around it. If the area of the lawn and the border are the same, find the width of the border.

17 A man bought a farm for $12,600. He then sold all but 40 acres at a profit of $20 per acre, receiving the entire cost of the farm. How many acres did he buy?

18   A group of women plan to pay equal amounts toward a gift that cost \$200. If two more women are added to the group, the cost to each will be reduced by \$5. Find the original number of women.

19   A department store bought a number of records for \$180 and sold all but 6 at a profit of \$2 per record. With the total amount received they could buy 30 more records than before. Find the cost per record.

20   Tom is 4 years older than Kim and the product of their present ages is three times what the product of their ages was 4 years ago. Find their present ages.

21   A rocket is fired upward from the ground at an initial velocity of 480 feet per second, and its distance s feet above the ground t seconds later is given by the equation $s = 480t - 16t^2$. In how many seconds will it reach a height of 2,000 feet?

22   A rocket is fired upward from the ground at an initial velocity of 120 meters per second, and its distance s meters above the ground t seconds later is given by the equation $s = 120t - 16t^2$. In how many seconds will it reach a height of 176 meters?

23   A girl riding a bicycle traveled a distance of 8 kilometers. She would have completed the trip 20 minutes earlier if she had ridden 2 kilometers per hour faster. Find her rate of speed.

24   A man travels for 20 miles at one speed, and then increases his speed by 20 miles per hour for the next 30 miles. How fast was he traveling originally if the total trip took 1 hour?

REVIEW PROBLEM SET

In problems 1–16, find the solution set of each equation by using the factoring method.

1 $x^2 - 5x - 84 = 0$

2 $x^2 - 14x - 15 = 0$

3 $t^2 + t - 12 = 0$

4 $6y^2 - 15y = 0$

5 $5x^2 - 10x = 0$

6 $5 - 14x - 3x^2 = 0$

7 $4t^2 - 4t = -1$

8 $16y^2 - 24y + 9 = 0$

9 $8 - 2x - x^2 = 0$

10 $6 - 5t - 6t^2 = 0$

11 $4 + 5y - 9y^2 = 0$

12 $(1 - 5x)^2 - 121 = 0$

13 $x^2 + 2ax = b^2 - a^2$;
a and b = constants

14 $(ax - bx)^2 = x(b - a)$;
a and b = constants

15 $\dfrac{3}{4x^2} + \dfrac{7}{8x} - \dfrac{5}{2} = 0$

16 $\dfrac{2}{x - 1} - \dfrac{3}{2x + 5} = \dfrac{5}{3}$

In problems 17–20, find a quadratic equation whose solution set is given.

17 $\{5,-1\}$ **18** $\{\frac{2}{3},-\frac{3}{2}\}$

19 $\{-\frac{1}{2},\frac{3}{5}\}$ **20** $\{-\frac{5}{6},-\frac{7}{8}\}$

In problems 21–28, find the solution set of each equation by using the extraction-of-roots method.

21 $x^2 = 144$ **22** $5x^2 - 75 = 0$

23 $3x^2 + 48 = 0$ **24** $2(x-1)^2 - 50 = 0$

25 $(x-\frac{3}{5})^2 = 0$ **26** $(x-1)^2 + 4 = 0$

27 $(3x-1)^2 - 256 = 0$ **28** $5(2x-1)^2 + 80 = 0$

In problems 29–40, find the solution set of each equation by using the completing-the-square method.

29 $x^2 - 6x + 7 = 0$ **30** $x^2 + 4x - 21 = 0$

31 $t^2 - 8t + 9 = 0$ **32** $x^2 - 4x + 5 = 0$

33 $9y^2 - 6y + 2 = 0$ **34** $9t^2 - 12t + 1 = 0$

35 $4x^2 + 13 = 12x$ **36** $16y^2 - 24y + 9 = 0$

37 $4x^2 + 4x - 3 = 0$ **38** $2w^2 + 3 = 8w$

39 $x^2 + cx = 6c^2$; $c > 0$ **40** $0.3x^2 - 0.06x - 0.144 = 0$

In problems 41–56, find the solution set of each equation by using the quadratic formula.

41 $x^2 + 10x - 13 = 0$ **42** $5 - 8x + x^2 = 0$

43 $x^2 + x + 3 = 0$ **44** $3 - x - x^2 = 0$

45 $4x^2 - x - 1 = 0$ **46** $3x^2 - 8x + 1 = 0$

47 $2t^2 + 5t - 17 = 0$ **48** $16 - 6y - 3y^2 = 0$

49 $3y^2 + 7y + 2 = 0$ **50** $5t^2 - 8t + 1 = 0$

51 $2x^2 - 5x + 4 = 0$ **52** $4y^2 + 19y + 10 = 0$

53 $(2x-1)(x+2) = 5$ **54** $(x-1)(3x+5) = 2$

55 $x^2 + x + k = kx$;   **56** $(x-a)(x-3) = 5a$;
    k = constant a = constant

In problems 57–60, find the solution set of each equation by squaring both sides of each equation twice. Check the solutions to reject any extraneous roots.

57 $\sqrt{x-3} - \sqrt{x} = -1$

58 $3\sqrt{x^2 - 8x + 32} = 20 - 5x$

59 $\sqrt{2t+1} + \sqrt{t} = 1$

60 $\sqrt{5t-4} - \sqrt{2t+1} = 1$

In problems 61–66, find the sum and the product of the roots of each equation without solving the equations.

61 $x^2 - 4x + 4 = 0$

62 $4x^2 - 11x + 6 = 0$

63 $x^2 + x + 1 = 0$

64 $x^2 - 14x - 15 = 0$

65 $(1 - 7x)^2 - 36 = 0$

66 $3x^2 + 7x + 1 = 0$

In problems 67–76, determine the kind of roots of each equation by using the discriminant.

67 $x^2 + x - 1 = 0$

68 $x^2 + 11 = 0$

69 $2x^2 - 3x + 1 = 0$

70 $3x^2 - 5x - 4 = 0$

71 $5x^2 - 2x + 1 = 0$

72 $7x^2 + 5x + 1 = 0$

73 $3x^2 - x + 10 = 0$

74 $4x^2 - 12x + 3 = 0$

75 $4x^2 - 12x + 9 = 0$

76 $25x^2 - 20x + 4 = 0$

In problems 77–88, find the solution set of each equation by reducing each of them to quadratic equations in form, using the suggested substitution for u.

77 $x^4 - 5x^2 + 4 = 0;\ u = x^2$

78 $y^4 - 8y^2 + 16 = 0;\ u = y^2$

79 $16x^4 - 17x^2 + 1 = 0,\ u = x^2$

80 $x^8 - 2x^4 + 1 = 0;\ u = x^4$

81 $4x^{-4} - 11x^{-2} - 3 = 0;\ u = x^{-2}$

82 $1 - 2x^{-2} - 3x^{-4} = 0;\ u = x^{-2}$

83 $x^{2/3} + 2x^{1/3} - 3 = 0;\ u = x^{1/3}$

84 $x^{-3/2} - 26x^{-3/4} - 27 = 0;\ u = x^{-3/4}$

85 $(x^2 + x)^2 - (8x^2 + 8x) + 12 = 0;\ u = x^2 + x$

86 $(x^2 + 4x)^2 - 17x^2 - 60 - 68x = 0;\ u = x^2 + 4x$

87 $x^2 - 6x - \sqrt{x^2 - 6x - 3} = 5;\ u = \sqrt{x^2 - 6x - 3}$

88 $3x^2 - 4x + \sqrt{3x^2 - 4x - 6} = 18;\ u = \sqrt{3x^2 - 4x - 6}$

In problems 89–96, find the solution set of each inequality and illustrate the solution on the number line.

89 $x^2 + 2x < 15$ **90** $x^2 + 2x \geq 3$

91 $x^2 - 3x \geq 10$ **92** $3x^2 - 2x < 5$

93 $\dfrac{3x + 1}{2x - 5} < 0$ **94** $\dfrac{x - 2}{x + 3} \geq 0$

95 $\dfrac{3x - 1}{5x - 7} \geq 0$ **96** $\dfrac{7x - 3}{9x + 1} \leq 0$

97 Find three positive consecutive integers such that the sum of their squares is 149.

98 Find a number such that the sum of the number and its reciprocal is 4.

99 A boy has mowed a strip of uniform width around a lawn that is 80 feet by 60 feet. He still has half of the lawn to mow. How wide is the strip?

100 A swimming pool 25 feet by 15 feet is bordered by a concrete walk of uniform width. The area of the walk is 329 square feet. How wide is it?

101 A group of students chartered a bus for $60. When four students withdrew from the group, the share of each of the other students was increased by $2.50. How many students were in the group originally?

102 If the distance s in feet that a bomb falls in t seconds is given by the formula $s = 16t^2/(1 + 0.06t)$, how many seconds is required for a bomb released at 20,000 feet to reach its target?

103 A girl asked her father: "How old are you?" The father answered: "I was 30 years old when you were born and the product of our present ages is 736." How old is the father?

CHAPTER 7

First-Degree Equations in More Than One Variable

In this chapter we extend the concept of equations to include first-degree equations in more than one variable. In particular, we consider first-degree equations in two variables and their graphs. To investigate these graphs, we introduce the *Cartesian coordinate system*. One of the attractive features of the Cartesian coordinate system is the relative ease with which the graphical solutions of systems of first-degree equations in two variables are found.

7.1 Cartesian Coordinates and the Distance Formula

We have seen in Section 1.2 that each point on a number line is associated with exactly one real number, and each real number is represented by exactly one point on a number line. In this section we shall see that each point in a plane is associated with exactly one pair of numbers, and each pair of numbers is represented by exactly one point in a plane. The pairs of numbers in question are called *ordered pairs*. An *ordered pair* is a pair of numbers (usually written within parentheses) in which the order of listing is important. For example, the ordered pair (5,6) is not the same as the ordered pair (6,5). Two ordered pairs (a,b) and (c,d) are *equal* if and only if $a = c$ and $b = d$. Also in this section we shall use the Pythagorean theorem to derive a formula for the distance between any two points represented by ordered pairs in a plane.

Cartesian Coordinates

We shall begin by representing the set of all ordered pairs of real numbers as the set of points in a *plane*, using a two-dimensional indexing system called the *Cartesian coordinate system*. This system is constructed as follows:

First, two perpendicular lines L_1 and L_2 are constructed (Figure 7.1a). The point of intersection of the two lines is called the *origin*. Next, L_1 and L_2 are scaled as real lines by using the origin as the 0 point for each of the two lines. Traditionally, the portion of L_2 above the origin is chosen to be the positive direction, while the portion

Figure 7.1

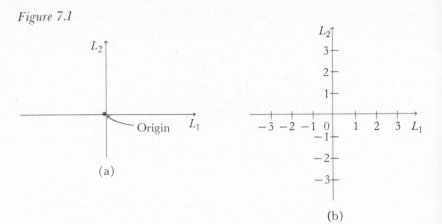

(a)

(b)

below the origin is chosen to be the negative direction of the number line; similarly, the portion of L_1 to the right of the origin is chosen to be the positive direction, while the portion to the left is chosen to be the negative direction (Figure 7.1b).

The resulting two real lines are called the *coordinate axes*. The coordinate axes, L_1 and L_2, are often referred to as the *horizontal axis* or the *x axis* and the *vertical axis* or the *y axis*, respectively. Given an ordered pair of real numbers (x,y) (the first member of the pair, x, is called the *abscissa;* the second member of the pair, y, is called the *ordinate; x* and *y* are called the *coordinates*), we can use the coordinate system to represent (x,y) as a point in the plane, as follows.

The abscissa x is located on the x axis. Then a line is drawn perpendicular to this axis at point x; the ordinate y is located on the y axis at point y and a line is drawn perpendicular to the axis at point y. The intersection of the two lines which have just been constructed is the point in the plane used to represent the ordered pair (x,y) (Figure 7.2).

Figure 7.2

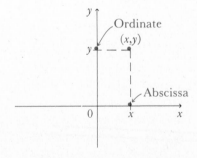

Thus for each ordered pair of real numbers we can associate a point in the plane. Conversely, for each point in the plane, we can associate an ordered pair of real numbers. Hence, there is a one-to-one correspondence between all possible pairs of real numbers and the points in the plane. For example, the ordered pair $(1, 1)$ is located by moving 1 unit to the right of 0, on the x axis, then 1 unit up from 0 on the y axis. Similarly, $(7,5)$ is located by moving 7 units to the right of 0 and 5 units up; $(-\frac{1}{2}, 0)$ is located by moving $\frac{1}{2}$ unit to the left of 0 and no units up from 0 (Figure 7.3).

Figure 7.3

y
(7,5)
5
4
3
2
(1, 1)
1
(−½,0)
−1 0 1 2 3 4 5 6 7
x

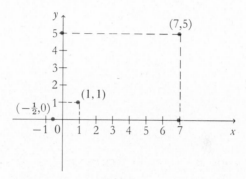

The set of all points in the plane whose coordinates correspond to the ordered pairs of a given set is called the *graph* of the set of ordered pairs. Locating these points in the plane by using the Cartesian coordinate system is called *graphing* (or *plotting*) the set of ordered pairs.

The coordinate axes divide the plane into four disjoint regions called *quadrants.* Thus, quadrant I includes all points (x, y) such that $x > 0$ and simultaneously $y > 0$. Quadrant II includes all points (x,y) such that $x < 0$ and simultaneously $y > 0$. Quadrant III includes all points (x,y) such that $x < 0$ and simultaneously $y < 0$. Quadrant IV includes all points (x,y) such that $x > 0$ and simultaneously $y < 0$ (Figure 7.4).

Figure 7.4

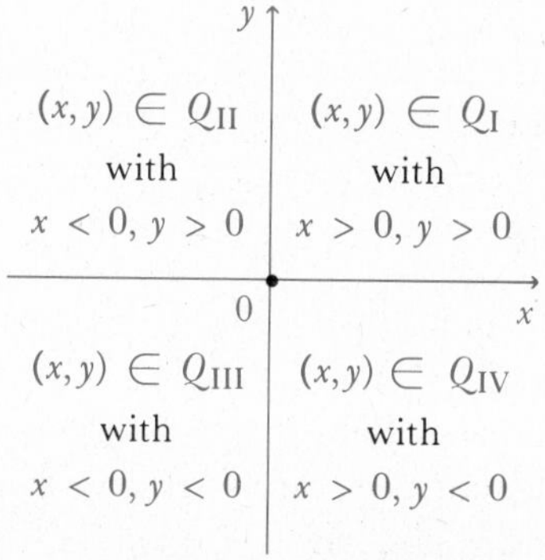

y
(x, y) ∈ Q_II
with
x < 0, y > 0
(x, y) ∈ Q_I
with
x > 0, y > 0
0
x
(x, y) ∈ Q_III
with
x < 0, y < 0
(x, y) ∈ Q_IV
with
x > 0, y < 0

EXAMPLES

1 What are the coordinates of a point 5 units to the left of the y axis
   and 3 units above the x axis?

   SOLUTION.   The abscissa is -5 and the ordinate is 3; therefore, the
   coordinates of the point are given by $(-5,3)$ (Figure 7.5).

Figure 7.5

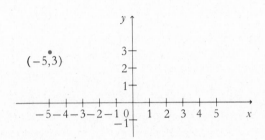

2 Plot each of the points $(1,-2)$, $(3,4)$, $(-2,-3)$, and $(3,0)$ and indicate
   which quadrant, if any, contains the points.

   SOLUTION.   These points are plotted in Figure 7.6. $(1,-2)$ lies in
   quadrant IV. $(3,4)$ lies in quadrant I. $(-2,-3)$ lies in quadrant III.
   $(3,0)$ does not lie in any quadrant.

Figure 7.6

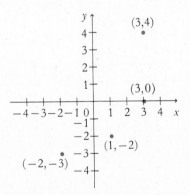

3 Plot the points $(-2,0)$, $(-1,0)$, $(0,0)$, $(1,0)$, and $(2,0)$.

   SOLUTION.   The y coordinate of each one of these points is 0. They
   all lie on the x axis. Every point on the x axis has y coordinate 0
   (Figure 7.7).

Figure 7.7

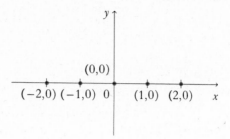

The Distance Formula

Suppose that a Cartesian coordinate system is established using the same scale for the x axis and the y axis. Then the distance between any two points, say P_1 and P_2, is the length of the line segment determined by the two points. (Figure 7.8). The following formula gives the distance between any two points P_1 and P_2 in the Cartesian plane when their Cartesian coordinates are known. The distance between P_1 and P_2 will be denoted by d or $|\overline{P_1P_2}|$.

Figure 7.8

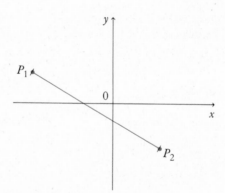

The *distance formula* is stated: Given any two points P_1 and P_2 with coordinates (x_1, y_1) and (x_2, y_2), respectively, the distance $d = |\overline{P_1P_2}|$ is given by

$$d = \sqrt{(x_1 - x_2)^2 + (y_1 - y_2)^2}$$

Using a geometric argument and considering the following cases, we derive the formula as follows:

a) If the two points lie on the same vertical line, that is, $x_1 = x_2$, then $|y_1 - y_2| = d$ (Figure 7.9a).

Figure 7.9

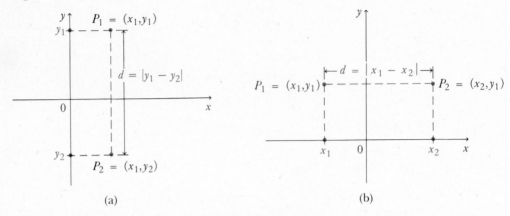

(a)   (b)

b) If the two points lie on the same horizontal line, that is, $y_1 = y_2$, then $|x_1 - x_2| = d$ (Figure 7.9b).

c) If the two points lie on a line that is neither horizontal nor vertical, then a right triangle (one with a 90° angle) is determined (Figure 7.10).

Figure 7.10

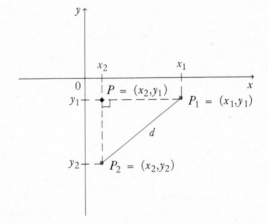

Now we use the Pythagorean theorem, which asserts that

$$|\overline{PP_1}|^2 + |\overline{PP_2}|^2 = |\overline{P_1P_2}|^2 = d^2$$

so that

$$|x_1 - x_2|^2 + |y_1 - y_2|^2 = d^2$$

or

$$(x_1 - x_2)^2 + (y_1 - y_2)^2 = d^2 \qquad \text{(Why?)}$$

Hence,

$$d = \sqrt{(x_1 - x_2)^2 + (y_1 - y_2)^2}$$

Notice that this formula is also applicable in the special cases where P_1 and P_2 are on the same vertical line or the same horizontal line. (Why?) Since $(a - b)^2 = (b - a)^2$, the order of subtracting the abscissas or the ordinates is irrelevant.

EXAMPLES

4   Plot the points $(2,-4)$ and $(-2,-1)$, and then find the distance d between them.

SOLUTION.   The points $P_1 = (2,-4)$ and $P_2 = (-2,-1)$ are plotted in Figure 7.11, and the distance d is given by

$$d = \sqrt{[2 - (-2)]^2 + [-4 - (-1)]^2}$$
$$= \sqrt{4^2 + (-3)^2}$$
$$= \sqrt{16 + 9} = \sqrt{25} = 5$$

Figure 7.11

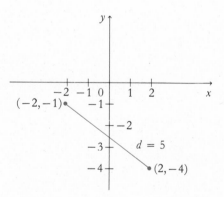

5   Plot the points $(-1,-2)$ and $(3,4)$, and then find the distance d between them.

SOLUTION.   The points $P_1 = (-1,-2)$ and $P_2 = (3,4)$ are plotted in Figure 7.12, and the distance d is given by

$$d = \sqrt{[3 - (-1)]^2 + [4 - (-2)]^2}$$
$$= \sqrt{4^2 + 6^2}$$
$$= \sqrt{16 + 36} = \sqrt{52} = 2\sqrt{13}$$

Figure 7.12

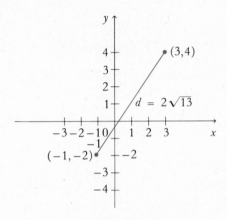

6   Find the distance between $(4, -7)$ and $(-2, 1)$.

SOLUTION

$$d = \sqrt{[4 - (-2)]^2 + (-7 - 1)^2}$$
$$= \sqrt{6^2 + (-8)^2}$$
$$= \sqrt{36 + 64} = \sqrt{100} = 10$$

7   Derive a formula for the distance between the origin and any point (x, y) in the plane.

SOLUTION.   The distance d between $(0, 0)$ and (x, y) (Figure 7.13) is given by

$$d = \sqrt{(x - 0)^2 + (y - 0)^2}$$

so that $d = \sqrt{x^2 + y^2}$.

Figure 7.13

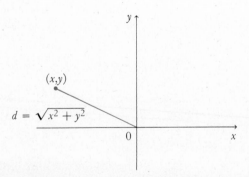

PROBLEM SET 7.1

In problems 1–12, plot the points on the Cartesian coordinate system.

1 $(1,2)$, $(3,5)$, $(1,-1)$, $(0,1)$

2 $(0,0)$, $(-6,0)$, $(0,-3)$, $(5,1)$

3 $(-1,-2)$, $(3,-2)$, $(3,-3)$, $(5,3)$

4 $(-2,2)$, $(-3,2)$, $(-4,2)$, $(5,2)$

5 $(-5,0)$, $(-4,0)$, $(5,0)$, $(4,0)$

6 $(7,0)$, $(0,7)$, $(-7,0)$, $(0,-7)$

7 $(\frac{1}{2},1)$, $(\frac{3}{2},2)$, $(\frac{2}{3},-1)$, $(-\frac{4}{5},2)$

8 $(6,2)$, $(-1,3)$, $(-\frac{3}{8},1)$, $(1,-\frac{1}{4})$

9 $(0,-\frac{1}{2})$, $(\frac{1}{2},3)$, $(-2,7)$, $(6,-\frac{5}{3})$

10 $(-1,\frac{5}{6})$, $(\frac{5}{6},-1)$, $(0,-\frac{2}{3})$, $(-\frac{2}{3},0)$, $(5,1)$

11 $(-1,-1)$, $(-2,-2)$, $(0,0)$, $(1,1)$, $(2,2)$

12 $(-5,0)$, $(-4,0)$, $(0,0)$, $(-\frac{1}{5},0)$

13   Give the coordinates of four points on the *x* axis.

14   Give the coordinates of four points on the *y* axis.

In problems 15–24, indicate which quadrant, if any, contains the points.

15 $(1,7)$ **16** $(3,2)$

17 $(-1,2)$ **18** $(-5,1)$

19 $(-3,-3)$ **20** $(-5,-1)$

21 $(5,0)$ **22** $(-7,0)$

23 $(0,2)$ **24** $(0,-3)$

In problems 25–34, find the distance between the two points.

25 $(1,2)$ and $(7,10)$ **26** $(-3,-4)$ and $(-5,-7)$

27 $(1,1)$ and $(-3,2)$ **28** $(-2,5)$ and $(3,-1)$

29 $(-4,-3)$ and $(0,0)$ **30** $(0,4)$ and $(4,0)$

31 $(1,5)$ and $(4,9)$ **32** $(-7,0)$ and $(0,-7)$

33 $(-\frac{1}{2},1)$ and $(2,3)$ **34** $(t,8)$ and $(t,7)$

In problems 35–40, find the length of the hypotenuse of the right triangle determined by the points.

35 $(0,0)$, $(-3,0)$, and $(-3,4)$ **36** $(-3,1)$, $(3,1)$, and $(3,10)$

37 $(-2,-2)$, $(0,0)$, and $(3,-3)$ **38** $(1,1)$, $(5,1)$, and $(5,7)$

39 $(0,0)$, $(0,-12)$, and $(5,0)$ **40** $(0,0)$, $(8,0)$, and $(8,-6)$

41 Given the points $P_1 = (-3,-2)$, $P_2 = (1,2)$ and $P_3 = (3,4)$:
     a) Find the lengths of segments $\overline{P_1P_2}$, $\overline{P_2P_3}$, and $\overline{P_1P_3}$, where $\overline{P_1P_2}$ means the line segment determined by P_1 and P_2, and so on.
     b) Are P_1, P_2, and P_3 *collinear?* That is, do the points P_1, P_2, and P_3 lie on a straight line?

42 Given points $P_1 = (a,b)$, $P_2 = (c,d)$, and $P_3 = \big((a + c)/2,\ (b + d)/2\big)$:
     a) Find the lengths of segments $\overline{P_1P_3}$ and $\overline{P_2P_3}$ in terms of a, b, c, and d by using the distance formula.
     b) How do these lengths compare?
     c) What can you conclude about the geometric position of P_3 with respect to P_1 and P_2?

43 Use part c in problem 42 to find the coordinates of the midpoints of the following pairs of points. Also use the distance formula to check the midpoints.
     a) $(5,6)$ and $(-7,8)$
     b) $(-4,7)$ and $(-3,0)$

44 Use the distance formula to determine if the triangle whose vertices are $(3,1)$, $(4,3)$, and $(6,2)$ is an isosceles triangle.

In problems 45–48, find the lengths of the sides of the triangle determined by the points.

45 $(5,-3)$, $(-2,4)$, and $(2,5)$ **46** $(-6,-3)$, $(2,1)$, and $(-2,-5)$

47 $(10,1)$, $(3,1)$, and $(5,9)$ **48** $(0,6)$, $(9,-6)$, and $(-3,0)$

7.2   Linear Equations and Their Graphs

Equations such as $2x + 3y = 6$, $y = 5 - \frac{1}{2}x$, and $\sqrt{3}x + 5y - 2 = 0$ are called *first-degree equations* or *linear equations* in two variables, x and y. The *solution* of a linear equation in two variables x and y is an ordered pair (x,y) which satisfies the equation — that is, which makes the equation true. The *solution set* of a linear equation is the set of ordered pairs that satisfies the equation. For example, some members of the solution set of the equation $2x + 3y = 6$ are $(0,2)$, $(3,0)$, $(-1,\frac{8}{3})$, and $(6,-2)$, because if we replace x in the given equation by $0, 3, -1$, and 6 and y by $2, 0, \frac{8}{3}$, and -2, respectively, the equation is satisfied. Indeed,

$$2(0) + 3(2) = 6$$

$$2(3) + 3(0) = 6$$

$$2(-1) + 3(\tfrac{8}{3}) = 6$$

$$2(6) + 3(-2) = 6$$

One should note that there are an unlimited number of elements in the solution set of linear equations in two variables. These solutions are obtained by arbitrarily replacing one of the variables by a specific real number and then solving the resulting equation for the remaining variable. For instance, in the equation $2x + 3y = 6$, arbitrarily replace x by 2. We then have $2(2) + 3y = 6$. Solving for y, we have $4 + 3y = 6$, so that

$$3y = 2 \qquad \text{or} \qquad y = \tfrac{2}{3}$$

Thus, the ordered pair $(2,\frac{2}{3})$ is an element of the solution set of the equation $2x + 3y = 6$. Each element of the solution set of a linear equation in two variables can be graphed on a Cartesian coordinate system. For example, the elements of the solution set of the equation $y = 3x$ consisting of the points $(-2,-6)$, $(-1,-3)$, $(0,0)$, $(1,3)$, and $(2,6)$ lie on the graph of the equation $y = 3x$, which, in turn, appear to lie on a straight line (Figure 7.14).

Figure 7.14

x	$y = 3x$
-2	-6
-1	-3
0	0
1	3
2	6

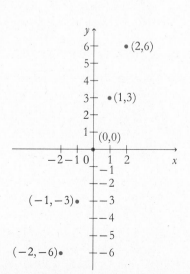

If other solutions of $y = 3x$ are graphed, they also will appear to lie on the same straight line (Figure 7.15a).

Now consider the graph of $y = 3x$ over the set of all real numbers x for which y is a real number. Each solution that is graphed will continue to lie on the same straight line; hence, by connecting these points, we get the graph of $y = 3x$, which appears to be a straight line (Figure 7.15b).

Figure 7.15

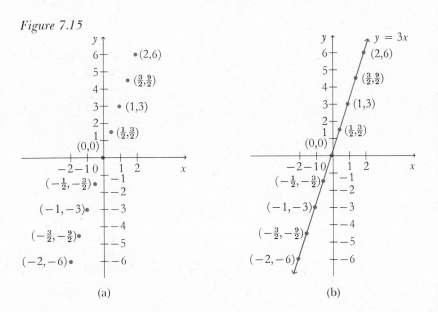

(a)   (b)

Consider the graph of the equation $y = -2x + 1$. Notice that if $x = 0$, then $y = -2(0) + 1 = 1$; thus, the ordered pair $(0, 1)$ is on the graph of $y = -2x + 1$; if $x = \frac{1}{2}$, then $y = -2(\frac{1}{2}) + 1 = 0$, so that the ordered pair $(\frac{1}{2}, 0)$ is on the graph of $y = -2x + 1$. Continuing in this manner, we can form a table.

Figure 7.16

x	y
0	1
$\frac{1}{2}$	0
1	-1
2	-3
-1	3

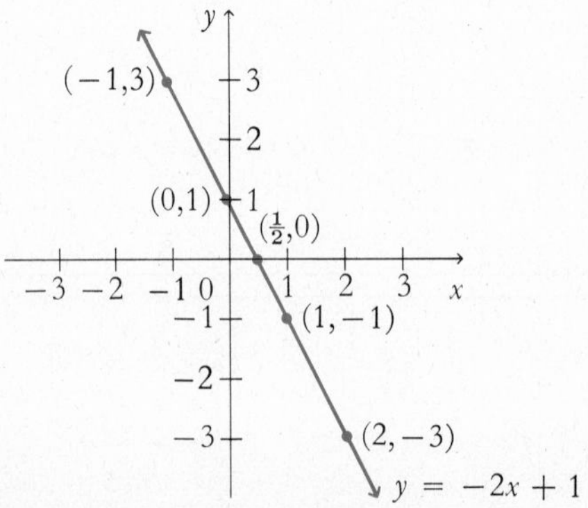

The graph of the equation $y = -2x + 1$ for any real number (Figure 7.16) also appears to be a straight line. It can be shown that the coordinates of each point on the line in Figure 7.16 satisfy the equation $y = -2x + 1$, and, conversely, every solution of the equation $y = -2x + 1$ corresponds to a point on the line.

In general, the graph of any first-degree equation in two variables is a straight line, and, conversely, a straight line is the graph of a first-degree equation in two variables. For instance, the graph of any equation of the form $ax + by = c$, where a, b, and c are real numbers and either a or b is nonzero, is a straight line.

Since we know from geometry that a straight line is completely determined by two distinct points, it will be enough to locate two points in order to graph the line. For example, to graph $y = 3x$, we know that $(0,0)$ and $(1,3)$ are solutions to the equation $y = 3x$; hence, if we were to plot $P_1 = (0,0)$ and $P_2 = (1,3)$, the graph of $y = 3x$ would be the straight line determined by the points P_1 and P_2 (Figure 7.17).

Figure 7.17

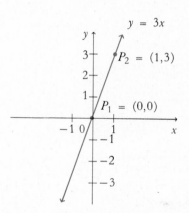

Similarly, the graph of the equation $y = -2x + 1$ is determined by $P_1 = (0,1)$ and $P_2 = (1,-1)$ (or any other two points) (Figure 7.18).

Figure 7.18

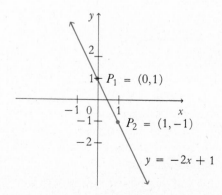

EXAMPLES

1   Graph $y = -2x$, where x is any real number.

SOLUTION.   The graph is a straight line (Figure 7.19). It can be drawn by locating any two points whose coordinates satisfy $y = -2x$. Thus, for $x = 0$, $y = 0$ and for $x = 1$, $y = -2$.

Figure 7.19

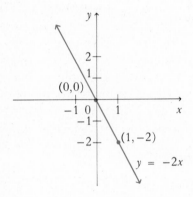

2   Graph $y = -2x + 4$.

SOLUTION.   Two solutions of $y = -2x + 4$ are $(0,4)$ and $(2,0)$. The graph is shown in Figure 7.20.

Figure 7.20

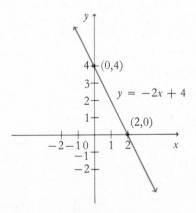

Since the straight line whose equation $ax + by + c = 0$, where a and b are not both zero is determined by two points, then we need to find at least two solutions of the equation to determine its graph. The solutions that are easy to find are $(x,0)$ and $(0,y)$. If we set $x = 0$ in the equation $ax + by + c = 0$, where $a \neq 0$ and $b \neq 0$, we get the equation

$by + c = 0$, so that $y = -c/b$. Therefore, $(0, -c/b)$ is a point on the graph. Letting $y = 0$, we have $ax + c = 0$, so that $x = -c/a$; therefore, $(-c/a, 0)$ is also a point on the graph. Hence, $(-c/a, 0)$ and $(0, -c/b)$ are the points where the graph of $ax + by + c = 0$ crosses the x axis and y axis, respectively. For this reason, we call the number $-c/a$ the x *intercept*, and the number $-c/b$ the y *intercept* of the graph. For example, the graph of $y = 3x + 5$ crosses the x axis at $(-\frac{5}{3}, 0)$ and the y axis at $(0, 5)$. Hence, the number $-\frac{5}{3}$ is the x intercept of the line and the number 5 is its y intercept (Figure 7.21).

Figure 7.21

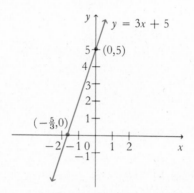

If either a or b is zero, the graph of the equation $ax + by + c = 0$ is either a horizontal or a vertical line. This illustrates the fact that if $a = 0$ and $b \neq 0$, the graph of $by + c = 0$ is a horizontal line whose y intercept is $-c/b$, whereas if $b = 0$ and $a \neq 0$, the graph of $ax + c = 0$ is a vertical line whose x intercept is $-c/a$. For example, the graph of the equation $3y - 12 = 0$ whose y intercept is 4 is a horizontal line (Figure 7.22a), while the graph of $2x - 6 = 0$ whose x intercept is 3 is a vertical line (Figure 7.22b).

Figure 7.22

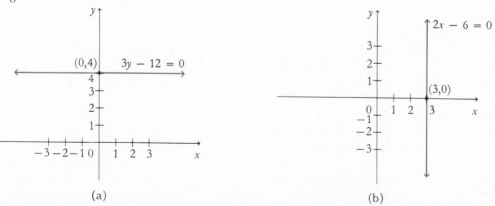

(a)

(b)

EXAMPLE

3   Find the x and y intercepts of the line $2x + 5y = 10$ and sketch the graph.

SOLUTION.   If $y = 0$, we see that $x = 5$. Thus, the x intercept is 5. Similarly, if $x = 0$, then $y = 2$ and the y intercept is 2 (Figure 7.23).

Figure 7.23

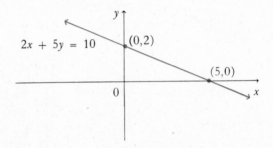

Slope of a Line

An important property of a straight-line segment joining two points is its *slope*, which is defined as follows. Suppose that $P_1 = (x_1, y_1)$ and $P_2 = (x_2, y_2)$ are two endpoints of a line segment (Figure 7.24). The number m that is defined by the equation

$$m = \frac{y_2 - y_1}{x_2 - x_1}$$

provided that $x_1 \neq x_2$, is called the *slope* of the line segment. Since

$$\frac{y_1 - y_2}{x_1 - x_2} = \frac{-(y_1 - y_2)}{-(x_1 - x_2)} = \frac{y_2 - y_1}{x_2 - x_1}$$

Figure 7.24

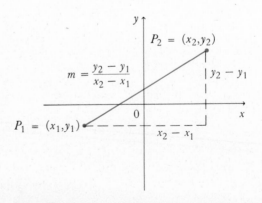

the order in which we take the two points when subtracting the coordinates does not change the value. For example, the slope of the line segment with endpoints $(-2, 5)$ and $(3, -4)$ is

$$m = \frac{-4 - 5}{3 - (-2)} = -\frac{9}{5}$$

or, equivalently,

$$m = \frac{5 - (-4)}{-2 - 3} = -\frac{9}{5}$$

Accordingly, we can speak of $(y_2 - y_1)/(x_2 - x_1)$, as the slope of the segment joining the two points (x_1, y_1) and (x_2, y_2), without specifying which comes first.

The slope of a line segment can also be interpreted geometrically. For example, consider the line segment joining the points $P_1 = (3, 3)$ and $P_2 = (5, 6)$. The slope of this segment is $\frac{3}{2}$, since

$$\frac{6 - 3}{5 - 3} = \frac{3}{2}$$

or, equivalently,

$$\frac{3 - 6}{3 - 5} = \frac{3}{2} \qquad \text{(Figure 7.25)}$$

Figure 7.25

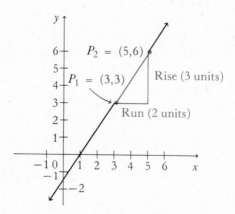

Note that $\frac{3}{2}$ is also the ratio of the vertical distance from P_1 to P_2 to the horizontal distance from P_1 to P_2. In more informal language, the *rise* divided by the *run* is called the slope, that is, $m = \text{rise/run}$.

If the line segment $\overline{P_1P_2}$ is slanted upward to the right, its rise will be considered to be positive, and so its slope $m = $ rise/run is positive (Figure 7.26a); whereas, if the line segment $\overline{P_1P_2}$ is slanted downward to the right, its rise will be considered to be negative; hence, its slope $m = $ rise/run is negative (Figure 7.26b).

Figure 7.26

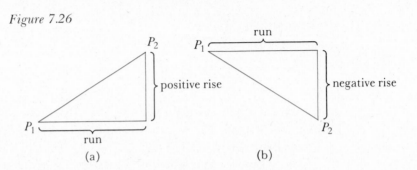

(a)   (b)

One can show that the slope m of the line segment is the same regardless of the choice of x_1 and x_2. Therefore, the slope is the same for every segment of the line. Hence, we may consider the slope to be a property of a *line* as a whole, rather than a particular line segment.

If $x_1 = x_2$ for each distinct pair of points on the line L, then $x_2 - x_1 = 0$, so that

$$m = \frac{y_2 - y_1}{x_2 - x_1} = \frac{y_2 - y_1}{0}$$

This has no meaning in the algebra of real numbers, and we say that the slope m of the line is undefined. However, if $y_1 = y_2$ for each distinct pair of points on the line L, then $y_2 - y_1 = 0$ and so

$$m = \frac{y_2 - y_1}{x_2 - x_1} = \frac{0}{x_2 - x_1} = 0$$

that is, the slope of the line L is 0. For example, the slope m of the line $2x - 6 = 0$ in Figure 7.22b is undefined since $x_1 = x_2 = 3$, whereas the slope of the line $3y - 12 = 0$ in Figure 7.22a is 0, since $y_1 = y_2 = 4$.

EXAMPLE

4   Determine the slope of the line containing the two points whose coordinates are given.

a) (6,2) and (3,7)   b) (3,−2) and (5,−6)

SOLUTION

a) The slope of the line is given by $m = (y_2 - y_1)/(x_2 - x_1)$, so that if $P_1 = (6,2)$ and $P_2 = (3,7)$, then

$$m = \frac{7-2}{3-6} = \frac{5}{-3} = -\frac{5}{3}$$

b) If $P_1 = (3,-2)$ and $P_2 = (5,-6)$, then

$$m = \frac{-6-(-2)}{5-3} = \frac{-6+2}{5-3} = \frac{-4}{2} = -2$$

Parallel and Perpendicular Lines

Two lines with slopes m_1 and m_2 are *parallel* if and only if $m_1 = m_2$. For example, the line containing the points $P_1 = (3,3)$ and $P_2 = (5,6)$ is parallel to the line containing the points $P_3 = (-1,1)$ and $P_4 = (1,4)$, since their slopes are identical (Figure 7.27). That is,

$$m_1 = \frac{6-3}{5-3} = \frac{3}{2} \quad \text{and} \quad m_2 = \frac{4-1}{1+1} = \frac{3}{2}$$

Figure 7.27

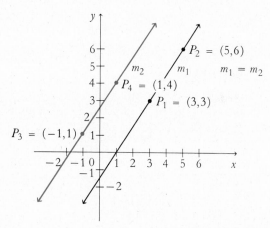

On the other hand, it can be shown that two lines with slopes m_1 and m_2 are *perpendicular* if and only if $m_1 m_2 = -1$.

For instance, the line containing the points $P_1 = (3,3)$, $P_2 = (5,6)$, with slope

$$m_1 = \frac{6-3}{5-3} = \frac{3}{2}$$

and the line containing the points $Q_1 = (1,4)$ and $Q_2 = (-2,6)$, with slope

$$m_2 = \frac{6-4}{-2-1} = -\frac{2}{3}$$

are perpendicular, since

$$m_1 m_2 = \left(\frac{3}{2}\right)\left(-\frac{2}{3}\right) = -1 \qquad \text{(Figure 7.28)}$$

Figure 7.28

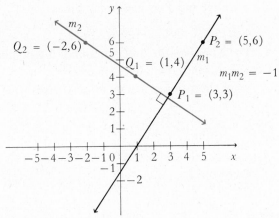

EXAMPLE

5   Given the points $P_1 = (3,2)$, $P_2 = (4,5)$, and $P_3 = (0,3)$, show that the line containing P_1 and P_2 is perpendicular to the line containing P_1 and P_3.

SOLUTION.   The slope m_1 of the line containing P_1 and P_2 is given by

$$m_1 = \frac{5-2}{4-3} = \frac{3}{1} = 3$$

The slope m_2 of the line containing P_1 and P_3 is given by

$$m_2 = \frac{3-2}{0-3} = \frac{1}{-3} = -\frac{1}{3}$$

Thus, $m_1 m_2 = (3)(-\frac{1}{3}) = -1$, and the two lines are perpendicular.

PROBLEM SET 7.2

In problems 1–6, indicate which equations in two variables are linear.

1 $x + 7y = 13$ 　　　　　　　**2** $\sqrt{3}x - \frac{1}{2}y = 17$

3 $\frac{1}{x} + 5y = 6$ 　　　　　　　**4** $x + 3y^2 - 7y = 1$

5 $x + \sqrt{2}y = 5$ 　　　　　　　**6** $\frac{1}{\sqrt{5}}x + \sqrt{17}y = 13$

In problems 7–12, indicate the ordered pairs which are elements of the solution set of each equation.

7 $y = 3x + 1$, $(-1,-2)$, $(0,1)$, $(3,2)$, $(4,5)$

8 $2x - 5y = -10$, $(-5,0)$, $(3,1)$, $(2,7)$, $(0,2)$

9 $2x - y = 0$, $(-5,1)$, $(1,2)$, $(-2,4)$, $(5,10)$

10 $2x + y = 3$, $(-2,7)$, $(3,1)$, $(5,-2)$, $(0,3)$

11 $3x + y - 7 = 0$, $(0,1)$, $(1,4)$, $(3,8)$, $(2,1)$

12 $5x - y = 6$, $(0,-6)$, $(4,3)$, $(1,-1)$, $(\frac{6}{5},0)$

In problems 13–22, sketch the graph of each equation.

13 $y = 2x$ for $x \in \{1,2\}$ 　　　　**14** $y = 2x$

15 $y = 3x + 1$ 　　　　　　　**16** $x - 3y = 5$

17 $y = -4x + 3$ 　　　　　　　**18** $\frac{x}{2} + y + 4 = 0$

19 $2x + 3y = -6$ 　　　　　　　**20** $4x - 3y - 12 = 0$

21 $-x + 5y = 10$ 　　　　　　　**22** $\frac{3x}{4} + \frac{4y}{5} = 1$

In problems 23–34, find the x intercept and the y intercept of each line, and sketch the graphs.

23 $y = -3x + 5$ 　　　　　　　**24** $3x - y + 1 = 0$

25 $2x - 3y - 9 = 0$ 　　　　　　**26** $-x + 2y + 1 = 0$

27 $y = -7x + 2$ 　　　　　　　**28** $y = 3$

29 $-4x + 3y = 0$ 　　　　　　　**30** $x = 4$

31 $\dfrac{x}{2} + \dfrac{y}{-1} = 1$ **32** $y - 2 = -5(x - 1)$

33 $y = 3 - 2x$ **34** $y = -\frac{3}{4}x + \frac{1}{2}$

In problems 35–40, find the slope of the line containing each pair of points.

35 $(2,-1)$ and $(-3,4)$ **36** $(1,-4)$ and $(2,3)$

37 $(1,5)$ and $(-2,3)$ **38** $(4,3)$ and $(-3,-4)$

39 $(6,-1)$ and $(0,2)$ **40** $(7,1)$ and $(-8,3)$

The points $P_1 = (x_1, y_1)$, $P_2 = (x_2, y_2)$, and $P_3 = (x_3, y_3)$ are collinear if the slope between P_1 and P_2 is the same as the slope between P_1 and P_3. In problems 41–44, use the concept of slope to determine whether or not the following points are collinear.

41 $(1,1)$, $(2,4)$, and $(3,2)$ **42** $(0,3)$, $(1,1)$, and $(2,-1)$
43 $(1,-3)$, $(-1,-11)$, and $(-2,-15)$ **44** $(1,5)$, $(-2,-1)$, and $(-3,-3)$

In problems 45–48, determine whether or not the lines determined by the pairs of points are parallel or perpendicular or neither.

45 $(2,-3)$, $(-1,3)$, and $(1,-\frac{1}{2})$, $(-4,-3)$

46 $(8,-7)$, $(-7,8)$, and $(10,-7)$, $(-4,6)$

47 $(2,4)$, $(3,8)$, and $(5,1)$, $(4,-3)$

48 $(-2,8)$, $(8,2)$, and $(-8,-2)$, $(2,-8)$

49 Show that the points $(2,4)$, $(3,8)$, $(5,1)$, and $(4,-3)$ determine a parallelogram.

50 Show that the four points $A = (1,9)$, $B = (4,0)$, $C = (0,6)$, and $D = (5,3)$ are vertices of a parallelogram.

51 Use the fact that two lines with slopes m_1 and m_2 are perpendicular if and only if $m_1 m_2 = -1$ to show that the triangle with vertices $A = (2,1)$, $B = (3,-1)$, and $C = (1,-2)$ is a right triangle.

7.3 Forms of Equations of a Line

We have seen that two points determine a straight line. A straight line is also determined by a given point on it and its slope. Suppose that L is a line with slope m that contains a point $P_1 = (x_1, y_1)$ (Figure 7.29). Then a point $P = (x, y)$, different from P_1, is on the line L if and only if the line containing P and P_1 also has slope m; that is, if and only if

$$m = \frac{y - y_1}{x - x_1} \qquad x \neq x_1$$

Figure 7.29

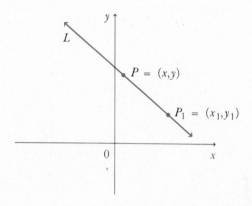

This equation can be written in the form

$$y - y_1 = m(x - x_1)$$

which is called the *point-slope form* of the equation of the line L containing the point (x_1, y_1), with slope m. Thus, the equation of the line containing the point $(-1, 2)$, with slope 3, is $y - 2 = 3(x + 1)$ (Figure 7.30).

Figure 7.30

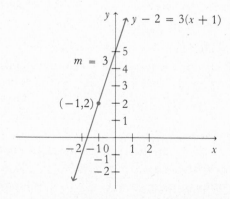

EXAMPLES

1   Find the equation of the line with slope $m = 5$ that contains the point $P_1 = (1,3)$. Sketch the graph.

SOLUTION.   Using the point-slope form $y - y_1 = m(x - x_1)$, we obtain $y - 3 = 5(x - 1)$. To sketch the graph, take $x = 0$, so that $y - 3 = 5(0 - 1)$ or $y = -2$. Hence, $(0,-2)$ is also a point on the graph (Figure 7.31).

Figure 7.31

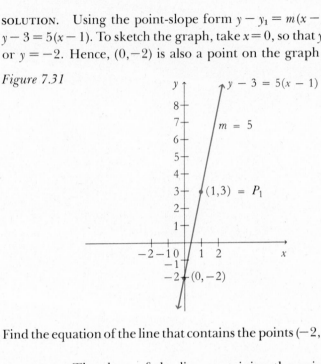

2   Find the equation of the line that contains the points $(-2,5)$ and $(3,-4)$.

SOLUTION.   The slope of the line containing the points $(-2,5)$ and $(3,-4)$ is given by

$$\frac{5 - (-4)}{-2 - 3} = \frac{9}{-5} = -\frac{9}{5}$$

Using $P_1 = (x_1, y_1) = (-2, 5)$, then the equation of the line is written as

$$y - 5 = -\frac{9}{5}(x + 2)$$

If we use $P_1 = (x_1, y_1) = (3, -4)$, the equation of the line becomes

$$y + 4 = -\frac{9}{5}(x - 3)$$

Both equations in Example 2 can be written in the same equivalent form. That is $y + 4 = -\frac{9}{5}(x - 3)$ and $y - 5 = -\frac{9}{5}(x + 2)$ can be written as $y = -\frac{9}{5}x + \frac{7}{5}$, where the slope is $-\frac{9}{5}$ and the y intercept is $\frac{7}{5}$ (Figure 7.32).

Figure 7.32

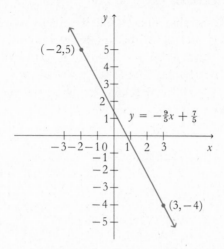

This equivalence procedure works in general. Let the line L with slope m contain the point $(0,b)$; then the equation $y - y_1 = m(x - x_1)$ can be written in the form $y - b = m(x - 0)$ or $y - b = mx$, so that

$$y = mx + b$$

This equation is called the *slope-intercept form* of the equation of L, since it involves only the slope m and the y intercept b of L (Figure 7.33).

Figure 7.33

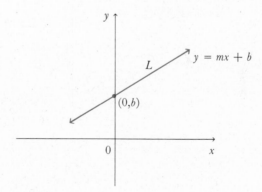

EXAMPLES

3   Find the equation of each of the following lines, with given slope and given y intercept. Sketch the graph.

a) $m = 2, b = -3$

b) $m = -4, b = 1$

SOLUTION

a) Using the slope-intercept form $y = mx + b$, we have $y = 2x - 3$.
To sketch the graph, we find another point on the line. If $y = 0$,
then $0 = 2x - 3$ or $x = \frac{3}{2}$, so that $(\frac{3}{2}, 0)$ is also a point on the graph
(Figure 7.34).

Figure 7.34

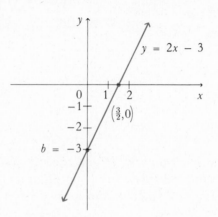

b) Using the slope-intercept form $y = mx + b$, we have $y = -4x + 1$.
Another point on the graph is found by letting $y = 0$; then $x = \frac{1}{4}$,
so that $(\frac{1}{4}, 0)$ is also a point on the graph (Figure 7.35).

Figure 7.35

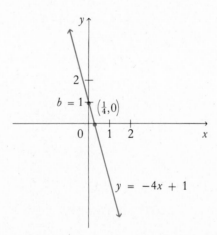

4   Determine the slope and the y intercept of each of the following lines.
Sketch each graph.

a) $y = -3x + 7$

b) $5x - y - 6 = 0$

SOLUTION

a) The slope of the line $y = -3x + 7$ is -3; and the y intercept is 7. (Why?) If $x = 1$, then $y = -3(1) + 7$ or $y = 4$, so that $(1, 4)$ is also a point on the graph (Figure 7.36).

Figure 7.36

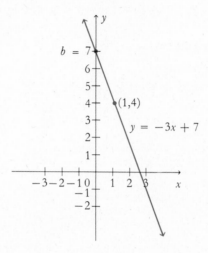

b) $5x - y - 6 = 0$ can be written as $y = 5x - 6$; therefore, the slope is 5 and the y intercept is -6. To find another point on the graph, let $x = 1$; then $y = 5(1) - 6 = -1$, so that $(1, -1)$ is also a point on the graph (Figure 7.37).

Figure 7.37

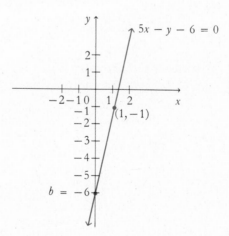

Now we summarize the forms of the equation of a line.

1 *Point-slope form:* $y - y_1 = m(x - x_1)$; slope is m and the line contains the point (x_1, y_1).

2 *Slope-intercept form:* $y = mx + b$; slope is m and the y intercept is b.

3 *Vertical line:* $x = h$; no slope, x intercept is h.

4 *Horizontal line:* $y = k$; slope is 0, y intercept is k.

5 *General form:* $ax + by + c = 0$; a and b not both 0, slope is $-a/b$ for $b \neq 0$, x intercept is $-c/a$, and y intercept is $-c/b$.

PROBLEM SET 7.3

In problems 1–18, find the equation of the line with the information given. Sketch the graph.

1   $m = -3$ and $(x_1, y_1) = (-1, 2)$ 2 $m = -7$ and $(x_1, y_1) = (0, 3)$

3   $m = 5$ and $(x_1, y_1) = (3, 1)$ 4 $m = \frac{22}{5}$ and $(x_1, y_1) = (-1, 4)$

5   $m = -\frac{13}{7}$ and $(x_1, y_1) = (0, 1)$ 6 $m = -\frac{3}{4}$ and $(x_1, y_1) = (0, 0)$

7   $m = 0$ and $(x_1, y_1) = (-1, -5)$ 8 $m = 0$ and $(x_1, y_1) = (-\frac{1}{2}, 0)$

9   $(x_1, y_1) = (-3, 2)$ and $(x_2, y_2) = (3, 5)$

10   $(x_1, y_1) = (-1, 7)$ and $(x_2, y_2) = (6, -1)$

11   $(x_1, y_1) = (-2, 4)$ and $(x_2, y_2) = (0, 1)$

12   $(x_1, y_1) = (0, 0)$ and $(x_2, y_2) = (-1, 2)$

13   $m = -3$ and y intercept 5   14 $m = -7$ and y intercept 2

15   no slope and x intercept -3 16   no slope and x intercept 2

17   $m = 0$ and y intercept 4   18 $m = 0$ and y intercept $-\frac{1}{5}$

In problems 19–26, express each equation in the slope-intercept form; find the x intercept and the y intercept. Also, sketch the graph.

19   $2x - 3y - 1 = 0$ 20 $2x + 3y + 12 = 0$

21   $y - 1 = -2(x - 2)$ 22 $5x - 7y - 8 = 0$

23   $4x - y + 5 = 0$ 24 $3x - 4y - 5 = 0$

25   $-2x + y = 0$ 26 $y - 3 = -4(x - 3)$

In problems 27–32, find the equation of the line that is parallel to the given line and contains the given point.

27   $y = -2x + 3$, $(x_1, y_1) = (3, -1)$ 28 $y = 7x - 13$, $(x_1, y_1) = (-2, 1)$

29   $x + 2y + 7 = 0$, $(x_1, y_1) = (-\frac{1}{2}, 1)$ 30 $y = 3$, $(x_1, y_1) = (7, -3)$

31 $x + y = 0$, $(x_1, y_1) = (0, 1)$ **32** $\dfrac{x}{7} + \dfrac{y}{3} = 1$, $(x_1, y_1) = (-1, 0)$

In problems 33-37, find the equation of the line that is perpendicular to the given line and contains the given point. Sketch the graph.

33 $y = 3x + 11$, $(x_1, y_1) = (1, 5)$

34 $2x + 3y - 1 = 0$, $(x_1, y_1) = (-1, 2)$

35 $x + 2y - 5 = 0$, $(x_1, y_1) = (-3, 2)$

36 $y = -2$, $(x_1, y_1) = (5, 3)$

37 $\dfrac{x}{3} + \dfrac{y}{2} = 1$, $(x_1, y_1) = (1, 1)$

38 Show that an equation of a line whose intercepts are $(a, 0)$ and $(0, b)$ with $a \neq 0$ and $b \neq 0$ is $\dfrac{x}{a} + \dfrac{y}{b} = 1$. This is called the *intercept form* of a line.

In problems 39–42, use the result of problem 38 to find the equation of the line whose intercepts are the given points. Sketch the graph.

39 $(5, 0)$ and $(0, 6)$ **40** $(-2, 0)$ and $(0, 7)$

41 $(-3, 0)$ and $(0, -11)$ **42** $(-1, 0)$ and $(0, -5)$

In problems 43–49, find an equation of the line that satisfies the given conditions.

43 The y intercept is 3 and it is parallel to the line $x - 2y = 5$.

44 The y intercept is 5 and it is perpendicular to the line $y = 3x$.

45 The x intercept is -1 and it is perpendicular to the line $3x - 2y = 5$.

46 It contains the origin and is parallel to the line containing the points $(5, 1)$ and $(-2, 3)$.

47 It contains the point $(1, -2)$ and is perpendicular to the line containing the points $(3, 0)$ and $(-3, 1)$.

48 It contains the point $(7, -3)$ and is parallel to the x axis.

49 It contains the point $(-1, 6)$ and is perpendicular to the y axis.

50 Find a value of k so that each of the following conditions will hold.
   a) The line $3x + ky + 2 = 0$ is parallel to the line $6x - 5y + 3 = 0$.
   b) The line $y = (2 - k)x + 2$ is perpendicular to the line $y = 3x - 1$.

7.4 Systems of Linear Equations in Two Variables

A set of equations such as

$$\begin{cases} 2x + 3y = 12 \\ 6x - 2y = 14 \end{cases}$$

is called a *system of linear equations* in two variables, x and y. The *solution set* of the system of two equations in two variables is the set of the ordered pairs whose coordinates satisfy both equations. For example, the ordered pair $(3,2)$ belongs to the solution set of the system, since

$$\begin{cases} 2(3) + 3(2) = 6 + 6 = 12 \\ 6(3) - 2(2) = 18 - 4 = 14 \end{cases}$$

Graphically, the solution of a system of linear equations in two variables is the set of all points lying on both lines represented by the equations. For example, since the ordered pair $(3,2)$ is the only common point on the graph of both lines in the above system (Figure 7.38), then $\{(3,2)\}$ is the solution set of the system.

Figure 7.38

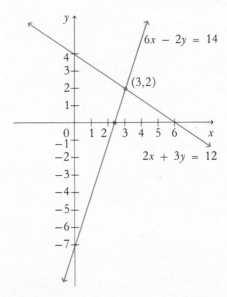

A system of two linear equations in the same two variables represents a pair of lines in the plane. The two lines in the same plane either *coincide* or *have no points in common (parallel lines)* or *intersect in exactly one point.* Each of these cases is illustrated by the following examples. In Figure 7.39a the two lines $2x - 3y + 7 = 0$ and $4x - 6y + 14 = 0$ *coincide*. In Figure 7.39b the two lines $2x - 3y + 7 = 0$ and $4x - 6y - 8 = 0$ are *parallel*, whereas the two lines $2x - 3y + 7 = 0$ and $x + 2y - 5 = 0$ in Figure 7.39c *intersect in exactly one point;* in this case, the point is $(\frac{1}{7}, \frac{17}{7})$.

Figure 7.39

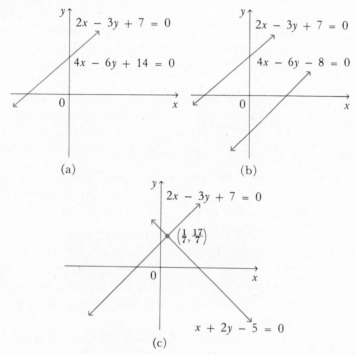

(a)

(b)

(c)

If each equation of a line in Figure 7.39a is expressed in the slope-intercept form, then

$$\begin{cases} y = \frac{2}{3}x + \frac{7}{3} \\ y = \frac{4}{6}x + \frac{14}{6} \end{cases}$$

have the same slope, $\frac{2}{3}$. The two lines in Figure 7.39b,

$$\begin{cases} y = \frac{2}{3}x + \frac{7}{3} \\ y = \frac{4}{6}x - \frac{8}{6} \end{cases}$$

also have the same slope, $\frac{2}{3}$, whereas the two lines in Figure 7.39c,

$$\begin{cases} y = \frac{2}{3}x + \frac{7}{3} \\ y = -\frac{1}{2}x + \frac{5}{2} \end{cases}$$

have different slopes, $\frac{2}{3}$ and $-\frac{1}{2}$, respectively. The solution set of the system of linear equations whose graph is shown in Figure 7.39a is the infinite set $\{(x,y)\,|\,y=\frac{2}{3}x+\frac{7}{3}\}$, in Figure 7.39b is the empty set $\varnothing$, and in Figure 7.39c consists of the ordered pair $(\frac{1}{7},\frac{17}{7})$ that represents the point of intersection of the two lines. The above results can be generalized as follows.

   When systems of two linear equations in two variables are graphed on the same coordinate plane, three general relationships between the two lines are evident: they coincide (Figure 7.39a); they are parallel lines (Figure 7.39b); or they are a pair of intersecting lines (Figure 7.39c). Hence, we have the following possibilities:

1   If the two lines coincide (Figure 7.39a), the coordinates of every point on the first line are solutions of the given system and all points on the first line are points on the second line. In this case we say that the system is *dependent*. (The slopes are equal.)

2   If the two lines are parallel (Figure 7.39b), they have no points in common. Hence, the solution set is the empty set and the system is called *inconsistent*. (The slopes are equal.)

3   If the two lines intersect at one point (Figure 7.39c), we call the system *independent*. (The slopes are unequal.)

Although the solution set of a linear system can be found by graphing, this method is often tedious and sometimes inaccurate, since it is difficult to get the exact measurements. For this reason, we introduce algebraic methods for solving independent systems called the substitution method and the addition–subtraction method.

Substitution Method

To solve a system of linear equations using this method, we express y in terms of x (or x in terms of y) in one of the equations, and then substitute this value in the remaining equation to obtain an equivalent equation in one variable. To illustrate this method, consider the system

$$\begin{cases} 2x - y = 1 \\ 3x + y = 1 \end{cases}$$

Solving the first equation for y produces $y = 2x - 1$. Replacing y with $2x - 1$ in the second equation, we obtain $3x + (2x - 1) = 1$ or $5x - 1 = 1$, so that $5x = 2$ or $x = \frac{2}{5}$. To find the corresponding value for y, we have

$$y = 2x - 1 = 2\left(\tfrac{2}{5}\right) - 1 = \tfrac{4}{5} - 1 = -\tfrac{1}{5}$$

Hence, the solution set is $\{(\tfrac{2}{5}, -\tfrac{1}{5})\}$.

EXAMPLES

Solve the following systems by the substitution method.

1 $\begin{cases} 3x - y = 0 \\ 5x - y = -1 \end{cases}$

SOLUTION. Solving the first equation for y produces $y = 3x$. Replacing y by $3x$ in the second equation, we have $5x - (3x) = -1$, so that $2x = -1$ or $x = -\tfrac{1}{2}$. To find the corresponding value of y, we have $y = 3x = 3(-\tfrac{1}{2}) = -\tfrac{3}{2}$.

Check: To check our solution, replace x by $-\tfrac{1}{2}$ and $y = -\tfrac{3}{2}$ in the original system, so that

$$\begin{cases} 3(-\tfrac{1}{2}) - (-\tfrac{3}{2}) = 0 \\ 5(-\tfrac{1}{2}) - (-\tfrac{3}{2}) = -1 \end{cases}$$

Hence, the solution set is $\{(-\tfrac{1}{2}, -\tfrac{3}{2})\}$.

2 $\begin{cases} x + y = 3 \\ 2x - 3y = 1 \end{cases}$

SOLUTION. Solving the first equation for x produces $x = 3 - y$. Replacing x by $3 - y$ in the second equation, we have $2(3 - y) - 3y = 1$, so that $6 - 2y - 3y = 1$ or $-5y = -5$ or $y = 1$. Thus, $x = 3 - y = 3 - 1 = 2$. The solution set is $\{(2, 1)\}$.

Addition–Subtraction Method

Systems of linear equations can also be solved by an appropriate term-by-term addition or subtraction of two equations in a given system. In performing this method, an equation will result in which one of the variables is present and the other variable is eliminated. For this reason we often refer to this procedure as the *elimination-of-variables method.* It should be noted that in applying this method, it is sometimes necessary to first multiply one or both equations by some constant (or constants) in order to eliminate a variable.

To illustrate this technique, let us consider the following system of linear equations:

$$\begin{cases} x + y = 5 \\ x - y = 1 \end{cases}$$

Adding the corresponding members of the given equations, we have $x + y + x - y = 5 + 1$, so that $2x = 6$ or $x = 3$. To find the value of y, we substitute $x = 3$ in either of the given equations. Thus, using $x + y = 5$, we have $3 + y = 5$ or $y = 2$.

To check this solution, we substitute $x = 3$ and $y = 2$ in both equations, so that $3 + 2 = 5$ and $3 - 2 = 1$. Hence, the solution set is $\{(3,2)\}$.

EXAMPLES

Solve each of the following systems of linear equations by the addition–subtraction method.

3    $$\begin{cases} 2x + y = 5 \\ x - y = 1 \end{cases}$$

SOLUTION.   Since the coefficients of the y terms are 1 and -1, we can eliminate the y terms by addition. Thus,

$$\begin{aligned} 2x + y &= 5 \\ \underline{x - y} &= \underline{1} \\ 3x \phantom{{}+y} &= 6 \end{aligned}$$

or $x = 2$. Substituting $x = 2$ in the first equation, we have $2(2) + y = 5$, so that $4 + y = 5$ or $y = 1$.
Check: Substituting $x = 2$ and $y = 1$ in both equations, we have $2(2) + 1 = 4 + 1 = 5$ and $2 - 1 = 1$. Hence, the solution set is $\{(2,1)\}$.

4    $$\begin{cases} 2x + 3y = 7 \\ x + y = 2 \end{cases}$$

SOLUTION.   In order to eliminate a variable by addition or subtraction, we must have the coefficients of one variable either equal or the negatives of each other. If we first multiply the second equation by 2, we have

$$\begin{cases} 2x + 3y = 7 \\ 2x + 2y = 4 \end{cases}$$

so that the x terms can now be eliminated by subtraction. Thus,

$$\begin{array}{l} 2x + 3y = 7 \\ \underline{2x + 2y = 4} \\ y = 3 \end{array}$$

Substituting $y = 3$ in the second equation of the given system, we have $x + 3 = 2$ or $x = -1$. Hence, the solution set is $\{(-1,3)\}$.

5 $\begin{cases} 3x + 2y = 8 \\ 2x - 3y = 14 \end{cases}$

SOLUTION. Multiplying the first equation by 3 and the second equation by 2, we obtain

$$\begin{cases} 9x + 6y = 24 \\ 4x - 6y = 28 \end{cases}$$

Adding these two equations, we can eliminate the y terms. Thus,

$$\begin{array}{l} 9x + 6y = 24 \\ \underline{4x - 6y = 28} \\ 13x = 52 \end{array}$$

or $x = 4$.

Substituting $x = 4$ in the first equation, we have $3(4) + 2y = 8$ or $12 + 2y = 8$, so that $2y = -4$ or $y = -2$. Hence, the solution set is $\{(4,-2)\}$.

6 $\begin{cases} 3x + y = a \\ x + y = b \end{cases}$ where a and b are constants

SOLUTION. Subtracting the second equation from the first equation, we have

$$\begin{array}{l} 3x + y = a \\ \underline{x + y = b} \\ 2x = a - b \end{array}$$

or $x = (a - b)/2$. Substituting this value of x in the second equation, we obtain $(a - b)/2 + y = b$, so that $y = b - (a - b)/2$ or $y = (3b - a)/2$. Hence, the solution set is $\{((a-b)/2, (3a-b)/2)\}$.

PROBLEM SET 7.4

In problems 1–6, sketch the graph of each system of linear equations. Use the graph to determine whether each system is dependent, inconsistent, or independent. If the system is independent, determine the coordinates of the solution graphically.

1 $\begin{cases} 3x - 2y = 1 \\ 6x - 8y = 2 \end{cases}$ **2** $\begin{cases} y = 2x - 3 \\ 4x - 2y = 6 \end{cases}$

3 $\begin{cases} x + 3y = 6 \\ 2x + 6y = 8 \end{cases}$ **4** $\begin{cases} 2x = y + 3 \\ 4x - 2y = 5 \end{cases}$

5 $\begin{cases} 2x - 3y = 1 \\ 5x + 2y = 12 \end{cases}$ **6** $\begin{cases} 3x + 2y = 11 \\ -2x + y = 2 \end{cases}$

In problems 7–16, find the solution set of each system by the method of substitution.

7 $\begin{cases} 2x - y = 5 \\ x + 3y = 13 \end{cases}$ **8** $\begin{cases} 3x + y = 4 \\ 7x - y = 6 \end{cases}$

9 $\begin{cases} y = x - 2 \\ 2y = x - 3 \end{cases}$ **10** $\begin{cases} 3x + 5y = 1 \\ x - 4y = -6 \end{cases}$

11 $\begin{cases} y = 3x + 2 \\ y = -x + 6 \end{cases}$ **12** $\begin{cases} 7y = -5x + 4 \\ 6y = x + 14 \end{cases}$

13 $\begin{cases} x + y = 1 \\ 5x - y = 13 \end{cases}$ **14** $\begin{cases} 3x = 4y - 5 \\ 2y = 3 - 2x \end{cases}$

15 $\begin{cases} 2x - y = 3 \\ 3x + y = 22 \end{cases}$ **16** $\begin{cases} \frac{1}{2}y = -\frac{1}{3}x + 4 \\ \frac{1}{4}x - \frac{1}{3}y = -1 \end{cases}$

In problems 17–36, find the solution set of each system by the addition–subtraction method.

17 $\begin{cases} 2x - y = 1 \\ x + y = 2 \end{cases}$ **18** $\begin{cases} x + y = 0 \\ x - y = 0 \end{cases}$

19 $\begin{cases} x + 3y = 9 \\ x - y = 1 \end{cases}$ **20** $\begin{cases} x + y = 1 \\ y - 2x = 0 \end{cases}$

21 $\begin{cases} 2x + 4y = 2 \\ -x + y = 8 \end{cases}$

22 $\begin{cases} 7x + y = 5 \\ y - 2x = 3 \end{cases}$

23 $\begin{cases} 3x + 2y = 4 \\ 5x + 3y = 7 \end{cases}$

24 $\begin{cases} x + y + 8 = 0 \\ 2x - 7y = 5 \end{cases}$

25 $\begin{cases} 7x + y = 3 \\ 5x + y = 6 \end{cases}$

26 $\begin{cases} 3y + 2z = 5 \\ 2y = -3z + 1 \end{cases}$

27 $\begin{cases} -3x + y = 3 \\ 4x + 2y = 10 \end{cases}$

28 $\begin{cases} 5x - 2y = 35 \\ x + 4y = 25 \end{cases}$

29 $\begin{cases} 4x - 3y = 1 \\ 3x + 4y = 6 \end{cases}$

30 $\begin{cases} 8 = x + 3y \\ x + 8y = 53 \end{cases}$

31 $\begin{cases} 2x - y = 5 \\ x + 2y = 25 \end{cases}$

32 $\begin{cases} 13y + 5z = 2 \\ 2 - 2z = 6y \end{cases}$

33 $\begin{cases} \frac{1}{2}x + \frac{1}{3}y = 13 \\ \frac{1}{5}x + \frac{1}{8}y = 5 \end{cases}$

34 $\begin{cases} \frac{1}{3}x - \frac{1}{4}y = 2 \\ \frac{1}{4}x - \frac{1}{2}y = 7 \end{cases}$

35 $\begin{cases} 3x + y = a \\ x - 3y = b; \end{cases}$
a and b = constants

36 $\begin{cases} 3ax + 2by = 6 \\ 2ax - 5by = 7; \end{cases}$
a and b = constants

In problems 37–42, find the solution set of each system by using the substitution $u = 1/x$ and $v = 1/y$.

37 $\begin{cases} \dfrac{2}{x} - \dfrac{1}{y} = 9 \\ \dfrac{5}{x} - \dfrac{3}{y} = 14 \end{cases}$

38 $\begin{cases} \dfrac{5}{x} - \dfrac{2}{y} = 1 \\ \dfrac{8}{x} + 11 = \dfrac{5}{y} \end{cases}$

39 $\begin{cases} \dfrac{3}{x} + \dfrac{2}{y} = 2 \\ \dfrac{1}{x} - \dfrac{1}{y} = 9 \end{cases}$

40 $\begin{cases} \dfrac{4}{x} - \dfrac{3}{y} = 1 \\ \dfrac{3}{x} - \dfrac{4}{y} = 6 \end{cases}$

41 $\begin{cases} \dfrac{5}{x} + \dfrac{2}{y} = 1 \\ \dfrac{13}{x} + \dfrac{8}{y} = 11 \end{cases}$

42 $\begin{cases} \dfrac{4}{x} + \dfrac{1}{y} = 16 \\ \dfrac{3}{x} + \dfrac{1}{y} = 11 \end{cases}$

7.5   Systems of Linear Equations in Three Variables

We have seen in Section 7.4 how to solve a system of two linear equations in two variables. Sometimes situations arise in which it is necessary to solve a system of *three linear equations in three variables*.

$$\begin{cases} x + y + z = 6 \\ 2x - y + z = 3 \\ 3x + y - z = 2 \end{cases}$$

is an example of such a system.

A solution of an equation of three variables, such as $2x - y + z = 5$, is called an *ordered triple* and is denoted by (x, y, z). For example, the ordered triples $(1, -1, 2)$ and $(-2, 1, 10)$ are solutions of the equation $2x - y + z = 5$. The *solution set* of a system of three linear equations in three variables is the set of ordered triples that satisfies the three equations. For instance, the solution set of the three linear equations in the example above is $\{(1, 2, 3)\}$, since the elements of the ordered triple satisfy the three equations. That is,

$$\begin{cases} 1 + 2 + 3 = 6 \\ 2(1) - 2 + 3 = 3 \\ 3(1) + 2 - 3 = 2 \end{cases}$$

The substitution method or the addition–subtraction method can be used to solve such systems. The procedure is illustrated in the following examples.

EXAMPLES

1   Use the substitution method to solve the following system.

$$\begin{cases} x + y + z = 6 \\ 2x - y - z = 0 \\ x - y + 2z = 7 \end{cases}$$

SOLUTION.   We are seeking an ordered triple of numbers (x, y, z) that satisfies all three equations simultaneously. First, rewrite the equation $x + y + z = 6$ as $z = 6 - x - y$. Then, substituting $6 - x - y$ for z in the remaining two equations, we get

$$\begin{cases} 2x - y - (6 - x - y) = 0 \\ x - y + 2(6 - x - y) = 7 \end{cases}$$

that is,

$$\begin{cases} 3x - 6 \qquad = 0 \\ -x - 3y + 12 = 7 \end{cases}$$

However, the latter two equations in the system can be solved as a linear system containing two variables. Here we get $x = 2$ and $y = 1$ (why?), so that $z = 6 - x - y = 6 - 2 - 1 = 3$.
Check:

$$\begin{cases} 2 + 1 + 3 = 6 \\ 2(2) - 1 - 3 = 0 \\ 2 - 1 + 2(3) = 7 \end{cases}$$

Hence, the solution set is $\{(2, 1, 3)\}$.

2 Use the addition–subtraction method to solve the following system.

$$\begin{cases} x + y + z = 2 \\ 2x + 3y - z = 3 \\ 3x + 5y + z = 8 \end{cases}$$

SOLUTION.   Adding the first and second equations, we obtain

$$\begin{array}{r} x + y + z = 2 \\ 2x + 3y - z = 3 \\ \hline 3x + 4y \qquad = 5 \end{array}$$

Adding the second and third equations, we obtain

$$\begin{array}{r} 2x + 3y - z = 3 \\ 3x + 5y + z = 8 \\ \hline 5x + 8y \qquad = 11 \end{array}$$

We now have a system of two equations in two variables, that is

$$\begin{cases} 3x + 4y = 5 \\ 5x + 8y = 11 \end{cases}$$

which can be written as

$$\begin{cases} 6x + 8y = 10 \\ 5x + 8y = 11 \end{cases}$$

Subtracting these two equations, we have

$$
\begin{array}{r}
6x + 8y = 10 \\
5x + 8y = 11 \\
\hline
x = -1
\end{array}
$$

Substituting $x = -1$ in $3x + 4y = 5$, we have $3(-1) + 4y = 5$, so that $4y = 8$ or $y = 2$. Substituting $x = -1$ and $y = 2$ in the equation $x + y + z = 2$, we have $-1 + 2 + z = 2$ or $z = 1$. Hence, the solution set is $\{(-1, 2, 1)\}$.

PROBLEM SET 7.5

In problems 1–6, find the solution set of each system by the substitution method.

1
$$\begin{cases} x + y = 5 \\ x + z = 1 \\ y + z = 2 \end{cases}$$

2
$$\begin{cases} 2x + 3y = 28 \\ 3y + 4z = 46 \\ 4z + 5x = 53 \end{cases}$$

3
$$\begin{cases} x + y + 2z = 11 \\ x - y + z = 3 \\ 2x + y + 3z = 17 \end{cases}$$

4
$$\begin{cases} x - 3y = -11 \\ 2y - 5z = 26 \\ 7x - 3z = -2 \end{cases}$$

5
$$\begin{cases} 2x - y + z = 8 \\ x + 2y + 3z = 9 \\ 4x + y - 2z = 1 \end{cases}$$

6
$$\begin{cases} x + 3y - z = 4 \\ 3x - 2y + 4z = 11 \\ 2x + y + 3z = 13 \end{cases}$$

In problems 7–18, find the solution set of each system by the addition–subtraction method.

7
$$\begin{cases} x + y + 2z = 4 \\ x + y - 2z = 0 \\ x - y = 0 \end{cases}$$

8
$$\begin{cases} x + y + z = 2 \\ x + 2y - z = 4 \\ 2x - y + z = 0 \end{cases}$$

9
$$\begin{cases} x + y + z = 6 \\ x - y + 2z = 12 \\ 2x + y - z = 1 \end{cases}$$

10
$$\begin{cases} x + y + 2z = 4 \\ x - 5y + z = 5 \\ 3x - 4y + 7z = 24 \end{cases}$$

11
$$\begin{cases} x + y = 4 \\ 3x - y + 3z = 7 \\ 5x - 7y + 2z = -2 \end{cases}$$

12
$$\begin{cases} 7x + y + 3z = -6 \\ 4x - 5y + 6z = -27 \\ x + 15y - 9z = 64 \end{cases}$$

(handwritten:) should be –? Ans $(3, -1, 4)$

13 $\begin{cases} 2x + y - 3z = 9 \\ x - 2y + 4z = 5 \\ 3x + y - 2z = 15 \end{cases}$

14 $\begin{cases} x + 3y - 2z = -21 \\ 7x - 5y + 4z = 31 \\ 2x + y + 3z = 17 \end{cases}$

15 $\begin{cases} 2x + 3y + z = 6 \\ x - 2y + 3z = 3 \\ 3x + y - z = 8 \end{cases}$

16 $\begin{cases} 8x + 3y - 18z = -76 \\ 10x + 6y - 6z = -50 \\ 4x + 9y + 12z = 10 \end{cases}$

17 $\begin{cases} x + y + z = a \\ x + 2y - z = b \\ 2x + 4y + z = c; \end{cases}$
$a, b,$ and c = constants

18 $\begin{cases} x + y - z = a \\ x - y + z = b \\ x + y + z = c; \end{cases}$
$a, b,$ and c = constants

In problems 19 and 20, find the solution set of each system by using the substitutions $u = 1/x$, $v = 1/y$, and $w = 1/z$.

19 $\begin{cases} \dfrac{3}{x} - \dfrac{4}{y} + \dfrac{6}{z} = 1 \\ \dfrac{9}{x} + \dfrac{8}{y} - \dfrac{12}{z} = 3 \\ \dfrac{9}{x} - \dfrac{4}{y} + \dfrac{12}{z} = 4 \end{cases}$

20 $\begin{cases} \dfrac{3}{x} + \dfrac{1}{y} - \dfrac{1}{z} = 5 \\ \dfrac{4}{x} - \dfrac{1}{y} + \dfrac{2}{z} = 13 \\ \dfrac{2}{x} + \dfrac{2}{y} + \dfrac{3}{z} = 22 \end{cases}$

7.6 Determinants

Solving the following system of two linear equations in two variables,

$$\begin{cases} a_1 x + b_1 y = c_1 \\ a_2 x + b_2 y = c_2 \end{cases}$$

By the methods of Section 7.5, we obtain the solution (x, y) where

$$x = \frac{b_2 c_1 - b_1 c_2}{a_1 b_2 - a_2 b_1} \quad \text{and} \quad y = \frac{a_1 c_2 - a_2 c_1}{a_1 b_2 - a_2 b_1}$$

provided that $a_1 b_2 - a_2 b_1 \neq 0$.

Since any system of two linear equations in two variables can be arranged in the form above, we observe that the solution appears as fractions with a common denominator. This denominator will be denoted by the symbol

$$\begin{vmatrix} a_1 & b_1 \\ a_2 & b_2 \end{vmatrix}$$

called a *determinant,* which is defined as another form for $a_1b_2 - a_2b_1$. We have, then, by definition,

$$\begin{vmatrix} a_1 & b_1 \\ a_2 & b_2 \end{vmatrix} = a_1b_2 - a_2b_1$$

The numbers a_1, b_1, a_2, and b_2 are called the *elements* of the determinant, with two *rows,* a_1, b_1 and a_2, b_2 and two *columns,* a_1, a_2 and b_1, b_2. For this reason, it is called a *two-by-two* (2×2) determinant, or a determinant of *order* 2.

EXAMPLES

Find the value of the following 2×2 determinants.

1   $\begin{vmatrix} 1 & -2 \\ 3 & 4 \end{vmatrix}$

SOLUTION.   Here we have $a_1 = 1$, $b_1 = -2$, $a_2 = 3$, and $b_2 = 4$, and so by substitution in the expression $a_1b_2 - a_2b_1$, we have

$$\begin{vmatrix} 1 & -2 \\ 3 & 4 \end{vmatrix} = 1(4) - 3(-2) = 4 + 6 = 10$$

Rather than substituting in the expression $a_1b_2 - a_2b_1$ each time, we can find the value of a 2×2 determinant by following the cross-multiplication scheme suggested by the definition. Thus,

$$\begin{vmatrix} 1 & -2 \\ 3 & 4 \end{vmatrix} = 1(4) - 3(-2) = 10$$

2   $\begin{vmatrix} 1 & 0 \\ 2 & -1 \end{vmatrix}$

SOLUTION. Using the cross-multiplication scheme above, we have

$$\begin{vmatrix} 1 & 0 \\ 2 & 1 \end{vmatrix} = 1(-1) - 2(0) = -1 - 0 = -1$$

The value of a 3×3 determinant is defined in terms of 2×2 determinants in the following way:

$$\begin{vmatrix} a_1 & b_1 & c_1 \\ a_2 & b_2 & c_2 \\ a_3 & b_3 & c_3 \end{vmatrix} = a_1 \begin{vmatrix} b_2 & c_2 \\ b_3 & c_3 \end{vmatrix} - a_2 \begin{vmatrix} b_1 & c_1 \\ b_3 & c_3 \end{vmatrix} + a_3 \begin{vmatrix} b_1 & c_1 \\ b_2 & c_2 \end{vmatrix}$$

By employing the definition of a 2×2 determinant, we also have

$$\begin{vmatrix} a_1 & b_1 & c_1 \\ a_2 & b_2 & c_2 \\ a_3 & b_3 & c_3 \end{vmatrix} = a_1 b_2 c_3 - a_1 b_3 c_2 - a_2 b_1 c_3 + a_2 b_3 c_1 + a_3 b_1 c_2 - a_3 b_2 c_1$$

The right-hand side of this equation is called the *expansion* of a 3×3 determinant. It should be noted that the values for both 2×2 and 3×3 determinants are expressed in terms of sums and differences of products of their elements.

EXAMPLES

Find the values of the following 3×3 determinants.

3
$$\begin{vmatrix} 1 & 0 & 2 \\ 4 & 6 & -1 \\ -1 & 0 & -1 \end{vmatrix}$$

SOLUTION. By definition,

$$\begin{vmatrix} 1 & 0 & 2 \\ 4 & 6 & -1 \\ -1 & 0 & -1 \end{vmatrix} = 1 \begin{vmatrix} 6 & -1 \\ 0 & -1 \end{vmatrix} - 4 \begin{vmatrix} 0 & 2 \\ 0 & -1 \end{vmatrix} + (-1) \begin{vmatrix} 0 & 2 \\ 6 & -1 \end{vmatrix}$$

$$= 1[6(-1) - 0(-1)] - 4[0(-1) - 0(2)]$$
$$- 1[0(-1) - 6(2)]$$

$$= 1(-6) - 4(0) - 1(-12) = -6 + 12 = 6$$

$$4 \quad \begin{vmatrix} 3 & 1 & -1 \\ 0 & 2 & 4 \\ -1 & 4 & 2 \end{vmatrix}$$

SOLUTION

$$\begin{vmatrix} 3 & 1 & -1 \\ 0 & 2 & 4 \\ -1 & 4 & 2 \end{vmatrix} = 3 \begin{vmatrix} 2 & 4 \\ 4 & 2 \end{vmatrix} - 0 \begin{vmatrix} 1 & -1 \\ 4 & 2 \end{vmatrix} + (-1) \begin{vmatrix} 1 & -1 \\ 2 & 4 \end{vmatrix}$$

$$= 3(-12) - 0 - 1(6) = -42$$

The 3×3 determinants can be expanded by a cross multiplication scheme that parallels the method illustrated for 2×2 determinants. First, we rewrite the given determinant as

$$\begin{vmatrix} a_1 & b_1 & c_1 \\ a_2 & b_2 & c_2 \\ a_3 & b_3 & c_3 \end{vmatrix} \begin{matrix} a_1 & b_1 \\ a_2 & b_2 \\ a_3 & b_3 \end{matrix}$$

Next, form the products of the diagonal elements as follows:

$$\begin{vmatrix} a_1 & b_1 & c_1 \\ a_2 & b_2 & c_2 \\ a_3 & b_3 & c_3 \end{vmatrix} \begin{matrix} a_1 & b_1 \\ a_2 & b_2 \\ a_3 & b_3 \end{matrix}$$

$$+ \quad + \quad +$$

(positive terms $= a_1b_2c_3 + a_3b_1c_2 + a_2b_3c_1$)

$$\begin{vmatrix} a_1 & b_1 & c_1 \\ a_2 & b_2 & c_2 \\ a_3 & b_3 & c_3 \end{vmatrix} \begin{matrix} a_1 & b_1 \\ a_2 & b_2 \\ a_3 & b_3 \end{matrix}$$

$$- \quad - \quad -$$

(negatives terms $= a_3b_2c_1 + a_1b_3c_2 + a_2b_1c_3$)

so that

$$\begin{vmatrix} a_1 & b_1 & c_1 \\ a_2 & b_2 & c_2 \\ a_3 & b_3 & c_3 \end{vmatrix} = (a_1b_2c_3 + a_3b_1c_2 + a_2b_3c_1) - (a_3b_2c_1 + a_1b_3c_2 + a_2b_1c_3)$$

EXAMPLES

Use the cross-multiplication scheme to evaluate the following 3×3 determinants.

5 $\quad \begin{vmatrix} 1 & 2 & 3 \\ 2 & 4 & 1 \\ 1 & 1 & 1 \end{vmatrix}$

SOLUTION

$$\begin{array}{ccc|cc} 1 & 2 & 3 & 1 & 2 \\ 2 & 4 & 1 & 2 & 4 \\ 1 & 1 & 1 & 1 & 1 \end{array} = [1(4)(1) + 2(1)(1) + 3(2)(1)]$$
$$- [1(4)(3) + 1(1)(1) + 1(2)(2)]$$
$$= (4 + 2 + 6) - (12 + 1 + 4)$$
$$= 12 - 17 = -5$$

6 $\quad \begin{vmatrix} 3 & 0 & -1 \\ -1 & 2 & 0 \\ 2 & -2 & 1 \end{vmatrix}$

SOLUTION

$$\begin{array}{ccc|cc} 3 & 0 & -1 & 3 & 0 \\ -1 & 2 & 0 & -1 & 2 \\ 2 & -2 & 1 & 2 & -2 \end{array}$$

$$= [3(2)(1) + 0(0)(2) + (-1)(-1)(-2)]$$
$$- [2(2)(-1) + (-2)(0)(3) + 1(-1)(0)]$$
$$= (6 + 0 - 2) - (-4 + 0 + 0)$$
$$= 4 - (-4) = 8$$

Although the expansion of higher-order determinants (that is, determinants with rows and columns containing more than three elements) can be accomplished in a similar manner to that of order 2 and order 3, we shall restrict our discussions to 2×2 and 3×3 determinants.

PROBLEM SET 7.6

In problems 1–20, evaluate each determinant.

1 $\quad \begin{vmatrix} 7 & 1 \\ -5 & 3 \end{vmatrix}$

2 $\quad \begin{vmatrix} 0 & 1 \\ 1 & 0 \end{vmatrix}$

3
$$\begin{vmatrix} 7 & 0 \\ -5 & 4 \end{vmatrix}$$

4
$$\begin{vmatrix} 1 & 2 \\ 3 & 5 \end{vmatrix}$$

5
$$\begin{vmatrix} -3 & -1 \\ -5 & \frac{1}{2} \end{vmatrix}$$

6
$$\begin{vmatrix} 2 & 1 \\ -10 & 4 \end{vmatrix}$$

7
$$\begin{vmatrix} 3 & 0 \\ 6 & 0 \end{vmatrix}$$

8
$$\begin{vmatrix} 3 & -2 \\ 3 & 2 \end{vmatrix}$$

9
$$\begin{vmatrix} 1 & 0 \\ 0 & 1 \end{vmatrix}$$

10
$$\begin{vmatrix} 7 & 14 \\ 3 & 6 \end{vmatrix}$$

11
$$\begin{vmatrix} -3 & 1 & 7 \\ 0 & 2 & 6 \\ -4 & 5 & 1 \end{vmatrix}$$

12
$$\begin{vmatrix} 2 & 1 & 1 \\ 9 & 3 & 6 \\ 0 & 0 & 1 \end{vmatrix}$$

13
$$\begin{vmatrix} 1 & 0 & 0 \\ 0 & 1 & 0 \\ 0 & 0 & 1 \end{vmatrix}$$

14
$$\begin{vmatrix} -10 & -1 & 5 \\ -7 & 8 & 2 \\ 3 & -6 & 0 \end{vmatrix}$$

15
$$\begin{vmatrix} -1 & 3 & 5 \\ -7 & 4 & 2 \\ -6 & 2 & 0 \end{vmatrix}$$

16
$$\begin{vmatrix} 3 & -1 & 2 \\ 0 & 1 & -5 \\ 6 & 7 & 4 \end{vmatrix}$$

17
$$\begin{vmatrix} 2 & 3 & 5 \\ 9 & 4 & 2 \\ 11 & -6 & 2 \end{vmatrix}$$

18
$$\begin{vmatrix} 2 & 2 & 2 \\ 3 & 3 & 3 \\ 4 & 4 & 4 \end{vmatrix}$$

19
$$\begin{vmatrix} 2 & -1 & 3 \\ 9 & -7 & 4 \\ 11 & -6 & 2 \end{vmatrix}$$

20
$$\begin{vmatrix} \frac{1}{2} & 4 & 7 \\ 1 & -1 & 2 \\ 3 & 2 & 5 \end{vmatrix}$$

In problems 21–25, solve each equation for x.

21
$$\begin{vmatrix} x & x \\ 5 & 3 \end{vmatrix} = 2$$

22
$$\begin{vmatrix} x+1 & x \\ x & x-2 \end{vmatrix} = -6$$

23
$$\begin{vmatrix} x & 4 & 5 \\ 0 & 1 & x \\ 5 & 2 & 0 \end{vmatrix} = 7$$

24
$$\begin{vmatrix} x & 0 & 1 \\ 4x & 1 & 2 \\ 3x & 1 & 3 \end{vmatrix} = 4$$

25
$$\begin{vmatrix} x & 5 \\ 4 & 2-x \end{vmatrix} = -x^2 + 3$$

26 Show that the equation

$$\begin{vmatrix} 0 & x-2 & x-3 \\ x+2 & 0 & x-4 \\ x+3 & x+4 & 0 \end{vmatrix} = 0 \text{ has 0 as a root.}$$

27 For what values of x is

$$\begin{vmatrix} x & 2 \\ 2 & x \end{vmatrix} > 0?$$

28 What kind of roots does $ax^2 + bx + c = 0$, $a \neq 0$ have if

a) $\begin{vmatrix} b & 4a \\ c & b \end{vmatrix} > 0?$

b) $\begin{vmatrix} b & 4a \\ c & b \end{vmatrix} = 0?$

c) $\begin{vmatrix} b & 4a \\ c & b \end{vmatrix} < 0?$

7.7 Cramer's Rule

The solution to the linear system in the form

$$\begin{cases} a_1 x + b_1 y = c_1 \\ a_2 x + b_2 y = c_2 \end{cases}$$

is the ordered pair (x, y) where

$$x = \frac{c_1 b_2 - c_2 b_1}{a_1 b_2 - a_2 b_1} \quad \text{and} \quad y = \frac{a_1 c_2 - a_2 c_1}{a_1 b_2 - a_2 b_1}$$

Now, by employing the definition of a determinant given in Section 7.6, the denominator $a_1 b_2 - a_2 b_1$ can be expressed as

$$\begin{vmatrix} a_1 & b_1 \\ a_2 & b_2 \end{vmatrix}$$

The numerators of the expressions in the solutions for x and y above can also be written as 2×2 determinants, so that

$$c_1 b_2 - c_2 b_1 = \begin{vmatrix} c_1 & b_1 \\ c_2 & b_2 \end{vmatrix} \quad \text{and} \quad a_1 c_2 - a_2 c_1 = \begin{vmatrix} a_1 & c_1 \\ a_2 & c_2 \end{vmatrix}$$

We may therefore write the solution of the linear system

$$\begin{cases} a_1 x + b_1 y = c_1 \\ a_2 x + b_2 y = c_2 \end{cases}$$

in the form

$$x = \frac{\begin{vmatrix} c_1 & b_1 \\ c_2 & b_2 \end{vmatrix}}{\begin{vmatrix} a_1 & b_1 \\ a_2 & b_2 \end{vmatrix}} \quad \text{and} \quad y = \frac{\begin{vmatrix} a_1 & c_1 \\ a_2 & c_2 \end{vmatrix}}{\begin{vmatrix} a_1 & b_1 \\ a_2 & b_2 \end{vmatrix}}$$

provided that the determinant (which we call D) in the denominator is not equal to zero. Notice that the determinant D is formed from the coefficients of the variables x and y of the system, when expressed in the above form. The determinants in the numerators can be obtained from D by replacing the first or second column by the constants. We denote the determinant in the numerator of x by D_x, and the determinant in the numerator of y by D_y. Hence, the solution of this linear system can be expressed as

$$x = \frac{D_x}{D} \quad \text{and} \quad y = \frac{D_y}{D}$$

where

$$D = \begin{vmatrix} a_1 & b_1 \\ a_2 & b_2 \end{vmatrix} \qquad D_x = \begin{vmatrix} c_1 & b_1 \\ c_2 & b_2 \end{vmatrix} \qquad D_y = \begin{vmatrix} a_1 & c_1 \\ a_2 & c_2 \end{vmatrix}$$

EXAMPLES

Solve the given systems by using determinants and check the results.

1 $\begin{cases} 2x - 3y = 3 \\ x + 4y = 7 \end{cases}$

SOLUTION

$$D = \begin{vmatrix} 2 & -3 \\ 1 & 4 \end{vmatrix} = 2(4) - 1(-3) = 11$$

$$D_x = \begin{vmatrix} 3 & -3 \\ 7 & 4 \end{vmatrix} = 3(4) - 7(-3) = 33$$

$$D_y = \begin{vmatrix} 2 & 3 \\ 1 & 7 \end{vmatrix} = 2(7) - 1(3) = 11$$

so that

$$x = \frac{D_x}{D} = \frac{33}{11} = 3 \quad \text{and} \quad y = \frac{D_y}{D} = \frac{11}{11} = 1$$

Check:

$$2(3) - 3(1) = 6 - 3 = 3$$

$$(3) + 4(1) = 3 + 4 = 7$$

The solution set is $\{(3, 1)\}$.

2 $$\begin{cases} 3x = 4y - 1 \\ y = 2x + 2 \end{cases}$$

SOLUTION. We first write the system in the form

$$\begin{cases} 3x - 4y = -1 \\ -2x + y = 2 \end{cases}$$

Thus,

$$D = \begin{vmatrix} 3 & -4 \\ -2 & 1 \end{vmatrix} = 3(1) - (-2)(-4) = -5$$

$$D_x = \begin{vmatrix} -1 & -4 \\ 2 & 1 \end{vmatrix} = -1(1) - 2(-4) = 7$$

$$D_y = \begin{vmatrix} 3 & -1 \\ -2 & 2 \end{vmatrix} = 3(2) - (-1)(-2) = 4$$

so that

$$x = \frac{D_x}{D} = -\frac{7}{5} \quad \text{and} \quad y = \frac{D_y}{D} = -\frac{4}{5}$$

The solution set is $\{(-\frac{7}{5}, -\frac{4}{5})\}$.

3    $\begin{cases} 2x + y = 1 \\ 4x + 2y = 3 \end{cases}$

SOLUTION

$$D = \begin{vmatrix} 2 & 1 \\ 4 & 2 \end{vmatrix} = 2(2) - 4(1) = 0$$

$$D_x = \begin{vmatrix} 1 & 1 \\ 3 & 2 \end{vmatrix} = 1(2) - 3(1) = -1$$

$$D_y = \begin{vmatrix} 2 & 1 \\ 4 & 3 \end{vmatrix} = 2(3) - 4(1) = 2$$

Since $x = D_x/D = -1/0$ and $y = D_y/D = 2/0$, the system is inconsistent and its solution set is empty.

Solving systems of linear equations by this method is known as the *Cramer's rule method*. It can be shown that this rule can be used to solve any system of n linear equations in n variables (n a natural number). However, we shall only extend this rule to include systems of three equations in three variables.

Consider the following system of linear equations:

$$\begin{cases} a_1x + b_1y + c_1z = d_1 \\ a_2x + b_2y + c_2z = d_2 \\ a_3x + b_3y + c_3z = d_3 \end{cases}$$

In this case D, D_x, D_y, and D_z are given by

$$D = \begin{vmatrix} a_1 & b_1 & c_1 \\ a_2 & b_2 & c_2 \\ a_3 & b_3 & c_3 \end{vmatrix} \qquad D_x = \begin{vmatrix} d_1 & b_1 & c_1 \\ d_2 & b_2 & c_2 \\ d_3 & b_3 & c_3 \end{vmatrix}$$

$$D_y = \begin{vmatrix} a_1 & d_1 & c_1 \\ a_2 & d_2 & c_2 \\ a_3 & d_3 & c_3 \end{vmatrix} \qquad D_z = \begin{vmatrix} a_1 & b_1 & d_1 \\ a_2 & b_2 & d_2 \\ a_3 & b_3 & d_3 \end{vmatrix}$$

so that

$$x = \frac{D_x}{D} \qquad y = \frac{D_y}{D} \qquad z = \frac{D_z}{D}$$

Again, note that the elements of the determinant D are the coefficients of the variables x, y, and z in a system of the form above. The determinant D_x is obtained from D by replacing the column of the coefficients of x by the corresponding constants on the opposite side of the equation; similarly, D_y and D_z can be obtained from the determinant D by replacing the column of coefficients of each variable by the corresponding constant terms on the side of the equations opposite the variables y and z, respectively.

EXAMPLE

4   Use Cramer's rule to find the solution set of the given system.

$$\begin{cases} x + y + z = 2 \\ 2x - y + z = 0 \\ x + 2y - z = 4 \end{cases}$$

SOLUTION

$$D = \begin{vmatrix} 1 & 1 & 1 \\ 2 & -1 & 1 \\ 1 & 2 & -1 \end{vmatrix}$$

$$= 1 \begin{vmatrix} -1 & 1 \\ 2 & -1 \end{vmatrix} - 2 \begin{vmatrix} 1 & 1 \\ 2 & -1 \end{vmatrix} + 1 \begin{vmatrix} 1 & 1 \\ -1 & 1 \end{vmatrix}$$

$$= 1(-1) - 2(-3) + 1(2) = 7$$

Since $D \neq 0$, we proceed to find D_x, D_y, and D_z as follows:

$$D_x = \begin{vmatrix} 2 & 1 & 1 \\ 0 & -1 & 1 \\ 4 & 2 & -1 \end{vmatrix}$$

$$= 2\begin{vmatrix} -1 & 1 \\ 2 & -1 \end{vmatrix} - 0\begin{vmatrix} 1 & 1 \\ 2 & -1 \end{vmatrix} + 4\begin{vmatrix} 1 & 1 \\ -1 & 1 \end{vmatrix}$$

$$= 2(-1) - 0(-3) + 4(2) = 6$$

$$D_y = \begin{vmatrix} 1 & 2 & 1 \\ 2 & 0 & 1 \\ 1 & 4 & -1 \end{vmatrix}$$

$$= 1\begin{vmatrix} 0 & 1 \\ 4 & -1 \end{vmatrix} - 2\begin{vmatrix} 2 & 1 \\ 4 & -1 \end{vmatrix} + 1\begin{vmatrix} 2 & 1 \\ 0 & 1 \end{vmatrix}$$

$$= 1(-4) - 2(-6) + 1(2) = 10$$

$$D_z = \begin{vmatrix} 1 & 1 & 2 \\ 2 & -1 & 0 \\ 1 & 2 & 4 \end{vmatrix}$$

$$= 1\begin{vmatrix} -1 & 0 \\ 2 & 4 \end{vmatrix} - 2\begin{vmatrix} 1 & 2 \\ 2 & 4 \end{vmatrix} + 1\begin{vmatrix} 1 & 2 \\ -1 & 0 \end{vmatrix}$$

$$= 1(-4) - 2(0) + 1(2) = -2$$

Then we have

$$x = \frac{D_x}{D} = \frac{6}{7}, \quad y = \frac{D_y}{D} = \frac{10}{7}, \quad z = \frac{D_z}{D} = -\frac{2}{7}$$

Hence, the solution set is $\{(\frac{6}{7}, \frac{10}{7}, -\frac{2}{7})\}$.

PROBLEM SET 7.7

In problems 1–22, use Cramer's rule to find the solution set of each system.

1 $\begin{cases} 2x - y = 0 \\ x + y = 1 \end{cases}$

2 $\begin{cases} -3x + y = 3 \\ -2x - y = -5 \end{cases}$

3 $\begin{cases} x + y = 0 \\ x - y = 0 \end{cases}$

4 $\begin{cases} 3x + y = 1 \\ 9x + 3y = -4 \end{cases}$

5
$$\begin{cases} x + y = 30 \\ 2x - 2y = 25 \end{cases}$$

6
$$\begin{cases} 7x - 9y = 13 \\ 5x + 2y = 10 \end{cases}$$

7
$$\begin{cases} 3x + 7y = 16 \\ 2x + 5y = 13 \end{cases}$$

8
$$\begin{cases} 7x + 4y = 1 \\ 9x + 4y = 3 \end{cases}$$

9
$$\begin{cases} 8x - 2y = 52 \\ 3x - 5y = 45 \end{cases}$$

10
$$\begin{cases} 5x + 11y = 102 \\ x - 3y + 16 = 0 \end{cases}$$

11
$$\begin{cases} x + y + z = 6 \\ 3x - y + 2z = 7 \\ 2x + 3y - z = 5 \end{cases}$$

12
$$\begin{cases} x + y + z = 9 \\ 27x + 9y + 3z = 93 \\ 8x + 4y + 2z = 36 \end{cases}$$

13
$$\begin{cases} 2x_1 - x_2 + x_3 = 3 \\ -x_1 + 2x_2 - x_3 = 1 \\ 3x_1 + x_2 + 2x_3 = -1 \end{cases}$$

14
$$\begin{cases} x + y + 2z = 4 \\ x + y - 2z = 0 \\ x - y = 0 \end{cases}$$

15
$$\begin{cases} 2x_1 - 3x_2 = 4 \\ x_1 + x_2 - 2x_3 = 1 \\ x_1 - x_2 - x_3 = 5 \end{cases}$$

16
$$\begin{cases} x + y + z = 4 \\ x - y + 2z = 8 \\ 2x + y - z = 3 \end{cases}$$

17
$$\begin{cases} 2x + 3y + z = 6 \\ x - 2y + 3z = -3 \\ 3x + y - z = 8 \end{cases}$$

18
$$\begin{cases} 3x + 2z = 6 - 2y \\ x - 5y + 6z = 2 \\ 6x - 8z = 12 \end{cases}$$

19
$$\begin{cases} x + y = 1 \\ y + z = 9 \\ x + z + 6 = 0 \end{cases}$$

20
$$\begin{cases} x_1 - x_2 + x_3 = 3 \\ 2x_1 + 3x_2 - 2x_3 = 5 \\ 3x_1 + x_2 - 4x_3 = 12 \end{cases}$$

21
$$\begin{cases} x + y + z = 6 \\ 3x - y + 2z = 7 \\ 2x + 3y - z = 5 \end{cases}$$

22
$$\begin{cases} 7x + y + 3z = 52 \\ 4x - 5y + 6z = 13 \\ x + 15y - 9z = 52 \end{cases}$$

7.8 Applications of Systems of Linear Equations

We have seen how to work applied problems containing one variable. In this section we shall see that certain word problems which contain more than one unknown quantity can be easily solved if more than one variable is introduced. However, before the problem can be completely solved, the number of equations formed must be equal to the number of variables used.

EXAMPLES

1   The sum of two numbers is 16 and their difference is 4. Find the numbers.

SOLUTION.   Let x and y represent the two numbers. Thus, the two equations which express the given conditions are

$$\begin{cases} x + y = 16 \\ x - y = 4 \end{cases}$$

Solving this system for x by the addition method, we have $2x = 20$ or $x = 10$, so that $10 + y = 16$ or $y = 6$. Hence, the two numbers are 10 and 6.

2   Jamie is 1 year more than twice Gus's age. The sum of their ages is 7. How old is each?

SOLUTION.   Let x years = Jamie's age and y years = Gus's age. The equations are

$$\begin{cases} x = 2y + 1 \\ x + y = 7 \end{cases}$$

Solving this system by substituting $2y + 1$ for x, in the second equation we have $2y + 1 + y = 7$ or $3y = 6$ or $y = 2$, so that $x = 2(2) + 1$ or $x = 5$. Hence, Jamie is 5 years old and Gus is 2 years old.

3   Charlotte has $1.95 in dimes and quarters. She has 2 more dimes than quarters. How many of each type of coin does she have?

SOLUTION.   Let x = number of dimes and y = number of quarters. We begin with $0.10x + 0.25y = 1.95$, so that $10x + 25y = 195$ and $x - y = 2$. Solving this system, we have

$$\begin{array}{r} 10x + 25y = 195 \\ 25x - 25y = 50 \\ \hline 35x = 245 \\ x = 7 \end{array}$$

and $7 - y = 2$ or $y = 5$. That is, there are 7 dimes and 5 quarters.

4   Roy invested $40,000, part of it at $4\frac{1}{2}$ percent interest and the other at 6 percent. The interest from the whole investment is $2,130. How much did he invest at each rate?

SOLUTION.   Let $x =$ amount invested at $4\frac{1}{2}$ percent and $y =$ amount invested at 6 percent. First, $0.045x + 0.06y = 2,130$, so that

$$\begin{cases} 45x + 60y = 2,130,000 \\ x + y = 40,000 \end{cases}$$

Using Cramer's rule, we have

$$D = \begin{vmatrix} 45 & 60 \\ 1 & 1 \end{vmatrix} = -15$$

$$D_x = \begin{vmatrix} 2,130,000 & 60 \\ 40,000 & 1 \end{vmatrix} = -270,000$$

$$D_y = \begin{vmatrix} 45 & 2,130,000 \\ 1 & 40,000 \end{vmatrix} = -330,000$$

so that

$$x = \frac{D_x}{D} = \frac{-270,000}{-15} = 18,000$$

and

$$y = \frac{D_y}{D} = \frac{-330,000}{-15} = 22,000$$

Hence, Roy invested $18,000 at $4\frac{1}{2}$ percent and $22,000 at 6 percent.

5   Suppose that 4 pounds of a mixture of three different types of candy is made. One of the three types costs $1.40 per pound, the second type costs $1.30 per pound, and the third type costs $1.50 per pound. Assume that the total cost of the 4-pound mixture is $5.50 and that the mixture contains twice as much of the $1.30 candy as of the $1.40 candy. How many pounds of each type is in the mixture?

SOLUTION.   Let x represent the number of pounds of the $1.40 candy, y the number of pounds of the $1.30 candy, and z the number of pounds of the $1.50 candy. The values of x, y, and z are found by solving the system

$$\begin{cases} x + y + z = 4 \\ 1.40x + 1.30y + 1.50z = 5.50 \\ y = 2x \end{cases}$$

which is equivalent to the system

$$\begin{cases} x + y + z = 4 \\ 14x + 13y + 15z = 55 \\ 2x - y = 0 \end{cases}$$

Using Cramer's rule, we have

$$D = \begin{vmatrix} 1 & 1 & 1 \\ 14 & 13 & 15 \\ 2 & -1 & 0 \end{vmatrix} = 5 \qquad D_x = \begin{vmatrix} 4 & 1 & 1 \\ 55 & 13 & 15 \\ 0 & -1 & 0 \end{vmatrix} = 5$$

$$D_y = \begin{vmatrix} 1 & 4 & 1 \\ 14 & 55 & 15 \\ 2 & 0 & 0 \end{vmatrix} = 10 \qquad D_z = \begin{vmatrix} 1 & 1 & 4 \\ 14 & 13 & 55 \\ 2 & -1 & 0 \end{vmatrix} = 5$$

so that

$$x = \frac{D_x}{D} = \frac{5}{5} = 1 \qquad y = \frac{D_y}{D} = \frac{10}{5} = 2 \qquad z = \frac{D_z}{D} = \frac{5}{5} = 1$$

Hence, the mixture contains 1 pound of $1.40 candy, 2 pounds of $1.30 candy, and 1 pound of $1.50 candy.

PROBLEM SET 7.8

In problems 1–20, use systems of linear equations to solve each problem.

1 The sum of two numbers is 12. If one of the numbers is multiplied by 5 and the other is multiplied by 8, the sum of the products is 75. Find the numbers.

2 Find two numbers such that twice the first plus 5 times the second is 20 and 4 times the first less 3 times the second is 14.

3 In a number the tens' digit is 5 greater than the units' digit and the number is 63 greater than the sum of its digits. Find the number.

4 The sum of the digits of a two-place number is 13 and the number is 1 less than 10 times the difference of the digits, the units' digit being greater. Find the number.

5 A farmer sold two grades of wheat, one grade at $2.25 per bushel and another grade at $2.50 per bushel. How many bushels of each kind did he sell if he received $7,200 for 3,000 bushels?

6 The rowing team of a certain school can row downstream at the rate of 7 miles per hour and upstream at the rate of 3 miles per hour. Find the team's rate in still water and the rate of the current.

7 A man invested a total of $4,000 in securities. Part was invested at 5 percent, and the rest was invested at 6 percent. His annual income from both investments was $230. What amount was invested at each rate?

8 Wilma has 11 coins, consisting of nickels and dimes. She has 3 more dimes than nickels. How many of each does she have if the total amount is 90 cents?

9 Joan is 3 years older than Doreen. Four years ago she was twice as old as Doreen. How old is each now?

10 Two cars start from the same point, at the same time, and travel in opposite directions. One car travels at a rate that is 10 miles per hour faster than the other. After 3 hours, they are 330 miles apart. How fast is each traveling?

11 A car that traveled at an average rate of speed of 40 miles per hour went 35 miles farther than one that went 55 miles per hour. The slower car traveled 2 hours longer than the faster one. How long did each car travel?

12 The perimeter of an isosceles triangle is 26 inches. The base is 2 inches longer than one of the equal sides of the triangle. Find the length of each of the sides of the triangle.

13 A rectangular field is 40 rods longer than it is wide and the length of the fence around it is 560 rods. What are the dimensions of the field?

14 Assume that the sum of three numbers is 5. The first number is twice the second number and the third number equals the sum of the other two numbers. Find the values of the three numbers.

15 A watch, chain, and ring together cost $225. The watch cost $50 more than the chain and the ring cost $25 more than the watch and chain together. What was the cost of each?

16   Mr. Brown, Mr. Jones, and Mr. Smith bought a grocery store for $100,000. If Mr. Smith invested twice as much as Mr. Brown, and Mr. Jones invested $20,000 more than Mr. Smith, how much did each invest?

17   Margaret bought three different kinds of common stocks for $20,000, one paying a 6 percent dividend, one paying a 7 percent dividend, and the other paying an 8 percent dividend. If the sum of the dividends from the 6 percent and the 7 percent stocks is $940, and the sum of the dividends from the 6 percent and the 8 percent stocks is $720, how much did she invest in each kind of stock?

18   How much cream containing 25 percent butterfat and milk containing 10 percent butterfat must be mixed to make 30 gallons of cream containing 20 percent butterfat?

19   The perimeter of a triangle is 45 inches. One side is twice as long as the shortest side, and the third side is 5 inches longer than the shortest side. Find the lengths of the three sides of the triangle.

20   A chemist has in his laboratory the same acid in two strengths. Six parts of the first mixed with four parts of the second gives a mixture 86 percent pure, and four parts of the first mixed with six parts of the second gives a mixture 84 percent pure. What is the percent of purity of each acid?

REVIEW PROBLEM SET

In problems 1–8, graph the point and indicate which quadrant, if any, contains the point.

1 $(3,2)$ **2** $(-2,-3)$

3 $(-4,3)$ **4** $(1,-1)$

5 $(0,2)$ **6** $(3,0)$

7 $(2,-3)$ **8** $(-2,2)$

In problems 9–12, find the distance between the two points.

9 $(3,1)$ and $(15,6)$ **10** $(4,-3)$ and $(-1,7)$

11 $(-2,1)$ and $(-5,5)$ **12** $(a,-b)$ and $(-a,b)$

In problems 13 and 14, use the distance formula to show that the triangles determined by the points are isosceles triangles.

13 $(2,1)$, $(9,3)$, and $(4,-6)$ **14** $(0,2)$, $(-1,4)$, and $(-3,3)$

In problems 15–18, sketch the graph of each linear equation.

15 $y = 2x - 3$ **16** $2y + 3 = 2x$

17 $2x - 5y = 10$ **18** $-3x + 6y + 5 = 0$

In problems 19–22, find the x intercept and the y intercept of each line and sketch the graphs.

19 $2x + y = 4$ **20** $y = -x + 2$

21 $-3x + 4y - 12 = 0$ **22** $\dfrac{x}{2} + \dfrac{y}{3} = 1$

In problems 23–26, find the slope of the line containing each pair of points.

23 $(7,3)$ and $(2,-2)$ **24** $(11,12)$ and $(7,4)$

25 $(-3,5)$ and $(1,3)$ **26** $(3,0)$ and $(0,-2)$

In problems 27–34, find the equation of the line that satisfies each condition.

27 Its slope is 3 and it contains the point $(1,1)$.

28 Its slope is -2 and it contains the point $(2,-3)$.

29 Its slope is 0 and it contains the point $(-3,-2)$.

30 It contains the points $(4,1)$ and $(2,3)$.

31 Its slope is 2 and its y intercept is 4.

32 Its slope is -3 and its y intercept is -2.

33 It is parallel to the line $3x + 2y + 2 = 0$ and it contains the point $(-3,-1)$.

34 It is perpendicular to the line $y = -\frac{1}{2}x + 7$ and contains the point $(2,1)$.

In problems 35–38, use the graph to determine whether each system is dependent, inconsistent, or independent. If the system is independent, find its solution set by the method of substitution.

35 $\begin{cases} y = -2x + 2 \\ y = x - 4 \end{cases}$ **36** $\begin{cases} y = 5x + 2 \\ 10x - 2y + 4 = 0 \end{cases}$

37 $\begin{cases} 3x + 2y = 1 \\ 3x + 2y = 3 \end{cases}$ **38** $\begin{cases} x + y = 4 \\ 2x - y = 8 \end{cases}$

In problems 39–44, find the solution set of each system of equations by the addition–subtraction method.

39 $\begin{cases} x - y = 3 \\ 2x + y = 3 \end{cases}$ 　　　　　　**40** $\begin{cases} 5x + 2y = 3 \\ 2x - 3y = 5 \end{cases}$

41 $\begin{cases} x - y + 2z = 0 \\ 3x + y + z = 2 \\ 2x - y + 5z = 5 \end{cases}$ 　　　　**42** $\begin{cases} 3x + 2y - z = -4 \\ x - y + 2z = 13 \\ 5x + 3y - 4z = -15 \end{cases}$

43 $\begin{cases} \dfrac{2}{x} + \dfrac{1}{y} = 8 \\[2mm] \dfrac{3}{x} - \dfrac{1}{y} = 7 \end{cases}$ 　　　　**44** $\begin{cases} \dfrac{1}{x} + \dfrac{2}{y} + \dfrac{3}{z} = 0 \\[2mm] \dfrac{2}{x} - \dfrac{3}{y} + \dfrac{1}{z} = 2 \\[2mm] \dfrac{5}{x} - \dfrac{4}{y} - \dfrac{2}{z} = -3 \end{cases}$

In problems 45–48, evaluate each determinant.

45 $\begin{vmatrix} 3 & 4 \\ 1 & 5 \end{vmatrix}$ 　　　　　　**46** $\begin{vmatrix} 4 & 2 \\ 1 & -1 \end{vmatrix}$

47 $\begin{vmatrix} 4 & 2 & 1 \\ 5 & 7 & 1 \\ 6 & 2 & 3 \end{vmatrix}$ 　　　　**48** $\begin{vmatrix} -1 & 4 & 3 \\ 7 & 1 & 4 \\ 1 & 3 & 5 \end{vmatrix}$

49　Use Cramer's rule to solve the systems of equations in problems 39–42.

50　Solve for x:

$$\begin{vmatrix} x - 1 & -3 \\ 2 & x + 3 \end{vmatrix} = 6$$

In problems 51–54, use systems of equations to solve each problem.

51　A college mailed 160 letters, some requiring 13 cents postage and the rest requiring 18 cents postage. If the total bill was \$25.80, find the number sent at each rate.

52　The specific gravity of an object is defined to be its weight in air divided by its loss of weight when submerged in water. An object made of gold, of specific gravity 16, and silver, of specific gravity 10.8, weighs 8 grams in air and 7.3 grams when submerged in water.

How many grams of gold and how many grams of silver are in the object?

53 A coin collection containing nickels, dimes, and quarters consisted of 35 coins. If there are twice as many nickels as quarters, and one fourth as many dimes as nickels, find the numbers of each kind of coin in the collection.

54 Find the values of a, b, and c of the equation $y = ax^2 + bx + c$ whose graph contains the points $(1,3)$, $(3,5)$, and $(-1,9)$.

CHAPTER 8

Logarithms

The exponents and their properties that were discussed in Chapter 5 play a key role in the study of *logarithms.* In this chapter we shall see that logarithms are essentially exponents. Logarithms provide an extremely useful means to simplify certain computations.

8.1 Logarithms

As we have said, logarithms are basically exponents and have the properties of exponents. To illustrate this idea, let us consider some examples. In the equation $2^3 = 8$, 2 is the base, 3 is the exponent, and 8 is the result of "raising 2 to the power 3." Suppose that it were desirable to describe this example by highlighting the role of the exponent 3 in $2^3 = 8$. We could say that "3 is the exponent to which the base 2 must be raised in order to get 8."

Similarly, for $5^{-2} = \frac{1}{25}$, we could say that "-2 is the exponent to which the base 5 must be raised in order to get $\frac{1}{25}$"; for $100^{3/2} = 1,000$, "$\frac{3}{2}$ is the exponent to which the base 100 must be raised in order to get 1,000."

In general, if $b^y = x$, for $b > 0$, $b \neq 1$, "y is the exponent to which the base b must be raised in order to get x." In this context we refer to y as "the logarithm of x to the base b."

These statements are written more briefly as follows:

$$\log_2 8 = \quad 3 \quad \text{is equivalent to} \quad 2^3 = 8$$

$$\log_5 \tfrac{1}{25} = -2 \quad \text{is equivalent to} \quad 5^{-2} = \tfrac{1}{25}$$

$$\log_{100} 1,000 = \quad \tfrac{3}{2} \quad \text{is equivalent to} \quad 100^{3/2} = 1,000$$

Hence, a logarithm is an exponent, and to formalize these statements, we introduce the following definition.

DEFINITION 1   LOGARITHMS

If $b > 0$, $b \neq 1$, then $y = \log_b x$ (reads "y equals the logarithm of x to the base b") is equivalent to $b^y = x$. b is called the *base* of the logarithm.

EXAMPLES

1 Write each of the following exponential equations as an equivalent logarithmic statement.

a) $3^2 = 9$ b) $10^0 = 1$ c) $\sqrt[3]{27} = 3$

SOLUTION. By Definition 1 of logarithms, we know that $b^y = x$ is equivalent to $y = \log_b x$; hence:

a) $3^2 = 9$ is equivalent to $2 = \log_3 9$
b) $10^0 = 1$ is equivalent to $0 = \log_{10} 1$
c) $\sqrt[3]{27} = 27^{1/3} = 3$ is equivalent to $\frac{1}{3} = \log_{27} 3$

2 Write each of the following logarithmic equations as an equivalent exponential statement.

a) $\log_{10} 10 = 1$ b) $\log_{1/2} 4 = -2$ c) $\log_{1/2} \frac{1}{4} = 2$

SOLUTION. By Definition 1 of logarithms, we have $y = \log_b x$ is equivalent to $b^y = x$; so that:

a) $\log_{10} 10 = 1$ is equivalent to $10^1 = 10$
b) $\log_{1/2} 4 = -2$ is equivalent to $(\frac{1}{2})^{-2} = 4$
c) $\log_{1/2} \frac{1}{4} = 2$ is equivalent to $(\frac{1}{2})^2 = \frac{1}{4}$

3 Use Definition 1 of logarithms to find the value of x in each of the following equations.

a) $\log_x 9 = 2$ b) $\log_8 x = -\frac{2}{3}$

SOLUTION

a) $\log_x 9 = 2$ is equivalent to $x^2 = 9$. Although the solution of this latter equation is $x = \pm 3$, we can only use $x = 3$ because, according to Definition 1, the base of a logarithm must be a positive number.
b) $\log_8 x = -\frac{2}{3}$ is equivalent to $8^{-2/3} = x$, so that $x = \frac{1}{4}$.

In order to find the values of logarithms of positive real numbers with positive bases, we need to solve exponential equations. To solve exponential equations, we make use of the following property:

$$\text{If } a^x = a^y \quad \text{then} \quad x = y \quad \text{for} \quad a \neq 0 \text{ or } a \neq 1$$

EXAMPLES

Find the solution sets of the following exponential equations.

4 $2^x = 16$

SOLUTION. Since $2^x = 16$, and $16 = 2^4$, then $2^x = 2^4$. This equation holds if $x = 4$. The solution set is $\{4\}$.

5 $5^{-x} = 25$

SOLUTION.   Since $5^{-x} = 25$ and $25 = 5^2$, then $5^{-x} = 5^2$. This equation holds if $-x = 2$, that is, if $x = -2$. The solution set is $\{-2\}$.

6 $11^x = (121)^{2/3}$

SOLUTION.   We can solve this equation if we can express the number $(121)^{2/3}$ as a power of 11. Since $(121)^{2/3} = (11^2)^{2/3} = 11^{4/3}$, we have $11^x = (121)^{2/3} = (11)^{4/3}$, so that $x = \frac{4}{3}$. The solution set is $\{\frac{4}{3}\}$.

The logarithm of a number to a given base can be found by solving the equivalent exponential equation. The following examples will illustrate this.

EXAMPLES

Find the values of the following logarithms.

7 $\log_3 81$

SOLUTION.   Let $\log_3 81 = x$, so that $3^x = 81$. Then, $3^x = 3^4$ or $x = 4$, and so $\log_3 81 = 4$.

8 $\log_8 32$

SOLUTION.   Let $\log_8 32 = x$, so that $8^x = 32$. Then $(2^3)^x = 2^5$, so that $3x = 5$ or $x = \frac{5}{3}$. Therefore, $\log_8 32 = \frac{5}{3}$.

9 $\log_{4/9} \frac{27}{8}$

SOLUTION.   Let $\log_{4/9} \frac{27}{8} = x$, so that $(\frac{4}{9})^x = \frac{27}{8}$. Thus, $[(\frac{2}{3})^2]^x = (\frac{2}{3})^{-3}$ or $(\frac{2}{3})^{2x} = (\frac{2}{3})^{-3}$, so that $2x = -3$ or $x = -\frac{3}{2}$. Therefore, $\log_{4/9} \frac{27}{8} = -\frac{3}{2}$.

10 $\log_b b$

SOLUTION.   Let $\log_b b = x$, so that $b^x = b$ or $x = 1$. Therefore, $\log_b b = 1$.

Some equations involving logarithms can be solved by changing them into equivalent polynomial equations by making use of the definition of logarithms and the following property:

If $\log_a x = \log_a y$ then $x = y$

EXAMPLE

11 Find the solution set of the equation $\log_2(5x - 3) = 5$.

SOLUTION.   Write the equation $\log_2(5x - 3) = 5$ in exponential form, so that $5x - 3 = 2^5 = 32$ or $5x = 35$ or $x = 7$. The solution set is $\{7\}$.

PROBLEM SET 8.1

In problems 1–16, write each equation in logarithmic form.

1   $5^3 = 125$

2   $4^4 = 256$

3   $10^5 = 100,000$

4   $49^{0.5} = 7$

5   $4^{-2} = \frac{1}{16}$

6   $(\frac{1}{3})^{-2} = 9$

7   $6^{-2} = \frac{1}{36}$

8   $2^{-3} = 0.125$

9   $\sqrt{9} = 3$

10   $\sqrt[5]{32} = 2$

11   $(100)^{-3/2} = 0.001$

12   $(\frac{1}{8})^{-2/3} = 4$

13   $7^0 = 1$

14   $15^0 = 1$

15   $x^3 = a$

16   $\pi^t = z$

In problems 17–30, write each equation in exponential form.

17   $\log_9 81 = 2$

18   $\log_6 36 = 2$

19   $\log_{27} 9 = \frac{2}{3}$

20   $\log_{27} \frac{1}{9} = -\frac{2}{3}$

21   $\log_{10} 0.001 = -3$

22   $\log_{10} \frac{1}{10} = -1$

23   $\log_{1/3} 9 = -2$

24   $\log_{36} 216 = \frac{3}{2}$

25   $\log_{10} 4.35 = 0.64$

26   $\log_{10} 9.14 = 0.96$

27   $\log_{\sqrt{16}} 2 = \frac{1}{2}$

28   $\log_{4/9} \frac{27}{8} = -\frac{3}{2}$

29   $\log_x 1 = 0$

30   $\log_x 2 = 4$

In problems 31–40, find the value of the variable in each equation.

31   $\log_5 N = 2$

32   $\log_b 36 = 2$

33   $\log_2 N = 5$

34   $\log_{10} N = -4$

35   $\log_b 81 = 4$

36   $\log_b 64 = -3$

37   $\log_{\sqrt{2}} N = -6$

38   $\log_b 16 = -\frac{4}{3}$

39   $\log_b 3 = \frac{1}{5}$

40   $\log_{\sqrt{3}} N = 4$

In problems 41–50, find the solution set of each equation.

41 $2^{2x} = 32$ **42** $5^{2x-1} = 25$

43 $3^{x-5} = 27$ **44** $3^{2x-1} = \frac{1}{81}$

45 $2^{3x} = 16$ **46** $5^{3x-1} = 125$

47 $3^{4x-5} = 81$ **48** $9^{-2-x} = 243$

49 $5^{2x} = 625$ **50** $3^{x-1} = 1$

In problems 51–70, find the value of each logarithm.

51 $\log_2 64$ **52** $\log_4 \frac{1}{16}$

53 $\log_9 3$ **54** $\log_4 8$

55 $\log_9 \frac{1}{3}$ **56** $\log_{1/2} \frac{1}{8}$

57 $\log_{1/9} \frac{1}{81}$ **58** $\log_2 \frac{1}{32}$

59 $\log_5 \frac{1}{125}$ **60** $\log_{10} 0.00001$

61 $\log_3 9\sqrt{3}$ **62** $\log_2 4\sqrt{2}$

63 $\log_{10} \frac{1}{10,000}$ **64** $\log_7 343$

65 $\log_3 729$ **66** $\log_5 5$

67 $\log_7 1$ **68** $\log_b 1$

69 $\log_6 \frac{1}{216}$ **70** $\log_b b^3$

In problems 71–78, find the solution set of each equation.

71 $\log_5 25 = x$ **72** $\log_6 216 = x$

73 $\log_{10} 10,000 = x$ **74** $\log_{49} 7 = x$

75 $\log_{10}(x + 1) = 1$ **76** $\log_5(2x - 7) = 0$

77 $\log_4(3x + 1) = 2$ **78** $\log_7(2x - 3) = 2$

8.2 Properties of Logarithms

Since logarithms are exponents, the properties of logarithms are derived from the properties of exponents. Using the definition of logarithm in Section 8.1, we can establish these properties. Each of these properties can be shown to follow the properties of exponents; however, we shall defer such proofs and simply state and accept them for now as being true.

THEOREM 1   PROPERTIES OF LOGARITHMS

Suppose that M, N, and b, $b \neq 1$, are positive numbers and that r is any real number. Then:

1    $\log_b MN = \log_b M + \log_b N$

2    $\log_b \dfrac{M}{N} = \log_b M - \log_b N$

3    $\log_b N^r = r \log_b N$

We illustrate these properties by the following examples.

EXAMPLES

Use the properties of logarithms in Theorem 1 to work the following examples.

1   Write each of the following expressions as a single logarithm.
    a) $\log_3 15 + \log_3 13$ b) $\log_5 216 - \log_5 54$
    c) $2 \log_6 7 - \log_6 14$ d) $3 \log_a x + 2 \log_a y,\ x > 0,\ y > 0$

SOLUTION

    a) $\log_3 15 + \log_3 13 = \log_3 (15)(13)$ (Property 1)
                         $= \log_3 195$

    b) $\log_5 216 - \log_5 54 = \log_5 \frac{216}{54}$ (Property 2)
                         $= \log_5 4$

    c) $2 \log_6 7 - \log_6 14 = \log_6 7^2 - \log_6 14$ (Property 3)
                         $= \log_6 49 - \log_6 14$
                         $= \log_6 \frac{49}{14}$ (Property 2)
                         $= \log_6 \frac{7}{2}$

    d) $3 \log_a x + 2 \log_a y = \log_a x^3 + \log_a y^2$ (Property 3)
                         $= \log_a x^3 y^2$ (Property 1)

2   Write each of the following expressions as a sum, difference, or multiple of logarithms of simpler quantities.
    a) $\log_3 5(7)$ b) $\log_8 \frac{17}{5}$

    c) $\log_6 \dfrac{2^5}{3^4}$ d) $\log_a (x^7 y^{11}),\ x > 0,\ y > 0$

SOLUTION

    a) $\log_3 5(7) = \log_3 5 + \log_3 7$ (Property 1)

    b) $\log_8 \frac{17}{5} = \log_8 17 - \log_8 5$ (Property 2)

c) $\log_6 \dfrac{2^5}{3^4} = \log_6 2^5 - \log_6 3^4$ \hspace{2cm} (Property 2)

$= 5 \log_6 2 - 4 \log_6 3$ \hspace{2cm} (Property 3)

d) $\log_a(x^7 y^{11}) = \log_a x^7 + \log_a y^{11}$ \hspace{1cm} (Property 1)

$= 7 \log_a x + 11 \log_a y$ \hspace{1cm} (Property 3)

3 Evaluate $\log_b b^p$.

SOLUTION

$\log_b b^p = p \log_b b$ \hspace{1cm} (Property 1)
$= p \cdot 1 = p$

4 Use $\log_b 2 = 0.35$, $\log_b 3 = 0.55$, and Theorem 1 to find the value of each expression.

a) $\log_b 6$ \hspace{4cm} b) $\log_b \frac{2}{3}$

c) $\log_b 8$ \hspace{4cm} d) $\log_b \sqrt{\frac{2}{3}}$

e) $\log_b 24$ \hspace{3.5cm} f) $\dfrac{\log_b 2}{\log_b 3}$

SOLUTION

a) $\log_b 6 = \log_b 2(3) = \log_b 2 + \log_b 3 = 0.35 + 0.55 = 0.90$

b) $\log_b \frac{2}{3} = \log_b 2 - \log_b 3 = 0.35 - 0.55 = -0.20$

c) $\log_b 8 = \log_b 2^3 = 3 \log_b 2 = 3(0.35) = 1.05$

d) $\log_b \sqrt{\frac{2}{3}} = \log_b (\frac{2}{3})^{1/2} = \frac{1}{2} \log_b \frac{2}{3} = \frac{1}{2}(\log_b 2 - \log_b 3)$
$= \frac{1}{2}(0.35 - 0.55) = -0.10$

e) $\log_b 24 = \log_b 8(3) = \log_b (2^3)(3) = \log_b 2^3 + \log_b 3$
$= 3 \log_b 2 + \log_b 3 = 3(0.35) + 0.55$
$= 1.05 + 0.55 = 1.60$

f) $\dfrac{\log_b 2}{\log_b 3} = \dfrac{0.35}{0.55} = \dfrac{7}{11} = 0.64$

The basic properties of logarithms along with Definition 1 of Section 8.1 can be used to solve equations involving logarithms. This is illustrated as follows.

EXAMPLES

Find the solution sets of the following equations.

5 $\log_3(x + 1) + \log_3(x + 3) = 1$

SOLUTION

$$\log_3(x + 1) + \log_3(x + 3) = \log_3[(x + 1)(x + 3)] = 1 \qquad \text{(Why?)}$$

Using Definition 1 of logarithms, we have $x^2 + 4x + 3 = 3^1 = 3$, which is equivalent to $x^2 + 4x = 0$. The solution set of $x^2 + 4x = 0$ is $\{0, -4\}$; however, -4 does not satisfy the original equation (why?), so that $x = 0$ is the only solution. Hence, the solution set is $\{0\}$.

6 $\log_4(x + 3) - \log_4 x = 1$

SOLUTION

$$\log_4(x + 3) - \log_4 x = \log_4 \frac{x + 3}{x} = 1$$

so that $(x + 3)/x = 4^1 = 4$, from which it follows that $x = 1$. Hence, the solution set is $\{1\}$.

Proofs of the Properties of Logarithms

Now we present the proofs of these properties in Theorem 1.

PROOF OF PROPERTY 1.   To prove $\log_b MN = \log_b M + \log_b N$, let $x = \log_b M$ and $y = \log_b N$, so that $b^x = M$ and $b^y = N$ or $MN = b^x \cdot b^y = b^{x+y}$. This statement is written in logarithmic form as $\log_b MN = x + y$. Therefore, $\log_b MN = \log_b M + \log_b N$.

PROOF OF PROPERTY 2.   To prove $\log_b(M/N) = \log_b M - \log_b N$, let $x = \log_b M$ and $y = \log_b N$, so that $b^x = M$ and $b^y = N$ or $M/N = b^x/b^y = b^{x-y}$. This statement is written in logarithmic form as $\log_b(M/N) = x - y$. Therefore, $\log_b(M/N) = \log_b M - \log_b N$.

PROOF OF PROPERTY 3.   To prove $\log_b N^r = r \log_b N$, let $y = \log_b N$; then $N = b^y$, so that $N^r = (b^y)^r = b^{yr}$. To write the statement in logarithmic form, we have $\log_b N^r = yr$. Therefore, $\log_b N^r = r \log_b N$.

PROBLEM SET 8.2

In problems 1–10, use the properties of logarithms to write each expression as a single logarithm. Assume that all variables represent positive real numbers.

1 $\log_5 \frac{3}{4} + \log_5 \frac{3}{2}$

2 $\log_2 \frac{5}{7} + \log_2 \frac{14}{70}$

3 $\log_7 \frac{3}{8} - \log_7 \frac{9}{4}$

4 $\log_3 \frac{3}{4} - \log_3 \frac{5}{8}$

5 $\log_5 \frac{24}{25} - \log_5 \frac{3}{50}$

6 $2\log_7 \frac{2}{3} - 3\log_7 \frac{4}{5}$

7 $2\log_3 \frac{4}{5} + 3\log_3 \frac{1}{2}$

8 $3\log_5 \frac{3}{4} + 2\log_5 \frac{1}{5}$

9 $\log_a \frac{x}{y} + \log_a \frac{y^2}{3x}$

10 $\log_b \frac{x^2}{y} - \log_b \frac{x^4}{y^2}$

In problems 11–20, use the properties of logarithms to write each expression as a sum, difference, or multiple of logarithms of simpler quantities. Assume that all variables represent positive real numbers.

11 $\log_2 5(7)$

12 $\log_3 xy$

13 $\log_4 3^4 \cdot 5^2$

14 $\log_5 3^7 \cdot 2^6$

15 $\log_2 \frac{x}{y}$

16 $\log_5 \frac{x^3}{y^2}$

17 $\log_4 \left(\frac{2^5}{5^3} \right)$

18 $\log_5 \sqrt[6]{3^2 \cdot 5^3}$

19 $\log_3 \sqrt[5]{xy}$

20 $\log_5 \sqrt[7]{x^2 y^3}$

In problems 21–25, evaluate each expression.

21 $\log_7 7^4$

22 $\log_x x^4$

23 $\log_{y^2} y^8$

24 $\log_{z^3} z^9$

25 $\log_{x^9} x^{27}$

26   Use Theorem 1 to show that $\log_b(xyz) = \log_b x + \log_b y + \log_b z$, where b, x, y, and z are positive numbers, with $b \neq 1$.

In problems 27–38, use $\log_{10} 2 = 0.3010$ and $\log_{10} 3 = 0.4771$ to find the value of each expression.

27 $\log_{10} 6$

28 $\log_{10} 12$

29 $\log_{10} 18$

30 $\log_{10} 24$

31 $\log_{10} \frac{3}{2}$

32 $\log_{10} \frac{1}{3}$

33 $\log_{10} 5$

34 $\log_{10} 32$

35 $\log_{10} 81$

36 $\log_{10} \sqrt[5]{2}$

37 $\log_{10} 0.5$

38 $\log_{10} 60$

In problems 39–42, find the solution set of each equation.

39 $\log_3 x + \log_3 (x - 6) = \log_3 7$

40 $\log_4 x + \log_4 (6x + 11) = 1$

41 $\log_{10}(x + 1) - \log_{10} x = 1$

42 $\log_{10}(x^2 - 9) - \log_{10}(x + 3) = 2$

8.3   Common Logarithms

The two logarithmic bases used most often for purposes of computation are base 10 and base e, where e, an irrational number, is approximately equal to 2.718. Logarithms with base 10 are called *common logarithms*. By convention, we usually do not write the base 10 when using logarithmic notation, and simply use $\log x$ as the abbreviated way of writing $\log_{10} x$. Logarithms with base e are called *natural logarithms*, and the natural logarithm, $\log_e x$, is usually written $\ln x$.

Logarithms — Base 10

Up to now, we have considered the definition and the properties of logarithms to any valid base. To make use of these properties in performing numerical calculations, we restrict ourselves to base 10. If we consider the positive values of x that are expressible in integral powers of 10, then $\log x$ can be evaluated without any difficulty. In particular, $\log 10 = 1$, $\log 100 = 2$, $\log 1{,}000 = 3$, $\log 0.1 = -1$, and $\log 0.01 = -2$. Indeed, $\log 10^n = n$ for any integer n. If x is a positive number that is not expressible in integral powers of 10, Table I in Appendix C will be used to find $\log x$. Such a procedure is discussed as follows.

If $x = 5{,}340$, then using scientific notation (Section 5.1), we can represent x as 5.34×10^3, so that

$$\log 5{,}340 = \log(5.34 \times 10^3) = \log 5.34 + \log 10^3 = \log 5.34 + 3$$

Hence, determining $\log 5{,}340$ has been reduced to finding $\log 5.34$.

Suppose that $x = 0.000234$; then

$$\log 0.000234 = \log(2.34 \times 10^{-4}) = \log 2.34 + \log 10^{-4}$$

$$= \log 2.34 + (-4)$$

and our problem is reduced to finding $\log 2.34$.

To generalize the procedure suggested by the two examples, let x be any positive number that can be represented in scientific notation as $x = s \times 10^n$, where $1 \le s < 10$ and n is an integer; then

$$\log x = \log(s \times 10^n) = \log s + \log 10^n = \log s + n$$

The form $\log x = \log s + n$ is called the *standard form* of $\log x$, where $\log s$ is called the *mantissa* of $\log x$ and n is called the *characteristic* of $\log x$. Notice that for $1 \le s < 10$, it follows that $\log 1 \le \log s < \log 10$; that is, $0 \le \log s < 1$. In other words, the mantissa is either 0 or a positive number between 0 and 1.

Hence, the task of determining the value of log x is reduced to determining log s, where s is always between 1 and 10. However, the approximate values of log s can be determined from the *common log tables* (Table 1 in Appendix C).

Since the value of the characteristic of the logarithm of a number x is the value of n (in scientific notation, substitute $s \times 10^n$ for x), the rule for finding the characteristic is that for finding n. This rule is equivalent to the following statement: The characteristic of the logarithm of a positive number x is equal to the number of places the decimal point of x is moved from its position when x is written in scientific notation. The characteristic is positive or zero for numbers greater than or equal to 1, and negative for numbers less than 1.

EXAMPLES

Use Table 1 in Appendix C to find the value of the given logarithms. Indicate the characteristic.

1 a) log 53,900 b) log 385 c) log 28.4

SOLUTION. To find the characteristics of the logarithms, we arrange them in tabular form:

x	Scientific notation: $s \times 10^n$	Places decimal point is moved	Characteristic of x
53,900	5.39×10^4	4	4
385	3.85×10^2	2	2
28.4	2.84×10^1	1	1

Thus, using Table I, we have
a) log 53,900 = log 5.39 + 4 = 0.7316 + 4 = 4.7316
b) log 385 = log 3.85 + 2 = 0.5855 + 2 = 2.5855
c) log 28.4 = log 2.84 + 1 = 0.4533 + 1 = 1.4533

2 a) log 4.06 b) log 0.628 c) log 0.0035

SOLUTION. The characteristics are illustrated in the following tabular form:

x	Scientific notation: $s \times 10^n$	Places decimal point is moved	Characteristic of x
4.06	4.06×10^0	0	0
0.628	6.28×10^{-1}	-1	-1
0.0035	3.5×10^{-3}	-3	-3

Therefore, using Table I, we have

a) $\log 4.06 = \log 4.06 + 0 = 0.6085 + 0 = 0.6085$
b) $\log 0.628 = \log 6.28 + (-1) = 0.7980 + (-1) = -0.2020$
c) $\log 0.0035 = \log 3.50 + (-3) = 0.5441 + (-3) = -2.4559$

Antilogarithms

The process of finding the number corresponding to a given logarithm is the reverse of the above process. The number found is called the *antilogarithm* of the given logarithm. For instance, given number r we can determine the value of x such that $\log x = r$. The number x is called the *antilogarithm of r* and is abbreviated by *antilog r*. Thus, if $\log x = r$, then $x = $ antilog r.

As was the case for finding values of the logarithm, it is easy to determine x for some values of r, but not for others. For example, if $\log x = -2$, then $x = 0.01$ (why?); if $\log x = 5$, then $x = 100,000$ (why?); however, if $\log x = 4.4969$, then the value of x is not as easy to determine.

The antilog of 4.4969, or the solution of the equation $\log x = 4.4969$, can be determined as follows. Write $\log x = 4.4969$ in standard form, that is, as the sum of a number between 0 and 1 and an integer — the given logarithm $4.4969 = 4 + 0.4969$ has characteristic 4 and mantissa 0.4969. Using Table I in Appendix C we can find a value s such that $\log s = 0.4969$, so that $s = 3.14$. Since the characteristic of $\log x$ is 4, then

$$\log x = 0.4969 + 4$$
$$= \log 3.14 + 4 = \log 3.14 + \log 10^4$$
$$= \log(3.14 \times 10^4) = \log 31,400$$

so that $x = 31,400$.

EXAMPLES

Find the values of the given antilogarithms.

3   Antilog 2.7210

SOLUTION.   Let $x = $ antilog 2.7210, so that $\log x = 2.7210$.

$$\log x = 0.7210 + 2$$

Using Table I, we find that $\log 5.26 = 0.7210$, so that

$$\log x = \log 5.26 + 2 = \log 5.26 + \log 10^2$$
$$= \log (5.26 \times 10^2) = \log 526$$

Therefore, $x = $ antilog $2.7210 = 526$.

4 Antilog $[0.5105 + (-3)]$

SOLUTION.   Let $x = $ antilog $[0.5105 + (-3)]$, so that

$$\log x = 0.5105 + (-3)$$

Using Table I, we find that $\log 3.24 = 0.5105$, so that

$$\log x = \log 3.24 + (-3) = \log 3.24 + \log 10^{-3}$$
$$= \log (3.24 \times 10^{-3}) = \log 0.00324$$

Therefore, $x = $ antilog $[0.5105 + (-3)] = 0.00324$.

5 Antilog (-2.0804)

SOLUTION.   Let $x = $ antilog(-2.0804), so that $\log x = -2.0804$. Since the mantissa must always be positive, then $-2.0804 = (-2.0804 + 3) - 3 = 0.9196 + (-3)$. Thus, $\log x = 0.9196 + (-3)$. Using Table I, we find that $\log 8.31 = 0.9196$, so that

$$\log x = \log 8.31 + (-3) = \log 8.31 + \log 10^{-3}$$
$$= \log (8.31 \times 10^{-3}) = \log 0.00831$$

Therefore, $x = $ antilog$(-2.0804) = 0.00831$.

Interpolation

The logarithms and antilogarithms that we have just computed were special in the sense that we were able to find the necessary numbers in the logarithm table (Table I, Appendix C). However, this will not always be the case. Suppose that we want to find log 1.234 or antilog 0.2217. We see that the corresponding values are not available in Table I. The problem can be resolved by locating a number that is between two consecutive tabulated values, or entries, of Table I. For this, a process called *interpolation* is frequently employed. For example, to determine log 1.234 by using interpolation, we find, from Table I,

$$\log 1.23 = 0.0899 \quad \text{and} \quad \log 1.24 = 0.0934$$

These numbers were selected because $1.23 < 1.234 < 1.24$. The use of interpolation depends on the following property: If x, y, and z are positive numbers such that $z < x < y$, then $\log z < \log x < \log y$. It also depends on the assumption that for small differences in numbers, the change in the mantissas are proportional to the change in numbers. (The assumption does not always hold, but the results are sufficiently accurate for our purposes, especially since the difference between consecutive tabulated mantissas are small.)

Using the property above, we conclude that $\log 1.23 < \log 1.234 < \log 1.24$, because $1.23 < 1.234 < 1.24$. The tabular difference (the difference between two consecutive tabulated mantissas) is $0.0934 - 0.0899 = 0.0035$; also, $1.234 - 1.23 = 0.004$ and $1.24 - 1.23 = 0.01$.

Since $\log 1.234 > \log 1.23$, there is a "correction" d such that $\log 1.234 = \log 1.23 + d$. d can be determined by using the above assumption, so that $d/0.0035 = 0.004/0.01$. Then,

$$d = (0.0035 \times 0.004)/0.01 = 0.0014$$

Therefore,

$$\log 1.234 = \log 1.23 + d$$
$$= 0.0899 + 0.0014 = 0.0913$$

Once the process is understood, it is possible to condense it further by using an abbreviated scheme, which is illustrated as follows:

EXAMPLE

6   Use interpolation to find $\log 52.33$.

SOLUTION.   Since $\log 52.3 < \log 52.33 < \log 52.4$, we have

$$0.1 \left[\begin{array}{c} \log 52.4 = 1.7193 \\ 0.03 \left[\begin{array}{c} \log 52.33 = \quad ? \\ \log 52.3 = 1.7185 \end{array} \right] d \end{array} \right] 0.0008$$

Now, $0.03/0.1 = d/0.0008$, so that $d = (0.03 \times 0.0008)/0.01 = 0.0002$. Therefore,

$$\log 52.33 = \log 52.3 + d$$
$$= 1.7185 + 0.0002$$
$$= 1.7187$$

This process can also be used to interpolate antilogarithms. For example, to find antilog 0.2217 by interpolation, we let $x =$ antilog 0.2217, so that $\log x = 0.2217$. The mantissa 0.2217 is not entered in Table I; hence, we select the two consecutive tabulated mantissas 0.2201 and 0.2227, since $0.2201 < 0.2217 < 0.2227$. Using Table I, we find that

$$\log 1.66 = 0.2201 \qquad \text{and} \qquad \log 1.67 = 0.2227$$

so that the number x must lie between 1.66 and 1.67. That is, $1.66 < x < 1.67$. The tabular difference is $0.2227 - 0.2201 = 0.0026$. Also, $0.2217 - 0.2201 = 0.0016$ and $1.67 - 1.66 = 0.01$. Then, $x =$ antilog $0.2217 = 1.66 + d$, where d is determined as follows:

$$\frac{d}{0.01} = \frac{0.0016}{0.0026}$$

so that $d = (0.01 \times 0.0016)/0.0026 = 0.006$. Therefore,

$$x = \text{antilog } 0.2217 = 1.66 + d$$
$$= 1.66 + 0.006 = 1.666$$

EXAMPLE

7 Use interpolation to find antilog (-1.7186).

SOLUTION. Let $x =$ antilog (-1.7186), so that $\log x = -1.7186$. Since the mantissa must be positive, we write

$$-1.7186 = (-1.7186 + 2) - 2$$
$$= 0.2814 + (-2)$$

Thus, $\log x = 0.2814 + (-2)$. Since $[0.2810 + (-2)] < [0.2814 + (-2)] < [0.2833 + (-2)]$,

$$0.0001 \left[d \left[\begin{array}{l} \text{—} \log 0.0192 = 0.2833 + (-2) \text{—} \\ \text{—} \log x \qquad = 0.2814 + (-2) \text{—} \\ \text{—} \log 0.0191 = 0.2810 + (-2) \text{—} \end{array} \right] 0.0004 \right] 0.0023$$

Now, $d/0.0001 = 0.0004/0.0023$, so that $d = (0.0001 \times 0.0004)/0.0023 = 0.00002$. Therefore,

$$x = \text{antilog}[0.2814 + (-2)]$$
$$= \text{antilog}[0.2810 + (-2)] + d$$
$$= 0.0191 + 0.00002 = 0.01912$$

PROBLEM SET 8.3

In problems 1–18, use Table I in Appendix C to find the value of each common logarithm.

1	log 317	2	log 3,910
3	log 53,400	4	log 348,000
5	log 17.1	6	log 5
7	log 6.81	8	log 7.59
9	log 1.18	10	log 9.81
11	log 0.315	12	log 0.712
13	log 0.0713	14	log 0.00512
15	log 0.000178	16	log 0.00081
17	log 0.000007	18	log 0.00000137

In problems 19–28, use Table I in Appendix C and interpolation to find the value of each common logarithm.

19	log 1,545	20	log 333.3
21	log 79.56	22	log 62.95
23	log 5.312	24	log 1.785
25	log 0.5725	26	log 0.7125
27	log 0.05342	28	log 0.006487

In problems 29–48, use Table I in Appendix C to find the value of each antilogarithm.

29	antilog 0.4133	30	antilog 0.4871
31	antilog 1.2945	32	antilog 1.7825
33	antilog 2.7427	34	antilog 2.9795
35	antilog 3.5514	36	antilog 3.8993
37	antilog[0.7348 + (−1)]	38	antilog[0.8082 + (−2)]
39	antilog[0.8993 + (−3)]	40	antilog[0.5922 + (−4)]
41	antilog(−1.6289)	42	antilog(−2.4157)
43	antilog(−3.4881)	44	antilog(−4.8153)
45	antilog(−0.1574)	46	antilog(−0.3251)

47 antilog(−2.1475)   **48** antilog(−3.1884)

In problems 49–56, use Table I in Appendix C and interpolation to find the value of each antilogarithm.

49 antilog 0.1452   **50** antilog 1.5375

51 antilog 1.5425   **52** antilog(−1.1275)

53 antilog[0.2259 + (−2)]   **54** antilog[0.4950 + (−2)]

55 antilog (−4.4625)   **56** antilog (−4.565)

8.4 Computation with Logarithms

Nowadays, computations of numerical expressions such as 65×0.32, $27.6/0.65$, or $(0.0532)^7$ can be accomplished with the aid of calculators or by the use of computers. However, since such devices are not always available, we use logarithms and their properties as a tool in performing such calculations. In this case we transform more complicated operations of multiplication, division, raising a number to a power (positive or negative), and extraction of roots of a number to simpler operations of addition, subtraction, and multiplication of their corresponding logarithms.

For convenience, we restate the properties of logarithms to base 10.

1 $\log MN = \log M + \log N$

2 $\log \dfrac{M}{N} = \log M - \log N$

3 $\log N^r = r \log N$

4 If $\log M = \log N$, then $M = N$.

These techniques are illustrated as follows.

EXAMPLES

In Examples 1–3, use the properties of logarithms above and Table I in Appendix C to evaluate the given logarithms.

1 $\log(65 \times 0.325)$

SOLUTION

$$\log(65 \times 0.325) = \log 65 + \log 0.325 \qquad \text{(Property 1)}$$

$$= 1.8129 + (0.5119 - 1) = 1.3248$$

2   $\log\dfrac{27.6}{0.652}$

SOLUTION

$$\log\frac{27.6}{0.652} = \log 27.6 - \log 0.652 \qquad\qquad\text{(Property 2)}$$

$$= 1.4409 - (0.8142 - 1) = 1.6267$$

3   $\log(0.0532)^7$

SOLUTION

$$\log(0.0532)^7 = 7\log(0.0532) \qquad\qquad\text{(Property 3)}$$

$$= 7(0.7259 - 2) = 5.0813 - 14 = -8.9187$$

In Examples 4–11, use the properties of logarithms and Table I in Appendix C to compute the following expressions.

4   53.7×0.83

SOLUTION.   Let $x = 53.7 \times 0.83$, so that

$$\log x = \log(53.7 \times 0.83) = \log 53.7 + \log 0.83$$
$$= 1.7300 + (0.9191 - 1) = 1.6491$$

Hence, $x = $ antilog $1.6491 = 44.6$.

5   $\dfrac{0.837}{0.00238}$

SOLUTION.   Let $x = 0.837/0.00238$, so that

$$\log x = \log\left(\frac{0.837}{0.00238}\right) = \log 0.837 - \log 0.00238$$

$$= (0.9227 - 1) - (0.3766 - 3) = 2.5461$$

Hence, $x = $ antilog $2.5461 = 351.7$.

6   $0.0742 \times 6.13 \times 0.312$

SOLUTION.   Let $x = 0.0742 \times 6.13 \times 0.312$, so that

$$\log x = \log(0.0742 \times 6.13 \times 0.312)$$
$$= \log 0.0742 + \log 6.13 + \log 0.312$$

From here on we use the following scheme as an aid to simplify the use of logarithms.

$$\begin{aligned}
\log 0.0742 &= 0.8704 - 2 \\
+ \log 6.13 \ \ &= 0.7875 \\
+ \log 0.312 \ &= 0.4942 - 1 \\
\hline
\log x &= 2.1521 - 3 = 0.1521 - 1
\end{aligned}$$

Hence, $x = $ antilog$[0.1521 + (-1)] = 1.419 \times 10^{-1} = 0.1419$.

7   $\dfrac{289 \times 3.47}{0.0987}$

SOLUTION.   Let $x = (289 \times 3.47)/0.0987$, so that

$$\log x = \log\left(\frac{289 \times 3.47}{0.0987}\right) = \log 289 + \log 3.47 - \log 0.0987$$

$$\begin{aligned}
\log 289 &= \ \ 2.4609 \\
+ \ \ \log 3.47 &= \ \ 0.5403 \\
\hline
&= \ \ 3.0012 \\
- \log 0.0987 &= -(0.9943 - 2) \\
\hline
\log x &= \ \ 2.0069 + 2 = 4.0069
\end{aligned}$$

Hence, $x = $ antilog $4.0069 = 10{,}160$.

8   $\dfrac{(134)^5 (0.35)^8}{(49)^3}$

SOLUTION.   Let $x = (134)^5 (0.35)^8/(49)^3$, so that

$$\log x = 5 \log 134 + 8 \log 0.35 - 3 \log 49 \qquad \text{(Why?)}$$

$$\begin{aligned}
5 \log 134 &= 5(2.1271) \ \ &= 10.6355 \\
+ \ 8 \log 0.35 &= 8(0.5441 - 1) = &\ \ 4.3528 - 8 \\
\hline
& &\ \ 6.9883 \\
- \ \ 3 \log 49 &= -3(1.6902) \ \ &= -5.0706 \\
\hline
\log x &= &\ \ 1.9177
\end{aligned}$$

Hence, $x = $ antilog $1.9177 = 82.74$.

9   $\sqrt[5]{17}$

SOLUTION.   Let $x = \sqrt[5]{17} = 17^{1/5}$, so that

$$\log x = \log 17^{1/5} = \tfrac{1}{5}\log 17$$

$$= \tfrac{1}{5}(1.2304) = 0.2461$$

Hence, $x =$ antilog $0.2461 = 1.762$.

10 $\sqrt[3]{\dfrac{83.5}{99.6}}$

SOLUTION. Let $x = (83.5/99.6)^{1/3}$, so that

$$\log x = \log\left(\frac{83.5}{99.6}\right)^{1/3} = \tfrac{1}{3}(\log 83.5 - \log 99.6)$$

$$\begin{array}{ll}\log 83.5 = & 1.9217 \\ -\ \log 99.6 = & -1.9983 \\ \hline & = -0.0766\end{array}$$

$\log x = \tfrac{1}{3}(-0.0766) = -0.0255$

To find antilog(-0.0255), we write $-0.0255 = (-0.0255 + 1) -$
$= 0.9745 - 1$, so that $x =$ antilog$[0.9745 + (-1)] = 9.43 \times 10^{-1} = 0.9\cdots$

11 $(39.1)^{-2/3} \cdot (0.642)^{-0.2}$

SOLUTION. Let $x = (39.1)^{-2/3} \cdot (0.642)^{-0.2}$, so that

$$\log x = \log[(39.1)^{-2/3} \cdot (0.642)^{-0.2}]$$

$$= -\tfrac{2}{3}\log 39.1 - 0.2\log 0.642$$

$$\begin{array}{lll}-\tfrac{2}{3}\log 39.1 = -\tfrac{2}{3}(1.5922) & = -1.0614 \\ -\ 0.2\log 0.642 = -0.2(0.8075 - 1) = & 0.0385 \\ \hline \log x = & -1.0229\end{array}$$

To find antilog(-1.0229), we write $-1.0229 = (-1.0229 + 2) -$
$= 0.9771 - 2$, so that $x =$ antilog$[0.9771 + (-2)] = 9.49 \times 10^{-2} = 0.09\cdots$

The properties of logarithms and Table I can also be used to so
exponential equations. This is illustrated as follows.

EXAMPLE

12 Use logarithms to solve $5^x = 7$.

SOLUTION. Since $5^x = 7$, we have $\log 5^x = \log 7$, so that $x\log 5 = \log$
Hence, $x = \log 7/\log 5 = 0.8451/0.6990 = 1.209$.

PROBLEM SET 8.4

In problems 1–12, use the properties of logarithms and Table I in Appendix C to evaluate each logarithm.

1 $\log(32.9 \times 0.372)$

2 $\log(6.92 \times 0.231)$

3 $\log(0.713 \times 0.00218)$

4 $\log(0.0127 \times 0.00318)$

5 $\log \dfrac{28.9}{0.657}$

6 $\log \dfrac{327}{0.00125}$

7 $\log \dfrac{0.00718}{0.712}$

8 $\log \dfrac{0.127}{0.00513}$

9 $\log(0.618)^3$

10 $\log(53.2)^4$

11 $\log \sqrt[3]{0.00521}$

12 $\log \sqrt[5]{41.2}$

In problems 13–50, use the properties of logarithms and Table I in Appendix C to compute each expression. Express answers to three significant digits.

13 5.72×32.9

14 0.681×8.97

15 92.3×3.37

16 3.75×9.66

17 0.00321×0.715

18 0.00279×0.315

19 $0.0391 \times 0.731 \times 0.219$

20 $36.5 \times 5.92 \times 8.47$

21 $98.3 \times 7.437 \times 0.00165$

22 $0.00819 \times 0.912 \times 785$

23 $45.6 \times 0.357 \times 0.932$

24 $0.00356 \times 22.7 \times 39.2$

25 $\dfrac{83.4}{20.7}$

26 $\dfrac{5.02}{81.6}$

27 $\dfrac{0.117}{6.39}$

28 $\dfrac{0.901}{1.03}$

29 $\dfrac{83.4 \times 2.09}{0.0642 \times 62.8}$

30 $\dfrac{32.9 \times 0.0472}{71.6 \times 1.25}$

31 $\dfrac{5.73 \times 0.927}{5.95 \times 3.91}$

32 $\dfrac{7.04 \times 1.06 \times 5.31}{3.04 \times 1.73}$

33 $(3.26)^4$

34 $(2.35)^5$

35 $(0.792)^6$

36 $(1.83)^7$

37 $(23.9)^{1/4}$

38 $(6.03)^{3/8}$

39 $\sqrt[3]{99.1}$

40 $\sqrt[7]{0.035}$

41 $\sqrt[5]{0.0763}$ **42** $\sqrt[4]{0.293}$

43 $\left(\dfrac{7.51}{3.27}\right)^3$ **44** $\sqrt[3]{\dfrac{85.6}{6.41}}$

45 $\sqrt[6]{\dfrac{7.16}{3.52}}$ **46** $\dfrac{\sqrt[5]{263} \times 0.781}{(1.53)^{3/5}}$

47 $\dfrac{(21.8)^3 \cdot (39.1)^2}{(85.1)^{0.3} \cdot (671)^{2.5}}$ **48** $0.684 \times \sqrt[5]{\dfrac{32.7}{9.15}}$

49 $(0.0824)^{-3/5} \cdot (1.47)^{-4}$ **50** $(0.706)^{-1/3} \cdot (2.46)^{-1.5}$

In problems 51–60, use the properties of logarithms and Table I in Appendix C to find the solution set of each equation.

51 $2^x = 7$ **52** $3^{2x} = 5$

53 $4^{-x} = 13$ **54** $3^{5x-1} = 29$

55 $7^{2x-1} = 6$ **56** $5^{7-3x} = 9$

57 $3^{5-4x} = 8^{x-4}$ **58** $4^{x+1} = 6^{8-3x}$

59 $13^{8x-1} = 5^{1-3x}$ **60** $5^{3-7x} = 7^{1-5x}$

8.5 Applications of Logarithms

Logarithms are a handy tool in the solution of problems involving geometric, engineering, and business applications. In this section we shall consider a few examples of these various applications.

EXAMPLES

1   Find the volume of a right cylindrical gas reservoir if its radius $r = 21.3$ feet and its height $h = 79.6$ feet.

SOLUTION.   From geometry the formula for the volume V of a cylinder of radius r and height h (see Appendix B, page 434) is given by $V = \pi r^2 h$ ($\pi = 3.14$), so that $V = (3.14) \cdot (21.3)^2 (79.6)$. Using logarithms to calculate the volume, we have

$$\log V = \log[(3.14) \cdot (21.3)^2 \cdot (79.6)]$$

$$= \log 3.14 + \log(21.3)^2 + \log 79.6$$

$$= \log 3.14 + 2 \log 21.3 + \log 79.6$$

$$= 0.4969 + 2(1.3284) + 1.9009$$

$$= 0.4969 + 2.6568 + 1.9009 = 5.0546$$

Therefore, $V = \text{antilog}(5.0546) = 113,400$ cubic feet.

2   Boyle's Law for adiabatic expansion of air is given by the equation $PV^{1.4} = C$, where P is the pressure, V is the volume, and C is a constant. At a certain instant the volume is 75.2 cubic inches and $C = 12,600$ pounds per inch. Find the pressure P.

SOLUTION. $PV^{1.4} = C$, so that $P = C/V^{1.4} = CV^{-1.4}$. Then

$$P = (12,600)(75.2)^{-1.4}$$

Using logarithms to calculate P, we have

$$\log P = \log[(12,600) \cdot (75.2)^{-1.4}] = \log 12,600 + \log(75.2)^{-1.4}$$
$$= \log 12,600 - 1.4 \log 75.2 = 4.1004 - 1.4(1.8762) = 1.4737$$

Therefore, $P = \text{antilog}(1.4737) = 29.76$ pounds per square inch.

If P dollars represents the amount invested at an annual interest rate r, then the amount A accumulated after n years when the interest is compounded t times per year is given by

$$A = P\left(1 + \frac{r}{t}\right)^{nt}$$

This formula is the basic formula of compound interest.

EXAMPLES

3   Karen invested \$1,000 in a bank at a yearly interest rate of 6 percent compounded every 4 months. How much money is accumulated in her account after 4 years?

SOLUTION.   Using the formula above, $A = P(1 + r/t)^{nt}$, where $P = 1,000$, $r = 0.06$, $n = 4$, and $t = 3$. We have $A = 1,000(1 + \frac{0.06}{3})^{4(3)} = 1,000(1.02)^{12}$, so that

$$\log A = \log[1,000(1.02)^{12}] = \log 1,000 + \log(1.02)^{12}$$
$$= \log 1,000 + 12 \log(1.02) = 3 + 12(0.0086) = 3 + 0.1032$$
$$= 3.1032$$

Therefore, $A = \text{antilog}(3.1032) = 1,268$. Hence, the money accumulated in Karen's account after 4 years is \$1,268.

4   In how many years would the \$1,000 in Example 3 double?

SOLUTION. Again, we use the formula $A = P(1 + r/t)^{nt}$, where A $= 2,000$, $P = 1,000$, $r = 0.06$, and $t = 3$, so that $2,000 = 1,000(1.02)^{3n}$ or $(1.02)^{3n} = 2$. Using logarithms, we have $\log(1.02)^{3n} = \log 2$, so that $3n \log(1.02) = \log 2$ or $3n = \log 2/\log 1.02$. Therefore,

$$n = \frac{1}{3}\left(\frac{\log 2}{\log 1.02}\right) = \frac{1}{3}\left(\frac{0.3010}{0.0086}\right) = 11.7 \qquad \text{(approximately)}$$

Therefore, the money will be doubled in 11.7 years.

An *annuity* is a series of equal payments separated by equal intervals of time. The interval of time between successive payments is called the *payment period*. One common type of annuity problem is to find the total value to which the series of payments will accumulate at the end of the term of the annuity, assuming that these payments can earn interest at a prescribed rate. An annuity is said to be *simple* if the interval between successive payments coincides with the conversion period at which interest is being paid. If the size of each payment is P dollars and the interest rate per conversion period is r, then the accumulated value S of all payments at the time of the nth payment is given by the equation

$$S = P\left[\frac{(1 + r)^n - 1}{r}\right]$$

EXAMPLE

5   A teacher deposits \$100 at the end of each month in the credit union. If the interest rate is 6 percent converted monthly, find the amount in the account after 4 years.

SOLUTION. The annuity consists of $n = 12(4) = 48$ payments. The interest rate per conversion period is $r = \frac{1}{12}(0.06) = \frac{1}{2}$ percent $= 0.005$. The periodic payment $P = 100$. Then,

$$S = 100\left[\frac{(1 + 0.005)^{48} - 1}{0.005}\right] = 20,000[(1.005)^{48} - 1]$$

$$= 20,000(1.005)^{48} - 20,000$$

To evaluate the expression $20,000(1.005)^{48}$, we use logarithms. Let $x = 20,000(1.005)^{48}$, so that

$$\log x = \log[20,000(1.005)^{48}] = \log 20,000 + \log(1.005)^{48}$$

$$= \log 20,000 + 48 \log(1.005) = 4.3010 + 48(0.0022)$$

$$= 4.3010 + 0.1056 = 4.4066$$

Therefore, $x =$ antilog $4.4066 = 25,510$, and so $S = 25,510.00 - 20,000 = 5,510$. The amount in the account after 4 years is $5,510.

The *present value A* of an annuity is the amount of the annuity discounted to the beginning of the first payment period and is given by the equation

$$A = P\left[\frac{1 - (1 + r)^{-n}}{r}\right]$$

where P, r, and n are the same as above.

EXAMPLE

6   A young couple bought a new home by making a down payment of $8,000 and contracting for monthly payments of $275 for 25 years. If the interest rate on the mortgage is 9 percent, what price did the young couple pay for the home? (Assume that the first mortgage payment occurs 1 month after the down payment has been made.)

SOLUTION.   Let A be the amount of mortgage on the home, $n = 25 \times 12 = 300$ and $r = \frac{1}{12}(0.09) = 0.0075$, so that

$$A = P\left[\frac{1 - (1 + r)^{-n}}{r}\right]$$

$$= 275\left[\frac{1 - (1 + 0.0075)^{-300}}{0.0075}\right]$$

$$= 36,666.67[1 - (1.0075)^{-300}]$$

$$= 36,666.67 - 36,666.67(1.0075)^{-300}$$

To evaluate the expression $36,666.67(1.0075)^{-300}$, we use logarithms. Let $x = 36,666.67(1.0075)^{-300}$, so that

$$\log x = \log[36,666.67(1.0075)^{-300}]$$

$$= \log(36,666.67) + \log(1.0075)^{-300}$$

$$= \log(36,666.67) - 300 \log(1.0075)$$

$$= 4.5643 - 300(0.0032) = 4.5643 - 0.9600 = 3.6043$$

Therefore, $x =$ antilog $3.6043 = 4,021.00$, and so $A = 36,666.67 - 4,021.00 = 32,645.67$. The price of the home was $32,645.67 + 8,000.00 = \$40,645.67$.

PROBLEM SET 8.5

In problems 1–18, use logarithms in solving each problem.

1   How long is a side of a square whose area is 2.13 square centimeters?

2   Find the volume of a right circular cone of radius 29.3 feet and height 67.8 feet. (*Hint:* $V = \frac{1}{3}\pi r^2 h$.)

3   Find the volume of a cylindrical coke can of radius 2.1 inches and height 7.4 inches.

4   The formula $A = \sqrt{s(s-a)(s-b)(s-c)}$ gives the area of a triangle, where a, b, and c denote the lengths of the sides and $s = (a+b+c)/2$. Find the area of the triangle whose sides are 42.3, 28.7, and 37.1 meters.

5   Find the volume of a sphere of radius 1.67 inches. (*Hint:* $V = \frac{4}{3}\pi r^3$.)

6   The area A of a cross section of a chimney is given by the formula $A = 0.06 p h^{-1}$, where p is the number of pounds of coal burned each hour and h is the height of the chimney in feet. What should be the area of the cross section of a chimney 72 feet high if 750 pounds of coal are burned each hour?

7   Suppose that \$1,500 is put into a savings plan that yields a yearly interest rate of $5\frac{1}{2}$ percent. How much money is accumulated after 5 years if the money is compounded
a)   Annually?
b)   Semiannually?
c)   Quarterly?
d)   Monthly?

8   Gus invests \$5,000 in a saving certificate at an interest rate of 8 percent compounded quarterly. How much money is accumulated after 4 years?

9   A retired couple bought a treasury bill for \$10,000 at 4 percent annual interest, compounded monthly. They redeemed the bill after 21 months. Find the amount received.

10   Find the amount and the interest if \$4,000 placed in a bank at an annual rate of $5\frac{1}{2}$ percent is compounded semiannually for 6 years.

11   \$2,000 is placed in a bank at a 6 percent annual rate compounded annually. In how many years would the money double at this rate?

12   To save for her college education, Mary's parents invested \$2,000 in a bank on the day she was born at 6 percent compounded annually. How much money has been accumulated by her eighteenth birthday?

13 A formula for the effective interest rate is given by $e = (1 + r/t)^t - 1$, where r is the interest rate compounded t times per year. A bank advertises as follows: Invest with us and earn $5\frac{1}{2}$ percent compounded monthly. What is the equivalent effective rate of interest?

14 A company invested \$50,000 at an interest rate of $5\frac{1}{2}$ percent compounded semiannually. It also invested another \$50,000 at an interest rate of 6 percent compounded annually. Which investment yields the higher return?

15 Penny deposits \$20 at the end of each month in a bank. If the interest rate is 5 percent converted monthly, find the amount in her account after 10 years.

16 If \$8,000 is deposited in a savings account at the end of every quarter at an interest rate of 6 percent converted quarterly, how much money is on deposit after 10 years?

17 Mr. and Mrs. Glass buy a home by making a down payment of \$10,000 and a monthly payment of \$250 for 30 years. If the interest rate on the mortgage is 9 percent, what price did Mr. and Mrs. Glass pay for the house? (Assume that the first mortgage payment occurs 1 month after the down payment has been made.)

18 A football player will receive a royalty payment of \$1,000 semiannually for the next 5 years for a commercial he made on television. Determine the present value of these payments if the money is worth 8 percent converted semiannually.

REVIEW PROBLEM SET

In problems 1–6, write each exponential equation in logarithmic form.

1 $2^5 = 32$

2 $27^{-2/3} = \frac{1}{9}$

3 $8^{5/3} = 32$

4 $(\frac{1}{32})^{4/5} = \frac{1}{16}$

5 $27^{2/3} = 9$

6 $13^0 = 1$

In problems 7–12, write each logarithmic equation in exponential form.

7 $\log_{10} 100 = 2$

8 $\log_2 64 = 6$

9 $\log_{17} 1 = 0$

10 $\log_\pi \pi^{100} = 100$

11 $\log_{10} \frac{1}{10} = -1$

12 $\log_{125} 625 = \frac{4}{3}$

In problems 13–18, find the value of x in each equation.

13 $\log_3 x = 2$ **14** $\log_x 16 = 4$

15 $\log_x 16 = -\frac{4}{3}$ **16** $\log_{\sqrt{2}} x = -3$

17 $\log_{10} x = -2$ **18** $\log_{\sqrt{5}} x = -2$

In problems 19–24, find the solution set of each exponential equation.

19 $2^{3x+6} = 32$ **20** $5^{x+2} = 625$

21 $27^{x-1} = 9$ **22** $6^{3x+7} = 216^{3-x}$

23 $(1.2)^{2x+1} = 1.44$ **24** $3^{2x+1} = 27^{x-1}$

In problems 25–36, find the value of each logarithm.

25 $\log_3 9$ **26** $\log_4 8$

27 $\log_6 1$ **28** $\log_5 0.04$

29 $\log_{100} 0.001$ **30** $\log_9 \frac{1}{3}$

31 $\log_{0.36} 0.6$ **32** $\log_2 \frac{1}{256}$

33 $\log_5 \frac{1}{625}$ **34** $\log_2 \frac{1}{1,024}$

35 $\log_4 8\sqrt{2}$ **36** $\log_{6/5} \frac{25}{36}$

In problems 37–42, find the solution set of each equation.

37 $\log_4 16 = x$ **38** $\log_3 9\sqrt{3} = x$

39 $\log_2(7x-1) = 4$ **40** $\log_5(3x-11) = 2$

41 $\log_9(17x-33) = 0$ **42** $\log_5\left(\frac{x}{2} - \frac{3}{2}\right) = 0$

In problems 43–48, use the properties of logarithms to write each expression as a single logarithm. Assume that all variables represent positive real numbers.

43 $\log_2 \frac{3}{7} + \log_2 \frac{14}{27}$

44 $\log_3 \frac{5}{12} + \log_3 \frac{4}{15}$

45 $\log_5 \frac{6}{7} - \log_5 \frac{27}{4} + \log_5 \frac{21}{16}$

46 $\log_9 \frac{11}{5} + \log_9 \frac{14}{3} - \log_9 \frac{22}{15}$

47 $5\log_a x - 3\log_a y$

48 $2\log_a x^3 + \log_a \frac{2}{x} - \log_a \frac{2}{x^4}$

In problems 49–54, use the properties of logarithms to write each expression as a sum, difference, or multiple of logarithms. Assume that all variables represent positive real numbers.

49 $\log_2 3^6 \cdot 4^7$

50 $\log_8 \dfrac{5^7}{9^3}$

51 $\log_2 xy^5$

52 $\log_2 \sqrt[7]{5^3 \cdot 5^6}$

53 $\log_a \dfrac{x^2}{y^4}$

54 $\log_5 (2^6 \cdot 3^7 \cdot 5^2)$

55 If $\log_b x = 3$, find $\log_{1/b} x$.

56 If $\log_b x = 3$, find $\log_b \dfrac{1}{x}$.

In problems 57–64, use the properties of logarithms and the fact that $\log_a 2 = 0.69$, $\log_a 3 = 1.10$, $\log_a 5 = 1.62$, and $\log_a 7 = 1.94$ to find the value of each expression.

57 $\log_a (3^5 \cdot 3^7)$

58 $\log_a \sqrt[5]{3^5 \cdot 3^7}$

59 $\log_a \dfrac{2^4}{3^4}$

60 $\sqrt[3]{\log_a \dfrac{2^4}{3^4}}$

61 $\log_a \dfrac{25}{27}$

62 $\log_a \sqrt[3]{\dfrac{25}{27}}$

63 $\log_a \dfrac{35}{a}$

64 $\log_a \sqrt[3]{16}$

In problems 65 and 66, use the properties of logarithms to find the solution set of each equation.

65 $\log_5 (2x - 1) + \log_5 (2x - 1) = 1$

66 $\log_{1/2} (4x^2 - 1) - \log_{1/2} (2x + 1) = 1$

In problems 67–84, use Table I in Appendix C to find the value of each common logarithm. Use interpolation when necessary.

67 $\log 844$

68 $\log 55.6$

69 $\log 77.2$

70 $\log 5.34$

71 $\log 3.18$

72 $\log 6.38$

73 $\log 0.473$

74 $\log 0.531$

75 $\log 0.0257$

76 $\log 0.0392$

77 $\log 0.00586$

78 $\log 0.000925$

79 log 876,000

80 log 92,600,000

81 log 11.32

82 log 475.3

83 log 6,153

84 log 1,995

In problems 85–98, use Table I in Appendix C to find each anti-logarithm. Use interpolation when necessary.

85 antilog 0.7466

86 antilog 0.5514

87 antilog 1.9533

88 antilog 2.9243

89 antilog 3.0347

90 antilog 4.1139

91 antilog[0.4518 + (−2)]

92 antilog[0.9289 + (−3)]

93 antilog(−1.0214)

94 antilog(−4.6972)

95 antilog(−3.4067)

96 antilog(−2.3487)

97 antilog(−1.1888)

98 antilog(−2.2612)

In problems 99–102, use the properties of logarithms and Table I in Appendix C to compute each expression.

99 $\log(73.4 \times 81.3)$

100 $\log \dfrac{0.721}{4.53}$

101 $\log(3.54)^3$

102 $\log \sqrt[3]{618}$

In problems 103–116, use the properties of logarithms and Table I in Appendix C to compute the value of each expression.

103 53.7×27.6

104 7.812×0.932

105 $98.3 \div 6.24$

106 $8.76 \div 0.00321$

107 $(34.5)^3$

108 $(4.61)^4$

109 $\sqrt[3]{0.346}$

110 $(0.682)^{2/3}$

111 $\sqrt{\dfrac{1.83 \times 16.7}{10.9}}$

112 $\sqrt[4]{\dfrac{24.5 \times 6.38}{45.7}}$

113 $\sqrt[5]{\dfrac{9.64 \times (1.83)^2}{0.0607}}$

114 $\sqrt[3]{\dfrac{(0.776)^2 \times (2.36)}{47.7}}$

115 $\sqrt[3]{98.4} \times \sqrt[4]{4.21}$

116 $\dfrac{97.6 \times \sqrt{47.3}}{(315)^2}$

In problems 117–120, use the properties of logarithms and Table I in Appendix C to find the solution set of each equation.

117 $3^x = 5$ **118** $5^{-2x} = 3$

119 $5^{2x} = 17$ **120** $2^{5x-1} = 11$

121 A sum of $5,000 is invested in a stock whose growth-rate average is 15 percent compounded annually. Assuming the rate of growth continues, find the value of the stock after 6 years.

122 Mike initially deposits $1,500 in a bank. At the end of each month for the next 18 months, he deposits an additional $150. The bank pays interest at 6 percent converted monthly. How much is in the account after 18 months?

CHAPTER 9

Relations and Functions

The concept of a function plays a fundamental role in the study of mathematics—indeed, many of the more sophisticated topics in mathematics depend on the understanding of functions. In this chapter we shall introduce and explore briefly a few kinds of functions. Then we shall discuss graphs of special relations and systems with second-degree equations.

9.1 Relations

The concept of a "relation" will be approached intuitively. The statements "Brian is married to Doreen," "Joan is a sister of Kathy's," "New York is larger than Michigan," and "5 is less than 8" involve what is commonly understood to be a relationship. Expressions of the type "is married to," "is a sister of," "is larger than," and "is less than" are also classified as relations. Hence, a relation suggests a correspondence or an association between the elements of two sets. For example, the table suggests a relation between the numbers in the x column with the numbers in the y column.

x	y
$1 \longrightarrow$	2
$2 \longrightarrow$	4
$3 \longrightarrow$	6
$4 \longrightarrow$	8
$5 \longrightarrow$	10
$\cdots \cdots \cdots$	
$n \longrightarrow$	$2n$

Here the correspondence between the numbers in the x column and the numbers in the y column is given by the formula $y = 2x$.

In the example above, there are three main elements: a first set, a second set, and a correspondence between the members of the two sets. This example suggests a relation between two sets. In order to describe a relation so that the corresponding members of the two sets are clearly identified, the ordered-pair notation is used (see Section 7.1).

DEFINITION 1 RELATION

A *relation* is a set of ordered pairs. The *domain* of a relation is the s
of all first members of the ordered pairs, and the *range* of a relatic
is the set of all second members of the ordered pairs.

Hence, $\{(x,y)\,|\,y=2x, x \text{ is a positive integer}\}$ is the relation suggeste
by the table on page 353. Since we will consider only those relatior
which are formed from real numbers, we can use the Cartesian cc
ordinate system to represent relations as points on a plane. This kir
of representation is called the *graph* of the relation; the graph pre
vides us with a "geometric picture" of the relation.

EXAMPLES

1   For relations R_1 and R_2, consider a set U to be $\{1,2,3\}$. Enumera
the members of the relation; indicate the domain and range, ar
graph the relation.
a) $R_1 = \{(x,y)\,|\,y=x, x \in U \text{ and } y \in U\}$
b) $R_2 = \{(x,y)\,|\,y>x, x \in U \text{ and } y \in U\}$

SOLUTION

a) $R_1 = \{(1,1),(2,2),(3,3)\}$, so that the domain and range are t
same set, that is, $\{1,2,3\}$. The graph is composed of three poin
(Figure 9.1a).
b) $R_2 = \{(x,y)\,|\,y>x\} = \{(1,2),(1,3),(2,3)\}$, so that the domain
R_2 is $\{1,2\}$ and the range of R_2 is $\{2,3\}$. The graph has thre
points (Figure 9.1b).

Figure 9.1

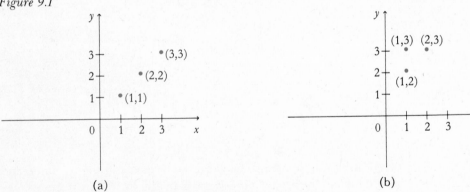

(a)   (b)

2   Graph the relation $\{(x,y)\,|\,y=3x, x \in R\}$.

SOLUTION.   The graph of the equation $y = 3x$ is a straight line (se
Chapter 7); hence it is enough to find two of its points and draw t
line joining them (Figure 9.2).

Figure 9.2

x	$y = 3x$
0	0
1	3

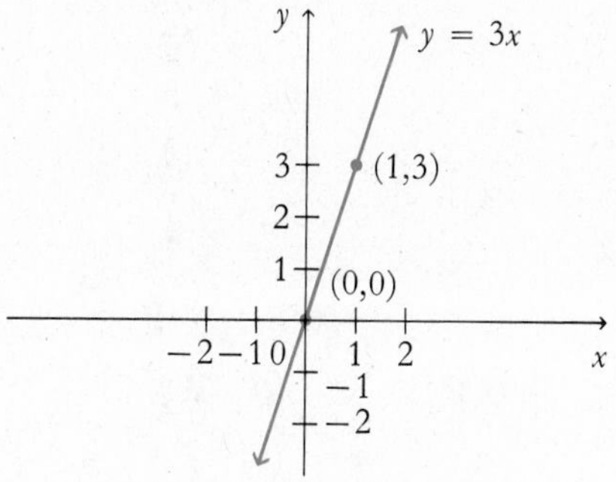

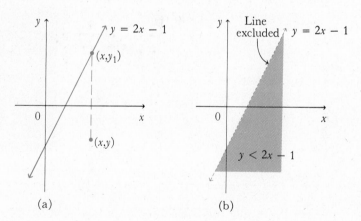

3 Graph the relation $\{(x,y) \mid y < 2x-1, x \in R\}$.

SOLUTION. The graph of $y < 2x - 1$ consists of all points (x, y) that lie below the point (x, y_1), where (x, y_1) is a point on the line $y = 2x - 1$ (Figure 9.3a). Thus, the graph of $y < 2x - 1$ is the shaded region below the graph of the line $y = 2x - 1$ (Figure 9.3b).

Figure 9.3

4 Graph the relation $\{(x,y) \mid y \geq 2x, x \in R\}$.

SOLUTION. The graph of $y \geq 2x$ consists of all points (x, y) that lie above the point (x, y_1), where (x, y_1) is a point on the line $y = 2x$ (Figure 9.4a). Thus, the graph of $y \geq 2x$ is the shaded region above, and including, the graph of the line $y = 2x$ (Figure 9.4b).

Figure 9.4

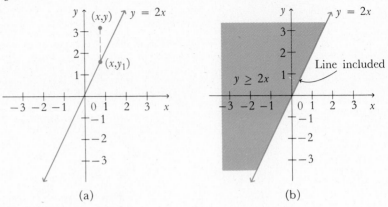

(a)   (b)

5   Graph the relation $\{(x, y) \mid y \leq -3x + 2, x \in R\}$.

SOLUTION.   The graph of $y < -3x + 2$ consists of all points (x, y) that
lie below the point (x, y_1), where (x, y_1) is a point on the line $y = -3x + 2$
(Figure 9.5a). Thus, the graph of $y \leq -3x + 2$ is the shaded region
below, and including, the graph of the line $y = -3x + 2$ (Figure 9.5b).

Figure 9.5

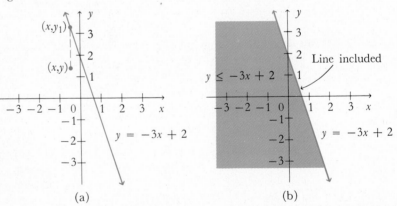

(a)   (b)

Quite often set notation is not used to describe a relation. When
this is the case, the set of ordered pairs is implied. For example,
$x < y$ is an abbreviated way of writing the relation $\{(x, y) \mid x < y, x \in R\}$.

Moreover, the domain and/or range of a relation is not always
given. In this case we determine the domain and/or range by inspection. Finally, if the universal set is not given, we assume the universal
set to be R, the set of real numbers.

Note that if (x, y) is a member of a relation, we can consider the real
numbers x and y from two viewpoints. On the one hand, x is a member

of the domain and y is the corresponding member of the range of the relation. On the other hand, x represents the abscissa and y represents the ordinate of a point on a plane. On the graph, then, the members of the domain are the abscissas and the members of the range are the ordinates (Figure 9.6). Any restriction on the domain is a restriction on the horizontal position of the graph, and any restriction on the range is a restriction on the vertical position of the graph.

Figure 9.6

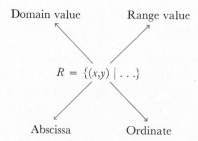

$$R = \{(x,y) \mid \ldots\}$$

Domain value   Range value

Abscissa   Ordinate

EXAMPLE

6   Graph $y = x^2$.

SOLUTION. $y = x^2$ implies the relation $\{(x, y) \mid y = x^2\}$. Since any real number can be squared and since the square of a real number is always nonnegative, the domain is R and the range is the set $\{y \mid y \geq 0\}$. This means that the graph "lies" in the region above and on the x axis (Figure 9.7). Finally, we can use a table to determine the pattern of the graph, which we complete "by inspection" (Figure 9.7).

Figure 9.7

x	$y = x^2$
0	0
1	1
-1	1
2	4
-2	4

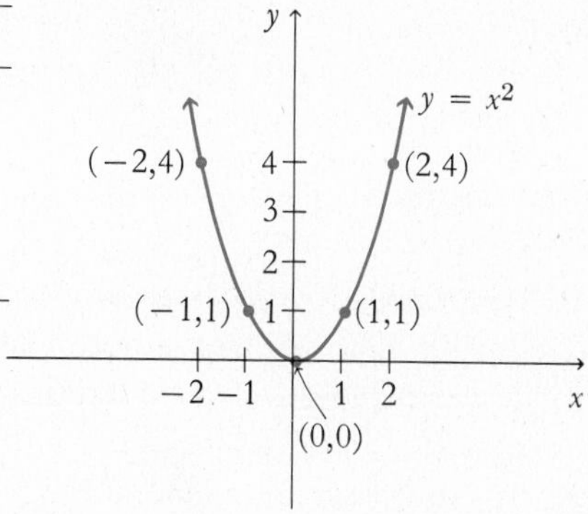

PROBLEM SET 9.1

In problems 1–10, identify the domain and range of each relation. Also, graph the relations.

1 $\{(1,2),(2,-3),(5,1),(2,4)\}$

2 $\{(3,5),(-6,8),(3,7),(4,8)\}$

3 $\{(0,6),(0,3),(0,-7)\}$

4 $\{(2,7),(-3,7),(0,7),(8,7),(-5,7)\}$

5 $\{(4,-1),(-3,-1),(-3,-2),(4,-2)\}$

6 $\{(1,2),(1,3),(2,2),(2,3),(3,3)\}$

7 $\{(x,y)\,|\,y^2=x,\ x\in\{7,9,16\}\}$

8 $\{(x,y)\,|\,x^2+y^2=1,\ y\in\{0,1\}\}$

9 $\{(x,y)\,|\,x^2+y^2=4,\ x\in\{0,2\}\}$

10 $\{(x,y)\,|\,y=|x|,\ y\in\{0,1,2\}\}$

11 Let $U=\{0,2,4,6,8\}$. Enumerate the members of the following relation, indicate the domain and range, and graph the relation:

$$\{(x,y)\,|\,y=x;\ x\in U \text{ and } y\in U\}$$

12 Let $U=\{-5,-4,-3,-2,-1,0\}$. Enumerate the members of the following relation, indicate the domain and range, and graph the relation:

$$\{(x,y)\,|\,y<x;\ x\in U \text{ and } y\in U\}$$

In problems 13–20, sketch the graph of each relation.

13 $y\geq -\frac{1}{3}x$ **14** $3y\leq 2x+1$

15 $2x-3y<0$ **16** $y>2-x$

17 $x-2y\geq 1$ **18** $y\geq -2x+5$

19 $2x<4-3y$ **20** $3x-4y+12>0$

9.2 Functions

We have already encountered the concept of functions in everyday living. For example, the amount of sales tax charged on a purchase of \$5 is a function of the sales tax rate; the number of books to be ordered for a course is a function of the number of students in the course; the number of members in the house of representatives for a particular state is a function of the population of the state.

Intuitively, a function suggests some kind of correspondence. In each of the examples above, there is an established correspondence between numbers — the amount of sales tax corresponds with the cost, the number of books with the number of students, and the number of representatives with the number of people. In general, we have the following definition.

DEFINITION 1 FUNCTION AS A CORRESPONDENCE

A *function* is a correspondence that assigns to each member in a certain set, called the *domain* of the function, exactly one member in a second set, called the *range* of the function.

For example, let us consider the situation in which as each student registers for classes at the beginning of the term, his tuition charge is recorded with his student account number as illustrated by the following partial table.

Student account number	Tuition charge
895	315.00
475	323.50
182	260.90
743	315.00
234	370.00

Here the set of student account numbers is the domain and the set of tuition charges is the range. The correspondence between the domain members and range members is suggested by the table.

Domain	"Corresponds to"	Range
895		315.00
475		323.50
182		260.90
743		315.00
234		370.00

Ordered-pair notation is well suited to representing functions: If x is a member of the domain of a function and y is the member of the range corresponding to x, we can represent the correspondence between x and y as the ordered pair (x, y). In fact, we can say that the function is the set of all such ordered pairs. In the example above, then, we can represent the function as the set of ordered pairs

$$\{. \ . \ . \ , (895, 315.00), (475, 323.50), (182, 260.90),$$
$$(743, 315.00), (234, 370.00), \ . \ . \ .\}$$

In this sense, we define a function as follows, recalling that a relation is a set of ordered pairs.

DEFINITION 2 FUNCTION AS A RELATION

A *function* is a relation in which no two different ordered pairs have the same first member. The set of all first members (of the ordered pairs) is called the *domain* of the function. The set of all second members (of the ordered pairs) is called the *range* of the function. For each ordered pair (x, y) of the function we say that x in the domain has y as the *corresponding* member in the range.

From Definition 2 we conclude that all functions are relations; however, not all relations are functions. For example, $\{(1,1), (2,2), (3,7), (3,5)\}$ is a relation, with domain $\{1,2,3\}$ and range $\{1,2,7,5\}$, but it is not a function because $(3,7)$ and $(3,5)$ have the same first member (see Definition 2). By contrast, $\{(1,2), (3,4), (4,4)\}$ is a relation that is a function with domain $\{1,3,4\}$ and range $\{2,4\}$. Note that in the latter example, two pairs have the same second member; this does not violate the definition of a function.

Hence, if at least two different ordered pairs of a relation have the same first member, the relation is not a function. In other words, if a domain member appears with more than one range member, the relation is not a function. Geometrically, this means that if the graph of a relation has more than one point with the same abscissa, the relation is not a function.

Consider the graphs of the relations in Figure 9.8a and b. The relation $\{(x, y) \mid y = \sqrt{25 - x^2}\}$ in part (a) represents a function, since the graph does not have two different points with the same x coordinate, whereas the relation $\{(x, y) \mid x = \sqrt{25 - y^2}\}$ in part (b) is not a function, since there are two different points with the same x coordinate—for example, the two points $(3,4)$ and $(3,-4)$.

Figure 9.8

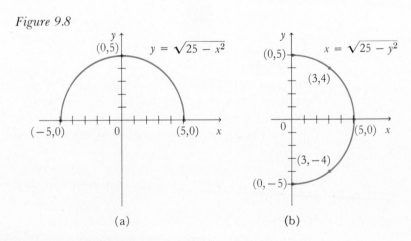

(a)   (b)

EXAMPLES

1 In each of the following parts, indicate whether or not the relation is a function. What is the domain and range? Graph the relation.
a) $\{(1,2),(2,3),(3,4),(4,4),(5,6)\}$
b) $y = 3$
c) $x = 1$
d) $y = 3x, x \in \{-1,0,2,3\}$
e) $y = 3x$

SOLUTION

a) $\{(1,2),(2,3),(3,4),(4,4),(5,6)\}$ is a function with domain $\{1,2,3,4,5\}$ and range $\{2,3,4,6\}$ (Figure 9.9).

Figure 9.9

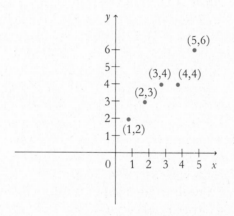

b) $y = 3$ is an abbreviated way of writing the relation $\{(x,y)|y=3\}$. The domain is the set of all real numbers because there is no restriction on x, whereas the range is $\{3\}$. The relation is a function because no two different ordered pairs have the same first member. The graph is the line parallel to the x axis (Figure 9.10).

Figure 9.10

x	y
1	3
$\frac{3}{2}$	3
-2	3
3	3
0	3

c) $x = 1$ is the relation $\{(x,y)\,|\,x = 1\}$. This relation is not a function because $(1,3)$ and $(1,4)$ are members of the set. The domain of the relation is $\{1\}$, and the range is the set of all real numbers. The graph is the line parallel to the y axis (Figure 9.11).

Figure 9.11

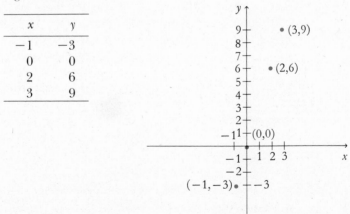

x	y
1	0
1	1
1	2
1	-1
1	-2

d) $y = 3x$, with $x \in \{-1,0,2,3\}$, represents the ordered pairs $\{(-1,-3), (0,0), (2,6), (3,9)\}$. It is a function with domain $\{-1,0,2,3\}$ and range $\{-3,0,6,9\}$, and the graph is composed of the four points (Figure 9.12).

Figure 9.12

x	y
-1	-3
0	0
2	6
3	9

e) $y = 3x$ is the set of ordered pairs $\{(x,y)\,|\,y = 3x\}$. It is a function and has as both its domain and range the set of all real numbers. It can be seen geometrically that this set of ordered pairs is a function, since no two points have the same abscissa (Figure 9.13).

Figure 9.13

x	y
0	0
1	3
−1	−3
2	6
3	9

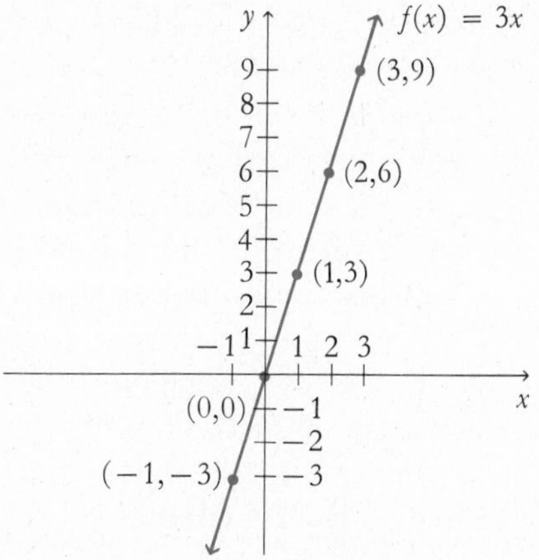

2   Examine the graphs of each of the following relations given in Figure 9.14a and b to decide whether or not each relation is a function.

Figure 9.14

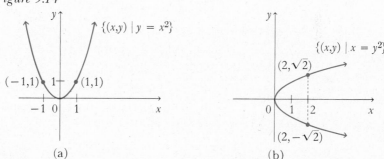

(a)   (b)

SOLUTION.   Since the graph in Figure 9.14a does not have two different points with equal abscissas, no two different ordered pairs of the relation have equal first members; thus the relation $\{(x,y)\,|\,y = x^2\}$ is a function. It can be seen from the graph in Figure 9.14b that there are two different points with abscissa 2, so that the relation $\{(x,y)\,|\,x = y^2\}$ is not a function.

If x represents members of the domain of $y = x^2$, then y is considered to be a function of x and we indicate this by writing $f(x) = x^2$, which reads "f of x equals x squared." Hence, $f(x) = x^2$ is another way of writing the function $\{(x,y)\,|\,y = x^2\}$.

In other words, $y = f(x)$ means that (x,y) is a member of the function f. For example, if $4x - 2y = 1$ and $y = f(x)$, then after solving for y in terms of x, we could write the function either as $\{(x,y)\,|\,y = 2x - \frac{1}{2}\}$ or as $f(x) = 2x - \frac{1}{2}$.

Now, it is important to realize that letters other than x, y, or f can be used to denote functions. For example, $h(r) = r^2 - 1$ is the function given by $\{(r,h(r)) \mid h(r) = r^2 - 1\}$; $3r + 5t = 3$, with $t = g(r)$ is the function given by $\{(r,t) \mid t = (3 - 3r)/5\}$; and $c(d) = \pi d$ is the function given by $\{(d,c(d)) \mid c(d) = \pi d\}$.

When functional notation is used, sometimes it is helpful to think of the variable which represents the members of the domain as a "missing blank." For example, $g(t) = t^2$ can be thought of as $g(\ ) = (\ )^2$; hence, if any expression (representing a real number) is to be used to represent a member of the domain, it is easy to see where this same expression is to be substituted in the equation describing the function. For example, using $g(t) = t^2$ again, $g(x + h)$ can be determined by first writing the function as $g(\ ) = (\ )^2$, so that, by substitution,

$$g(x + h) = (x + h)^2 = x^2 + 2xh + h^2$$

Similarly,

$$g(3 - 5x) = (3 - 5x)^2 = 9 - 30x + 25x^2$$

EXAMPLES

3   Let $f(x) = x + 1$. Find $f(2)$, $f(3)$, $f(5)$, $f(a - 6)$, and $f(a) - f(6)$. What is the domain of f?

SOLUTION.   The domain of f is the set of real numbers R. Since $f(x) = x + 1$, we have

$$f(2) = (2) + 1 = 3$$
$$f(3) = (3) + 1 = 4$$
$$f(5) = (5) + 1 = 6$$
$$f(a - 6) = (a - 6) + 1 = a - 5$$
$$f(a) - f(6) = [(a) + 1] - [(6) + 1] = a - 6$$

4   Let $f(x) = \sqrt{25 - x^2}$. Find $f(0)$, $f(3)$, $f(4)$, and $f(5)$.

SOLUTION.   Since $f(x) = \sqrt{25 - x^2}$, we have

$$f(0) = \sqrt{25 - (0)^2} = 5$$
$$f(3) = \sqrt{25 - (3)^2} = \sqrt{25 - 9} = \sqrt{16} = 4$$
$$f(4) = \sqrt{25 - (4)^2} = \sqrt{25 - 16} = \sqrt{9} = 3$$
$$f(5) = \sqrt{25 - (5)^2} = \sqrt{25 - 25} = 0$$

5   Given that $f(x) = x^2 - 1$, what is the domain and the range of f, and what is the value of each of the following expressions?
a) $f(3) + f(5)$ b) $f(x + 1)$
c) $f(2a + 4)$ d) $2f(a - 2)$

SOLUTION.   The domain of the function f is the set of real numbers R and the range of f is the set $\{x \mid x \geq -1\}$.
a) $f(3) = (3)^2 - 1 = 8$ and $f(5) = (5)^2 - 1 = 24$, so that
    $f(3) + f(5) = 8 + 24 = 32$
b) $f(x + 1) = (x + 1)^2 - 1 = x^2 + 2x$
c) $f(2a + 4) = (2a + 4)^2 - 1 = 4a^2 + 16a + 15$
d) $2f(a - 2) = 2[(a - 2)^2 - 1] = 2[a^2 - 4a + 3] = 2a^2 - 8a + 6$

6   Given the *identity function* $f(x) = x$, find $f(-2), f(-1), f(0), f(1)$, and $f(2)$. Describe the domain and range of f. Sketch the graph of f.

SOLUTION. $f(-2) = -2, f(-1) = -1, f(0) = 0, f(1) = 1$, and $f(2) = 2$. The domain is the set of real numbers R and the range is also the set of real numbers R (Figure 9.15).

Figure 9.15

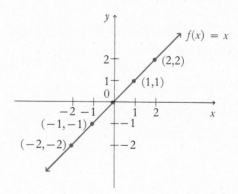

7   Given the *absolute value function* $f(x) = |x|$, find $f(-2), f(-1), f(0)$, $f(1)$, and $f(2)$. Describe the domain and range of f. Sketch the graph of f.

SOLUTION. $f(-2) = 2, f(-1) = 1, f(0) = 0, f(1) = 1$, and $f(2) = 2$. The domain is the set of real numbers R, and the range is the set of all nonnegative real numbers (Figure 9.16).

Figure 9.16

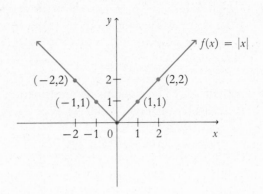

PROBLEM SET 9.2

In problems 1–10, determine if each set of ordered pairs is a function.

1 $\{(2,4),(3,6),(7,2),(9,-3)\}$ **2** $\{(1,1),(2,1),(3,1)\}$

3 $\{(-8,0),(-6,2),(5,3),(7,0)\}$ **4** $\{(1,1),(1,2),(1,3)\}$

5 $\{(x,y)\,|\,y=2x-1\}$ **6** $\{(x,y)\,|\,y<2x-1\}$

7 $\{(x,y)\,|\,y\leq x\}$ **8** $\{(x,y)\,|\,y^2-x^2=1\}$

9 $\{(x,y)\,|\,y=x^2-1\}$ **10** $\{(x,y)\,|\,y=\sqrt{x^2-1}\}$

In problems 11–20, find the domain of the given function. Also, find $f(1), f(3),$ and $f(5)$ for each function.

11 $f(x)=3x+1$ **12** $f=\{(1,2),(3,4),(5,7)\}$

13 $f=\{(3,8),(5,0),(1,-1)\}$ **14** $f(x)=2-3x$

15 $f(x)=\sqrt{x-1}$ **16** $f(x)=\dfrac{1}{2x-4}$

17 $f(x)=\dfrac{1}{x}$ **18** $f(x)=x^2+3$

19 $f(x)=\dfrac{1}{\sqrt{x+2}}$ **20** $f(x)=x^3-1$

In problems 21–26, find $f(0),$ $f(-1),$ $f(1),$ $f(a),$ $f(a+h),$ and $[f(a+h)-f(a)]/h.$ Also, sketch the graph of $f(x)$ and identify the domain and range.

21 $f(x)=3x-7$ **22** $f(x)=-2x+3$

23 $f(x)=2$ **24** $f(x)=3x^2+1$

25 $f(x) = 2x^2$ **26** $f(x) = x^3 + 2$

27 If x is the length of one side of a square, express the perimeter P as a function of x.

28 Define $f(x)$ so that $10^{f(x)} = x$. Find $f(1)$, $f(10)$, $f(100)$, $f(\frac{1}{10})$, and $f(\frac{1}{100})$.

29 If x is the length of one edge of a cube, express the volume V as a function of x.

30 If $f(x) = |x-2|$, find $f(0)$, $f(1)$, $f(2)$, $f(3)$, and $f(4)$. Sketch the graph of $f(x)$ and identify the domain and range.

9.3   Direct and Inverse Variations

Two important types of functions that are widely used in many laws in the sciences will be introduced in this section. We say that y *is directly proportional to x* or y *varies directly as x* if there is a constant number k such that $y = kx$ for every ordered pair (x, y). The number k is called the *constant of proportionality*. Hence, any function defined by the relation $y = kx$ is an example of direct variation. That is, if y is a function of x and y is directly proportional to x, then $y = f(x) = kx$. If $y = 4$ when $x = 1$, then $4 = k \cdot 1$ or $k = 4$. Thus, $y = f(x) = 4x$. The graph of this function is a straight line (Figure 9.17).

Figure 9.17

x	y
1	4
-1	-4
0	0

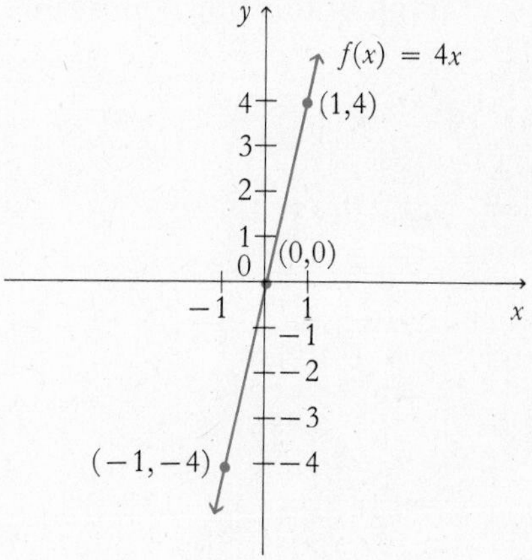

EXAMPLES

1   Express y as a function of x if y is directly proportional to x^2 and if $y = 98$ when $x = 7$.

SOLUTION.   Since y is directly proportional to x^2, there is a constant number k such that $y = kx^2$. The fact that $y = 98$ when $x = 7$ tells us that $98 = 49k$, or $k = 2$, so that the equation is $y = 2x^2$.

2   If y is directly proportional to x and $y = f(x)$, show that $f(ax) = af(x)$.

SOLUTION.   Since y is directly proportional to x, there is a constant k such that $y = kx$, but $y = f(x)$, so that $f(x) = kx$; finally, $f(ax) = k(ax) = a(kx) = af(x)$.

3   The area of a sphere is directly proportional to the square of the radius. If a sphere of radius of 4 inches has an area of 64π square inches, express the area of a sphere as a function of its radius.

SOLUTION.   Let A represent the area of the sphere and r represent its radius, in inches. Since A is directly proportional to r^2, there is a constant k such that $A = kr^2$. Since $A = 64\pi$ when $r = 4$, then $64\pi = 16k$, so that $k = \frac{64}{16}\pi = 4\pi$. Hence, $A = 4\pi r^2$.

Another type of variation that we shall consider here is called *inverse variation,* which is stated as follows: y is *inversely proportional* to x, or y *varies inversely as* x, if there is a number k such that $y = k/x$ for every ordered pair (x, y), $x \neq 0$. For example, if y is inversely proportional to x and $y = 2$ when $x = 6$, then y can be expressed as a function of x by $y = k/x$. Since $y = 2$ when $x = 6$, then $k = 12$, so that $y = 12/x$. The graph of this function is shown in Figure 9.18.

Figure 9.18

x	y
-2	-6
-1	-12
1	12
2	6

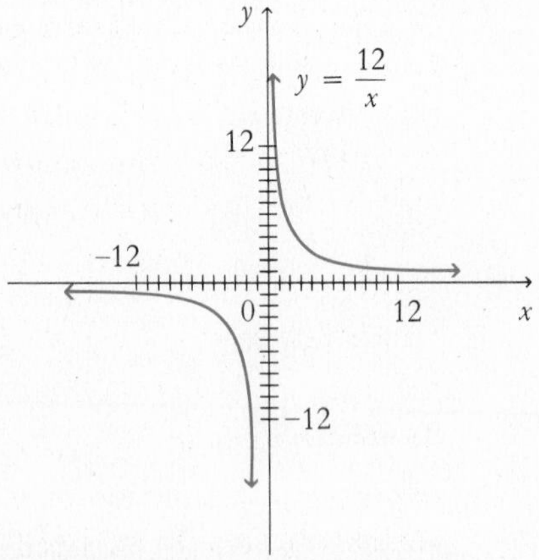

EXAMPLES

4 Express y as a function of x if y is inversely proportional to x^2 and $y = 12$ when $x = 2$.

SOLUTION. Since y is inversely proportional to x^2, there is a number k such that $y = k/x^2$. Given $y = 12$ when $x = 2$, then $12 = k/2^2$, so that $k = 48$. Thus, $y = 48/x^2$.

5 The intensity of a floodlight is inversely proportional to the square of the distance from the floodlight. How far from an object will the floodlight have to be placed for the object to receive three times the intensity when placed at 6 feet?

SOLUTION. Since the intensity I is inversely proportional to the square of the distance x, there is some constant k such that $I = k/x^2$. When $x = 6$ feet, we have $I = k/6^2$ or $I = k/36$. At some unknown distance x, the intensity will be three times the intensity at 6 feet; that is, $3I = k/x^2$. Solving these two equations for k, we have $k = 36I$ and $k = 3Ix^2$. Thus, $3Ix^2 = 36I$, so that $x^2 = 12$, or $x = \sqrt{12} = 2\sqrt{3}$. Hence, the required distance is $2\sqrt{3}$ feet.

6 Boyle's law states that the pressure P is inversely proportional to the volume of an ideal gas. Find the constant of variation if the pressure P is 30 pounds per square inch when the volume of the gas is 100 cubic inches.

SOLUTION. Since the pressure P is inversely proportional to V, there is a number k such that $P = k/V$ or $k = PV$. At $P = 30$ pounds per square inch, $V = 100$ cubic inches, so that $k = 30(100) = 3,000$ pounds per inch.

7 If z varies directly as the product of two variables x and y, the relationship between z and the product xy is known as *joint variation*. Thus, z varies jointly as x and y if there is a constant real number k such that $z = kxy$. The volume V of a right circular cone varies jointly as its altitude h and the square of its base radius r. If $V = 12\pi$ cubic inches when $r = 3$ inches and $h = 4$ inches, find V in terms of r and h.

SOLUTION. Since V varies jointly as r^2 and h, there is a real number k such that $V = kr^2h$. Substituting for r, h, and V, we have

$$12\pi = k(9)(4) \qquad \text{or} \qquad k = \frac{12\pi}{36} = \frac{\pi}{3}$$

Hence, the required formula is $V = \frac{1}{3}\pi r^2 h$.

8   If z varies directly as x and inversely as y^2, find the equation connecting x, y, and z if $z = 12$, when $x = 2$ and $y = 3$.

SOLUTION.   Since z varies directly as x and inversely as y^2, there is a constant real number k such that $z = kx/y^2$. Substituting for x, y, and z, we have

$$12 = \frac{k(2)}{3^2} \quad \text{or} \quad k = 54$$

Hence, the required equation is $z = 54x/y^2$.

PROBLEM SET 9.3

1   Let y be directly proportional to x, and $y = f(x)$. If $y = 8$ when $x = 4$, find a formula for f and sketch the graph of f. Also find $f(x + 2)$, $f(2) + f(3)$, and $[f(x + h) - f(x)]/h$, where $h \neq 0$.

2   If y is directly proportional to x^2 and $y = f(x)$, does $f(ax) = af(x)$?

In problems 3–8, if y is directly proportional to x^3, express y as a function of x in each case.

3   $y = 4$ when $x = 2$ 4 $y = 12$ when $x = -2$

5   $y = 3$ when $x = 1$ 6 $y = -2$ when $x = 3$

7   $y = 14$ when $x = 11$ 8 $y = 10$ when $x = -3$

9   If y is inversely proportional to x^2 and $x = 9$ when $y = 2$, find x when $y = 105$.

10 If y is inversely proportional to $\sqrt[3]{x}$ and $y = 9$ when $x = 8$, find y when $x = 216$.

11 If T is directly proportional to x and inversely proportional to y, and $T = 0.01$ when $x = 20$ and $y = 20$, express T as a function of x and y.

12 If y is inversely proportional to x^2 and $y = 8$ when $x = 10$, find y when $x = 2$.

13 If y is inversely proportional to x^3 and $y = 3$ when $x = 4$, express y as a function of x and sketch the graph.

14 If V varies directly as T and inversely as P, and $V = 40$ when $T = 300$ and $P = 30$, find V when $T = 324$ and $P = 24$.

15 The surfaces of two spheres have the ratio $9:4$. What is the ratio of their radii? Their volumes? (*Hint:* $S = 4\pi r^2$ and $V = \frac{4}{3}\pi r^3$.)

16 Coulomb's law states that the magnitude of the force F that acts on two charges q_1 and q_2 varies directly as the product of the magnitude of q_1 and q_2 and inversely as the square of the distance r between them. If the force on two charges, each of 1 coulomb, that are separated in air by a distance of 1,000 meters is 9,000 newtons, find the force when the two charges are separated by 2,000 meters.

17 The total surface area S of a cube is directly proportional to the square of the edge x. If a cube whose edge is 3 inches has a surface area of 54 square inches, express the surface area S as a function of x. Then find the surface area of a cube whose edge is 12 inches.

18 Newton's law of gravitational attraction states that the force F with which two particles of mass m_1 and m_2, respectively, attract each other varies directly as the product of the masses and inversely as the square of the distance r between them. If one of the masses is tripled, and the distance between the masses is also tripled, what happens to the force?

9.4 Polynomial Functions

Functions that can be expressed in polynomial forms (see Chapter 2) are called *polynomial functions.* Some examples that define such functions are $f(x) = 5$, $g(x) = 3x + 2$, and $h(x) = 5x^2 + 13x + 7$. The simplest polynomial functions to graph are polynomials of degree 0 and of degree 1.

Linear Functions

Polynomial functions whose graphs are straight lines are called *linear functions.* More formally, a *linear function* is a first-degree polynomial function of the form $f(x) = mx + b$, where m and b are constant real numbers. For example, $f(x) = 2x - 1$ and $g(x) = -3x + 7$ define linear functions. If $m = 0$, then $f(x) = b$ is called a *constant function.* The graph of such a function is the set of all points with an ordinate of b; it is a line parallel to the x axis through the point $(0,b)$. If, for example, $b = 3$, then $f(x) = 3$ has the set representation $f = \{(x,y) \mid y = 3\}$ (Figure 9.19).

Figure 9.19

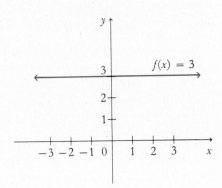

$f(x) = 3$

EXAMPLES

1   Find the domain and the range of the following linear function and sketch the graph.

$$f(x) = 2x + 5$$

SOLUTION.   The domain of the function f is the set of real numbers R, and the range of f is also the set of real numbers R. Since the graph of a linear function is a straight line, it is enough to determine the graph by two points (Figure 9.20).

Figure 9.20

x	$f(x)$
0	5
$-\frac{5}{2}$	0

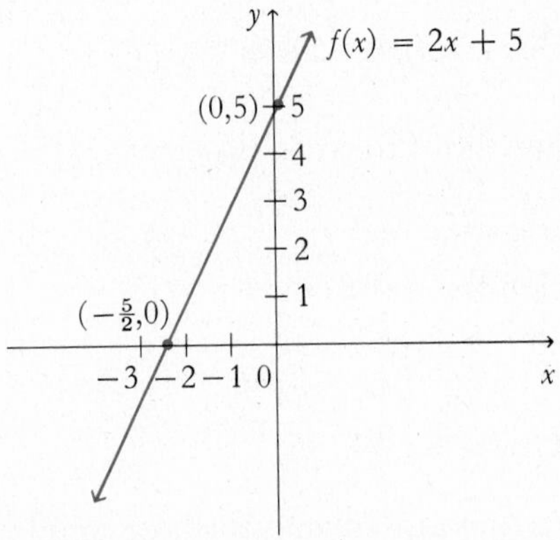

$f(x) = 2x + 5$

$(0,5)$

$\left(-\frac{5}{2},0\right)$

2   Suppose that $y = f(x)$ defines a linear function whose graph contains $(3,-4)$ and $(-2,5)$. Find $f(x)$ in equation form and sketch the graph of f.

SOLUTION.   Since $(3,-4)$ and $(-2,5)$ both lie on the same line, they must both satisfy the functional relationship $f(x) = mx + b$. We have

$$m = \frac{y_2 - y_1}{x_2 - x_1} = \frac{5 - (-4)}{-2 - 3} = -\frac{9}{5}$$

Substituting in $f(x) = mx + b$, we have $f(x) = -\frac{9}{5}x + b$. Using the point $(3, -4)$, we have $-4 = -\frac{9}{5}(3) + b$, so that $b = \frac{7}{5}$. Hence, $f(x) = -\frac{9}{5}x + \frac{7}{5}$ (Figure 9.21).

Figure 9.21

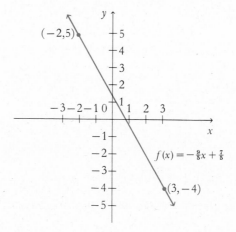

Quadratic Functions

Polynomial functions of degree two are called *quadratic functions*. Thus, $f(x) = ax^2 + bx + c$, where a, b, and c are real numbers, $a \neq 0$, is the general representation of a quadratic function. The graph of a quadratic function can be determined by plotting some of its points on the coordinate axes, and then drawing the curve connecting these points. For example, the graph of $f(x) = x^2$ is found by plotting some points (Figure 9.22a) and then drawing the curve suggested by these points (Figure 9.22b).

Figure 9.22

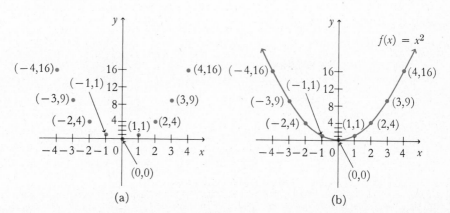

The graph of any quadratic function will have the same general shape as that illustrated in Figure 9.22, although the location of the graph will vary, depending upon the specific values of a, b, and c. Such graphs are called *parabolas*. Specific examples of quadratic functions are shown in Figure 9.23. Notice that the curve opens upward when the coefficient of the x^2 term is positive (Figure 9.23a and c), and downward when the coefficient of the x^2 term is negative (Figure 9.23b and d). In general, the graph of $f(x) = ax^2 + bx + c$ opens upward when $a > 0$ and downward when $a < 0$. It should be evident in the examples below that the graph crosses the y axis, although this is not always true for the x axis. In short, to graph a quadratic function, locate the x intercepts (if they exist) and the y intercept, and then sketch a parabola that contains these points. Another point on the graph of special interest is the "low" point (Figure 9.23a and c) or the "high" point (Figure 9.23b and d), which is obtained when $x = -b/2a$ and $y = c - (b^2/4a)$. The point $\left(-\dfrac{b}{2a}, f\left(-\dfrac{b}{2a}\right)\right)$ is referred to as the *extreme point*, or the *vertex*, of the parabola. Locating this point will

Figure 9.23

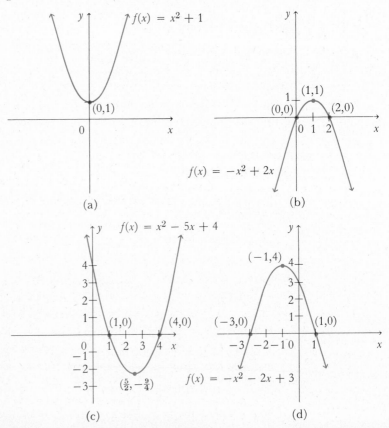

help in sketching the graphs of quadratic functions at a glance. For example, to graph $y = x^2 - 3x + 2$, notice that the solution of the quadratic equation $x^2 - 3x + 2 = 0$ is 1 or 2, so that the x intercepts are 1 and 2. By completing the square in the right-hand member of $y = x^2 - 3x + 2$, we obtain

$$y = (x^2 - 3x + \tfrac{9}{4}) - \tfrac{9}{4} + 2 \quad \text{or} \quad y = (x - \tfrac{3}{2})^2 - \tfrac{1}{4}$$

From this equation we can see that the extreme point occurs when the term $(x - \tfrac{3}{2})^2$ is zero, that is, when $x - \tfrac{3}{2} = 0$ or $x = \tfrac{3}{2}$. Thus, the extreme point is $(\tfrac{3}{2}, -\tfrac{1}{4})$. Also, the y intercept is found by setting x equal to zero, so that $y = 2$. By connecting these points, we are able to sketch the parabola (Figure 9.24).

Figure 9.24

x	y
0	2
$\tfrac{3}{2}$	$-\tfrac{1}{4}$
1	0
2	0

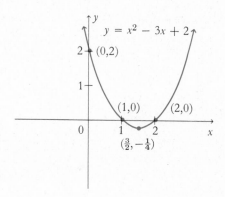

EXAMPLES

Find the domain, the range, the x intercept, the y intercept, and the extreme point of each of the following quadratic functions and sketch each graph.

3 $f(x) = -x^2 + 2x$

SOLUTION. The domain of f is the set of real numbers. In locating the x intercepts of the graph of the function f, let $f(x) = 0$, and find values of x for which $-x^2 + 2x = 0$. This can be written as $-x(x - 2) = 0$, so that $x = 0$ or 2. The y intercept is found by assigning the value 0 to x, so that the y intercept is 0. We obtain the coordinates of the extreme point by completing the square. Thus, $y = -x^2 + 2x$ can be written as $y = -(x^2 - 2x + 1) + 1$ (Why?), or $y = -(x - 1)^2 + 1$. For $x = 1$, $(x - 1)^2$ is zero, so that the extreme point is $(1, 1)$ (Figure 9.25). Hence, the range of f is $\{y \mid y \le 1\}$.

Figure 9.25

x	y
−1	−3
0	0
1	1
2	0

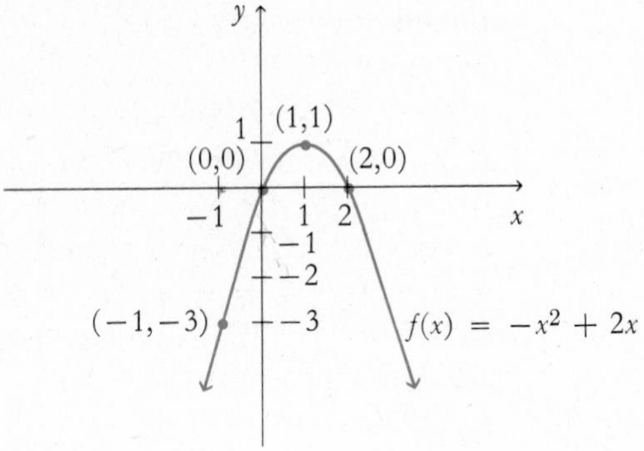

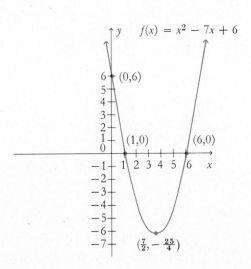

4   $f(x) = x^2 - 7x + 6$

SOLUTION.   The domain of f is the set of real numbers. The x intercepts of f are the values of x such that $x^2 - 7x + 6 = 0$; that is, $(x - 6)(x - 1) = 0$, so that $x = 1$ or 6. The y intercept of f is $f(0) = 6$. To obtain the coordinates of the extreme point write $y = f(x) = x^2 - 7x + 6$ as

$$y = x^2 - 7x + (\tfrac{7}{2})^2 - (\tfrac{7}{2})^2 + 6 \qquad \text{or} \qquad y = (x - \tfrac{7}{2})^2 - \tfrac{25}{4}$$

For $x = \tfrac{7}{2}$, y has an extreme value of $-\tfrac{25}{4}$. (Why?) Thus, the extreme point is at $(\tfrac{7}{2}, -\tfrac{25}{4})$ (Figure 9.26). Hence, the range of f is $\{y \mid y \geq -\tfrac{25}{4}\}$.

Figure 9.26

x	y
0	6
$\tfrac{7}{2}$	$-\tfrac{25}{4}$
1	0
6	0

The graphs of quadratic functions can be used to illustrate the kind of roots of associated quadratic equations $ax^2 + bx + c = 0$, $a \neq 0$. The roots of quadratic equation $ax^2 + bx + c = 0$, $a \neq 0$, can then be determined by inspecting the x intercepts of the associated function defined by $f(x) = ax^2 + bx + c$. Let us consider the graphs of $f(x) = x^2 - 4x + 4$, $f(x) = x^2 - 5x + 4$, and $f(x) = x^2 + 1$. The quadratic equation $x^2 - 4x + 4 = (x - 2)(x - 2) = 0$ has two equal roots, namely, $x = 2$ and $x = 2$, so that the graph of the associated function $f(x) = x^2 - 4x + 4$ intercepts the x axis at one point $x = 2$ (Figure 9.27a). The equation $x^2 - 5x + 4 = (x - 1)(x - 4) = 0$ has two different roots, namely, $x = 1$ and $x = 4$, so that the graph of the associated function $f(x) = x^2 - 5x + 4$ intercepts the x axis at two points, 1 and 4 (Figure 9.27b). Finally, the equation $x^2 + 1 = 0$ has no real roots; accordingly, the graph of the associated function $f(x) = x^2 + 1$ does not cross the x axis (Figure 9.27c).

Figure 9.27

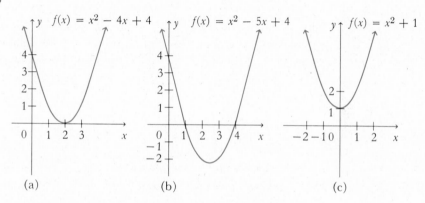

(a) (b) (c)

At this point we can restate the results of Section 6.3 as follows:

1   If the discriminant $b^2 - 4ac$ of a quadratic equation $ax^2 + bx + c = 0$, $a \neq 0$, is zero, then the associated function $f(x) = ax^2 + bx + c$ intercepts the x axis at one point (**Figure 9.27a**).

2   If the discriminant of a quadratic equation is positive, that is, if $b^2 - 4ac > 0$, then the associated function $f(x) = ax^2 + bx + c$, $a \neq 0$, intercepts the x axis at two different points (**Figure 9.27b**).

3   If the discriminant $b^2 - 4ac < 0$, then the associated function $f(x) = ax^2 + bx + c$, $a \neq 0$, does not intercept the x axis (**Figure 9.27c**).

PROBLEM SET 9.4

In problems 1–10, find the domain, range, and slope of each linear function and sketch each graph.

1   $f(x) = -3x + 5$ 2 $f(x) = 5x + 1$

3   $f(x) = -\frac{3}{4}x + 1$ 4 $f(x) = \frac{1}{4}x + 3$

5 $f(x) = -1$ **6** $f(x) = -3x$

7 $f(x) = 2(x - 2) + 1$ **8** $f(x) = 7$

9 $f(x) = 4x$ **10** $f(x) = 3 - 2(1 - x)$

In problems 11–16, find a linear function f such that each condition is satisfied.

11 $f(1) = 3$ and $f(2) = 5$ **12** $f(-2) = 7$ and $f(3) = -5$

13 $f(0) = 4$ and $f(3) = 0$ **14** $f(1) = -8$ and $f(\frac{1}{2}) = -6$

15 $f(-7) = 3$ and $f(5) = 3$ **16** $f(2) = -4$ and $f(-10) = -4$

17 Find a linear function f such that $2f(x) = f(2x)$ for every real number x.

18 Find a linear function f such that $f(3x + 2) = f(3x) + 2$.

19 Find a linear function f such that $f(1) = 3$ and such that the graph of f is parallel to the graph of the line determined by the points $(-2, 1)$ and $(3, 2)$.

20 Find a linear function f such that $f(1) = 3$ and such that the graph of f is perpendicular to the graph of the line determined by the points $(-2, 1)$ and $(3, 2)$.

In problems 21–30, determine the domain, range, extreme point, and x and y intercepts for each quadratic function. Also, sketch each graph.

21 $f(x) = 2x^2 - 3$ **22** $f(x) = x^2 - 3$

23 $f(x) = -x^2 - 2x - 1$ **24** $f(x) = (x - 5)^2$

$0 - \dfrac{9}{8}$

25 $f(x) = x^2 + 5x + 6$ **26** $f(x) = -x^2 - 1$

27 $f(x) = 2x^2 - 3x$ **28** $f(x) = -(x + 1)^2$

$b = -3$

29 $f(x) = x^2 + 4x + 3$ $a = 2$ **30** $f(x) = -x^2 + x - 5$

31 Sketch the graph of each of the following functions on the same coordinate system: $f(x) = x^2 - 1$, $f(x) = x^2 + 1$, $f(x) = x^2 - 2$, and $f(x) = x^2 + 2$.

32 Show that the coordinates of the extreme point of $y = ax^2 + bx + c$, $a \neq 0$, are

$$\left(-\frac{b}{2a}, \frac{4ac - b^2}{4a} \right)$$

(*Hint:* Use the method of completing the square.)

33 Apply the results of problem 32 to find the extreme points of the functions given in problems 21–30.

9.5 Exponential and Logarithmic Functions

The properties of exponents that have been used so far were developed for the set of rational numbers. We mentioned in Section 5.2 that the properties of exponents previously discussed for rational numbers apply also to irrational numbers. With this in mind, we will assume from here on that expressions such as $3^{\sqrt{2}}$ and $4^{\sqrt[3]{5}}$ are also real numbers and that the properties of exponents apply to the set of all real numbers R.

Exponential Functions

Computations in mathematical applications frequently require the use of functional relationships in which the exponent is a variable. One particular type of function of importance is called an *exponential function*. For example, $f(x) = 2^x$, and $f(x) = (\frac{1}{3})^x$ define exponential functions. In general: If b is a positive number, then the function $f(x) = b^x$, $b \neq 1$, is called an *exponential function* with *base b*.

The domain of $f(x) = b^x$ is the set of all real numbers R. The range of $f(x) = b^x$ is the set of all positive real numbers. In order to investigate the properties of exponential functions, we shall discuss the domain, the range, and the graph of such functions in the following examples.

EXAMPLES

Find the domain and range and sketch the graph of each of the following exponential functions.

1 $f(x) = 3^x$

SOLUTION. $f(x) = 3^x$ has base 3. The domain of $f(x) = 3^x$, as with all exponential functions, is the set of the real numbers R and the range is the set of positive real numbers. Using the table on page 380, we can locate some specific points and graph the function (Figure 9.28). Notice that the graph of $f(x) = 3^x$ goes up to the right as x gets larger. For this reason we say that $f(x)$ is an *increasing function*.

Figure 9.28

x	$f(x)$
-3	$\frac{1}{27}$
-2	$\frac{1}{9}$
-1	$\frac{1}{3}$
0	1
1	3
2	9
3	27

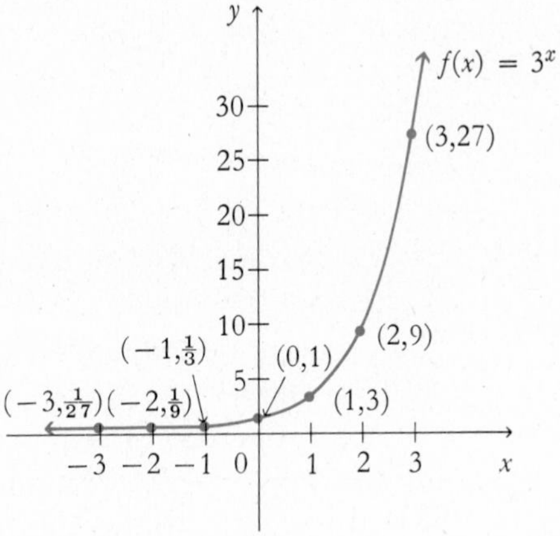

2   $f(x) = (\frac{1}{4})^x$

SOLUTION. $f(x) = (\frac{1}{4})^x$ has base $\frac{1}{4}$. The domain is the set of the real numbers R, and the range is the set of positive real numbers. The graph of $f(x) = (\frac{1}{4})^x$ goes down to the right as x gets larger; therefore, we say that the function $f(x) = (\frac{1}{4})^x$ is a *decreasing function* (Figure 9.29).

Figure 9.29

x	$f(x)$
-2	16
-1	4
0	1
1	$\frac{1}{4}$
2	$\frac{1}{16}$

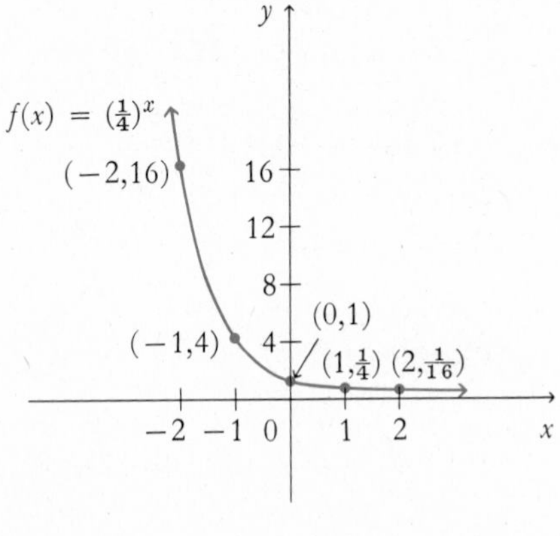

3   Let f be an exponential function with base b. Show that $f(u + v) = f(u) \cdot f(v)$ for any real numbers u and v.

SOLUTION.   Since f is an exponential function with base b, $f(x) = b^x$, $f(u + v) = b^{u+v}$. By the rules of exponents, $b^{u+v} = b^u b^v$, so that $f(u + v) = b^{u+v} = b^u \cdot b^v = f(u) \cdot f(v)$.

Logarithmic Functions

We saw in Section 8.1 that if $y = b^x$, $b > 0$, $b \neq 1$, then the exponent x is called the *logarithm of the number y to the base b*. Thus, $y = \log_b x$ is equivalent to $x = b^y$. The domain of $x = b^y$ is the set of real numbers R, and the range is the set of positive real numbers. That is, the domain is

$\{y \mid y$ is a real number$\}$

and the range is

$\{x \mid x$ is a positive number$\}$

By contrast, the domain of the function defined by $y = f(x) = \log_b x$ is the set of positive real numbers and the range is the set of real numbers R. Such a function is called a *logarithmic function*. In order to graph a logarithmic function $f(x) = \log_b x$, we can use the equivalent equation $x = b^{f(x)}$ to locate some points on the graph. For example, to graph $f(x) = \log_3 x$, we can use the equivalent equation $x = 3^{f(x)}$ $= 3^y$. Thus, for $f(x) = 0$, $x = 1$; for $f(x) = 1$, $x = 3$; and for $f(x) = 2$, $x = 9$. One should note that we are reversing the usual technique in finding points on the graph. Here we have selected a value of $y = f(x)$ first and then determined the corresponding value of x (Figure 9.30).

Figure 9.30

$x = 3^y$	y
$\frac{1}{9}$	-2
$\frac{1}{3}$	-1
1	0
3	1
9	2
27	3

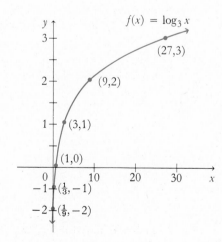

EXAMPLES

Find the domain and the range of each of the following functions, and sketch the graph.

4   $f(x) = \log_2 x$

SOLUTION.   The domain of f is the set of positive real numbers, and the range is the set of real numbers R. Notice that $f(x) = \log_2 x$ is equivalent to $2^{f(x)} = x$; hence, the following table can be determined by the latter exponential equation (Figure 9.31). The graph of $f(x) = \log_2 x$ goes up to the right as x gets larger (Figure 9.31); hence, we say that $f(x) = \log_2 x$ is an increasing function.

Figure 9.31

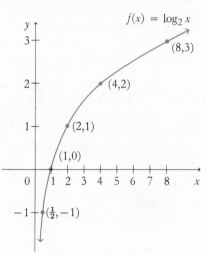

$x = 2^{f(x)}$	$f(x)$
$\frac{1}{2}$	-1
1	0
2	1
4	2
8	3

5   $f(x) = \log_{1/4} x$

SOLUTION.   The domain of f is the set of positive real numbers, and the range is the set of real numbers R. $f(x) = \log_{1/4} x$ is equivalent to $(\frac{1}{4})^{f(x)} = x$. The table is determined by using the exponential equation $x = (\frac{1}{4})^{f(x)}$. The graph of $f(x) = \log_{1/4} x$ goes down to the right as x gets larger. Therefore, the function f is a decreasing function (Figure 9.32).

Figure 9.32

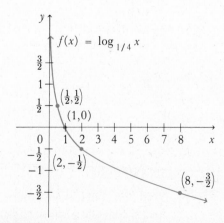

$x = (\frac{1}{4})^{f(x)}$	$f(x)$
$\frac{1}{2}$	$\frac{1}{2}$
1	0
2	$-\frac{1}{2}$
8	$-\frac{3}{2}$

PROBLEM SET 9.5

In problems 1–10, discuss the properties of each function. Indicate the domain and range, and graph the function. Is the function increasing or decreasing?

1 $f(x) = 2^x$

2 $f(x) = 1^x$

3 $f(x) = 3^{x+1}$

4 $f(x) = -2^x$

5 $f(x) = -(\frac{1}{3})^x$

6 $f(x) = 2^{-x}$

7 $f(x) = (\frac{1}{5})^{-x}$

8 $f(x) = (0.1)^x$

9 $f(x) = 5(3^x)$

10 $f(x) = -(4)^x$

In problems 11–16, determine the base of the exponential function $f(x) = b^x$ whose graph contains the given points.

11 $(2,9)$

12 $(3,27)$

13 $(2,16)$

14 $(5,3{,}125)$

15 $(0,1)$

16 $(\frac{1}{2}, \sqrt{10})$

17 Let f be an exponential function with base b. Show that $f(u - v) = f(u) \div f(v)$ for any real numbers u and v.

18 Use the graph in Figure 9.28 to approximate the value of $3^{3/2}$.

In problems 19–24, discuss the properties of each logarithmic function. Indicate the domain and range, and graph the function. Is the function increasing or decreasing?

19 $f(x) = -\log_2 x$

20 $f(x) = \log_5 x$

21 $f(x) = \log_{1/2} x$

22 $f(x) = \log_4 x$

23 $f(x) = \log_6 x$

24 $f(x) = \log_{1/3} x$

In problems 25–30, determine the base of the logarithmic function $f(x) = \log_b x$ whose graph contains the given points.

25 $(8,3)$

26 $(125,3)$

27 $(\frac{1}{16}, -2)$

28 $(8, \frac{3}{2})$

29 $(3, \frac{1}{2})$

30 $(b^{3/2}, 3)$

31 Let f be a logarithmic function with base b. Show that $f(u) + f(v) = f(uv)$ for any positive real numbers u and v.

32 Let f be a logarithmic function with base b. Show that $f(u) - f(v) = f(u/v)$ for any positive real numbers u and v.

9.6   Graphs of Special Relations—The Conics

In Section 9.4 we considered the graph of the general quadratic function of the form $y = f(x) = ax^2 + bx + c$. We shall now consider the relation represented by the general equation $ax^2 + bxy + cy^2 + dx + ey + f = 0$, where a, b, c, d, e, and f are constant real numbers with a and b both $\neq 0$. Notice that the quadratic function $y = ax^2 + bx + c$ is a special case of the relation above obtained by a suitable choice of values of the coefficients. The equation $ax^2 + bxy + cy^2 + dx + ey + f = 0$, for a and b both not equal to zero (with a few exceptional cases), produces one of the following four curves, called the *circle*, the *ellipse*, the *parabola*, and the *hyperbola* (Figure 9.33).

Figure 9.33

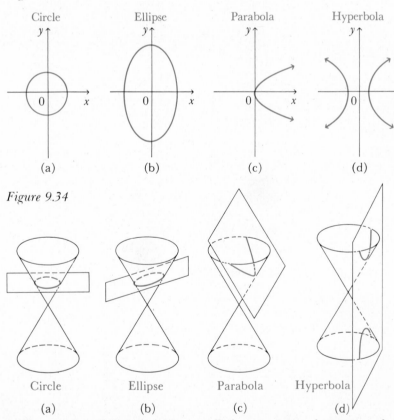

Circle	Ellipse	Parabola	Hyperbola
(a)	(b)	(c)	(d)

Figure 9.34

Circle	Ellipse	Parabola	Hyperbola
(a)	(b)	(c)	(d)

The graphs of Figure 9.33 are called *conic sections*, because each is formed by intersecting a right circular cone and a plane (Figure 9.34). The equations of the four conic sections can be obtained from the general second-degree equation $ax^2 + bxy + cy^2 + dx + ey + f = 0$ by placing conditions on the coefficients a, b, c, d, and e, and the constant term f. In particular, we have the following special cases.

1   The graph of any second-degree equation $Ax^2 + By^2 = C$, in which $A = B$ and A, B, and C have the same signs, represents a *circle*. The center of the circle is at the origin of the coordinate axes, and the radius r is determined by the equation $r = \sqrt{C/A}$. In particular, the graph of the equation $x^2 + y^2 = 16$ is a circle with radius 4 and with its center at the origin (Figure 9.35). Solving this equation for y, we have $y = \pm\sqrt{16 - x^2}$. Using the table, we can get a few points on the graph.

Figure 9.35

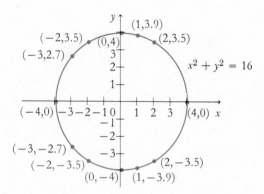

x	y
-4	0
-3	± 2.7
-2	± 3.5
-1	± 3.9
0	± 4
1	± 3.9
2	± 3.5
3	± 2.7
4	0

2   The graph of any second-degree equation of the form $Ax^2 + By^2 = C$, in which $A \neq B$ and A, B, and C have the same signs, is an *ellipse*. Its center is at the origin of the coordinate axes, and it has x intercepts at $x = \pm\sqrt{C/A}$ and y intercepts at $y = \pm\sqrt{C/B}$. For example, the graph of $16x^2 + 25y^2 = 400$ is an ellipse, with x intercepts -5 and 5 and y intercepts -4 and 4 (Figure 9.36). Solving for y, we have $y = \pm\frac{4}{5}\sqrt{25 - x^2}$, from which we can get the points on the graph.

$$y = \pm\sqrt{\frac{A}{B}}\sqrt{B - x^2}$$

Figure 9.36

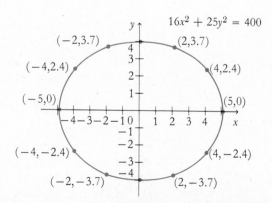

x	y
-5	0
-4	± 2.4
-2	± 3.7
0	± 4
2	± 3.7
4	± 2.4
5	0

3   The graph of the second-degree equation $y = Ax^2 + Bx + C,\, A \neq 0$, is a *parabola* with a vertical axis (see Section 9.4). Interchange of x and y in the equation gives an equation $x = Ay^2 + By + C$, which is also a parabola but with a horizontal axis. For example, the graph of $y^2 = 4x + 4$ is a parabola. The equation $y = \pm\sqrt{4x + 4}$, together with the table, will determine the graph (Figure 9.37).

Figure 9.37

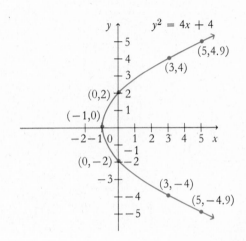

x	y
0	± 2
-1	0
3	± 4
5	± 4.9

4   The graph of a second-degree equation of the form $Ax^2 - By^2 = C$, in which A, B, and C are positive, is a *hyperbola*, with its center at the origin of the coordinate axes and with x intercepts at $x = \pm\sqrt{C/A}$.

Figure 9.38

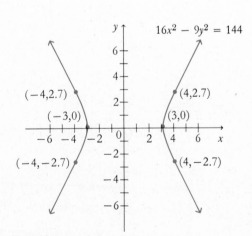

x	y
-4	± 2.7
-3	0
3	0
4	± 2.7

If x and y are interchanged, the resulting equation will be in the form $By^2 - Ax^2 = C$, $A, B,$ and C positive, with center at the origin and y intercepts at $y = \pm\sqrt{C/B}$. For example, the graph of the equation $16x^2 - 9y^2 = 144$ is a hyperbola with x intercepts at 3 and -3. Notice that there are no y intercepts. (Why?) Solving for y, we have $y = \pm\frac{4}{3}\sqrt{x^2 - 9}$, which can be used to find the points given in the table (Figure 9.38).

EXAMPLES

Graph each of the following relations. State the domain of each relation, find the x and y intercepts (if possible), and decide if the relation is a function. Identify the graph.

1 $x^2 + y^2 = 64$

SOLUTION. Solving the equation for y, we get $y = \pm\sqrt{64 - x^2}$. The domain of this relation is the set $\{x \mid -8 \le x \le 8\}$. The x intercepts are ±8 and the y intercepts are ±8. This relation is not a function and the graph is a circle (Figure 9.39).

Figure 9.39

x	y
±8	0
±6	±5.3
±4	±6.9
±2	±7.8
0	±8

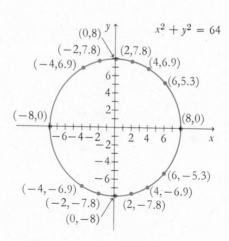

2 $9x^2 + 25y^2 = 225$

SOLUTION. Solving for y, we have $y = \pm\frac{3}{5}\sqrt{25 - x^2}$. The domain of the relation is the set $\{x \mid -5 \le x \le 5\}$. The x intercepts are ±5 and the y intercepts are ±3. The relation is not a function and the graph is an ellipse (Figure 9.40).

Figure 9.40

x	y
0	±3
−3	±2.4
3	±2.4
−5	0
5	0

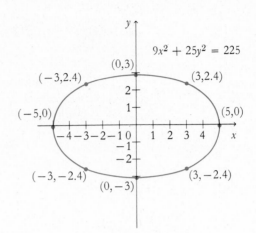

3   $y^2 = x + 1$

SOLUTION.   Solving for y, we have $y = \pm \sqrt{x + 1}$. The domain of the relation is the set $\{x \mid x \geq -1\}$. The y intercepts are ± 1 and there is one x intercept at -1. The relation is not a function and the graph is a parabola (Figure 9.41).

Figure 9.41

x	y
−1	0
0	±1
1	±1.4
2	±1.7
5	±2.4

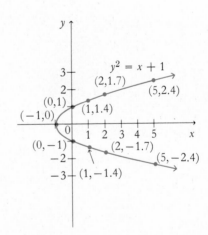

4   $9x^2 - 25y^2 = 225$

SOLUTION.   Solving for y, we have $y = \pm \frac{3}{5} \sqrt{x^2 - 25}$. The domain of the relation is the set $\{x \mid x \leq -5\} \cup \{x \mid x \geq 5\}$. The x intercepts are ± 5, and there are no y intercepts. (Why?) The relation is not a function and the graph is a hyperbola (Figure 9.42).

Figure 9.42

x	y
±5	0
±6	±1.9
±8	±3.8
±10	±5.2

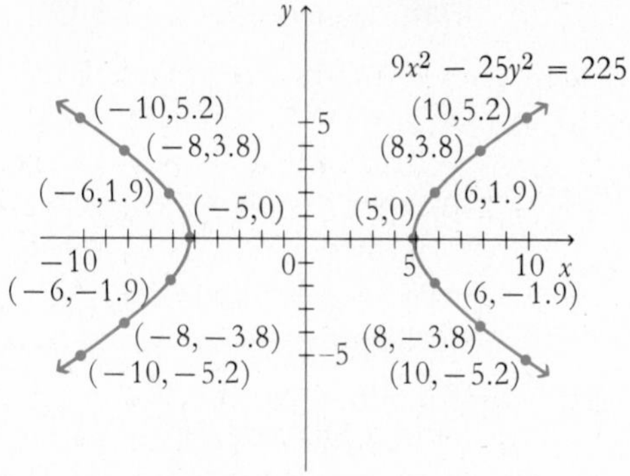

PROBLEM SET 9.6

In problems 1–20, graph each relation, state the domain and range of each relation, locate the x and y intercepts (if possible), and decide if the relation is a function. Identify the graph.

1 $x^2 - 4y = 0$ **2** $y^2 + 2x = 0$

3 $3x^2 + 2y = 0$ **4** $-2x^2 + 5y = 0$

5 $x^2 + y^2 = 4$ **6** $x^2 + y^2 = 81$

7 $4x^2 + 4y^2 = 64$ **8** $3x^2 + 3y^2 = 27$

9 $4x^2 + 9y^2 = 36$ **10** $9x^2 + 16y^2 = 144$

11 $16y^2 + 25x^2 = 400$ **12** $y^2 + 4x^2 = 16$

13 $4x^2 + 16y^2 = 64$ **14** $25x^2 + 9y^2 = 1$

15 $36x^2 - 9y^2 = 1$ **16** $16y^2 - 4x^2 = 48$

17 $y^2 - x^2 = 1$ **18** $\dfrac{x^2}{49} - \dfrac{y^2}{36} = 1$

19 $x^2 - 9y^2 = 9$ **20** $9x^2 - y^2 = 9$

9.7 Systems with Second-Degree Equations

We shall now use the methods introduced in Chapter 7 to solve systems with second-degree equations. One useful technique for solving such systems is the method of substitution. For example, to solve the system of equations

$$\begin{cases} x + y = 7 \\ x^2 + y^2 = 25 \end{cases}$$

we solve the first equation for y and obtain $y = 7 - x$, then substitute in the equation $x^2 + y^2 = 25$. Thus, we have

$$x^2 + (7 - x)^2 = 25 \qquad \text{or} \qquad x^2 - 7x + 12 = 0$$

so that $x = 3$ or $x = 4$. Hence, $y = 7 - 3 = 4$ when $x = 3$ and $y = 7 - 4 = 3$ when $x = 4$, so that the solution set is $\{(3,4),(4,3)\}$. If we had used the equation $x^2 + y^2 = 25$ to find values of y corresponding to $x = 3$ and $x = 4$, we would have obtained the four ordered pairs $(3,4)$, $(3,-4)$, $(4,3)$, and $(4,-3)$, respectively. However, only two ordered pairs, $(3,4)$ and $(4,3)$, satisfy both equations of the system. Hence, in solving systems involving one first-degree equation and one second-degree equation, the final substitution should be made in the first-degree equation.

The system of equations above has two solutions that can be illustrated graphically. The graph of $x^2 + y^2 = 25$ is a circle and the graph of $x + y = 7$ is a straight line. By graphing both equations on the same coordinate axes, we find that they intersect in exactly two points (Figure 9.43). From geometry, we observe that a straight line can intersect a circle in two, one, or no points.

Figure 9.43

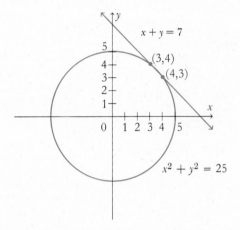

Solving systems with second-degree equations by graphical methods will usually produce only approximations to real solutions, if they exist. However, if the solutions are complex numbers, the solutions cannot be located on the graphs. It is therefore more practical to concentrate on algebraic methods, because the results are exact and solutions can be obtained whether real or complex numbers are involved. The graphical method is suggested only as a rough check of algebraic solutions. Geometric observations of the graphs of the different equations in a system with second-degree equations indicate that:

1   The curves cannot intersect in more than four points.

2   If the graphs are tangent to each other at one point, there is only one solution.

3   If the graphs do not intersect, the solutions are complex.

EXAMPLES

Solve each of the following systems of equations and check the solutions by sketching the graphs of the equations and approximating the points of intersection.

1   $$\begin{cases} 2x + 3y = 8 \\ 2x^2 - 3y^2 = -10 \end{cases}$$

SOLUTION.   From the graphs of the two given equations (Figure 9.44), we see that the solutions lie in the first and second quadrants and are approximately $(1,2)$ and $(-17,14)$. Solving the linear equation for x, we obtain $x = (8 - 3y)/2$. Then substituting the expression for x into the quadratic equation, we have

$$2\left(\frac{8 - 3y}{2}\right)^2 - 3y^2 = -10$$

so that

$$3y^2 - 48y + 84 = 0 \quad \text{or} \quad y^2 - 16y + 28 = 0$$

By factoring, we obtain $(y - 2)(y - 14) = 0$, so that $y = 2$ or $y = 14$. When $y = 2$, $x = [8 - 3(2)]/2 = 1$, and when $y = 14$, $x = [8 - 3(14)]/2 = -17$. Hence, the solution set of the system is $\{(1,2), (-17,14)\}$.

Figure 9.44

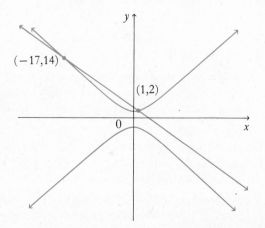

2 $\quad\begin{cases} 4x^2 + 7y^2 = 32 \\ -3x^2 + 11y^2 = 41 \end{cases}$

SOLUTION. From the graphs of the equations $4x^2 + 7y^2 = 32$ and $-3x^2 + 11y^2 = 41$, it is evident that we have four points of intersection (Figure 9.45). Thus, we conclude that the solutions are real. Multiplying the first equation by 3 and the second by 4, so that

$$\begin{cases} 12x^2 + 21y^2 = 96 \\ -12x^2 + 44y^2 = 164 \end{cases}$$

Using the addition–subtraction method, we have

$$\begin{array}{r} 12x^2 + 21y^2 = 96 \\ -12x^2 + 44y^2 = 164 \\ \hline 65y^2 = 260 \end{array}$$

so that $y^2 = 4$. That is, $y = -2$ or $y = 2$. When $y = -2$, $4x^2 + 7(-2)^2 = 32$, so that $x = -1$ or $x = 1$. When $y = 2$, $4x^2 + 7(2)^2 = 32$, so that $x = -1$ or $x = 1$. Hence, the solution set is $\{(1,2), (-1,2), (1,-2), (-1,-2)\}$. Examining the graph (Figure 9.45), we find that the solutions are $(1,2)$, $(-1,2)$, $(1,-2)$, and $(-1,-2)$, since these are the points of intersection of the two curves.

Figure 9.45

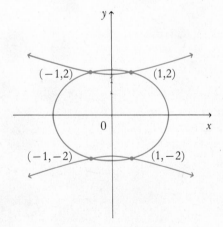

3 $\quad\begin{cases} x^2 + 2y^2 = 22 \\ 2x^2 + y^2 = 17 \end{cases}$

SOLUTION. From the graphs of the equations $x^2 + 2y^2 = 22$ and

$2x^2 + y^2 = 17$, it is clear that we have four points of intersection (Figure 9.46).

Figure 9.46

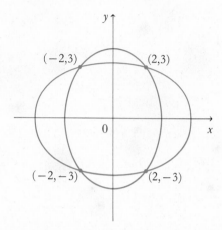

Multiplying the first equation by 2,

$$\begin{cases} 2x^2 + 4y^2 = 44 \\ 2x^2 + y^2 = 17 \end{cases}$$

Subtracting, we have

$$\begin{aligned} 2x^2 + 4y^2 &= 44 \\ 2x^2 + y^2 &= 17 \\ \hline 3y^2 &= 27 \end{aligned}$$

or

$$y^2 = 9$$

so that $y = 3$ or $y = -3$. When $y = 3$, $x^2 + 2(3)^2 = 22$, so that $x = 2$ or $x = -2$. When $y = -3$, $x^2 + 2(-3)^2 = 22$, so that $x = 2$ or $x = -2$. Hence, the solution set is $\{(2,3), (-2,3), (2,-3), (-2,-3)\}$. Examining the graph (Figure 9.46), we can see that the points of intersection of the two ellipses are $(2,3)$, $(2,-3)$, $(-2,3)$, and $(-2,-3)$.

PROBLEM SET 9.7

In problems 1–30, solve each system of equations and check the solution by sketching graphs of the equations (if possible) for approximate points of intersection.

1 $\begin{cases} x - y = 1 \\ x^2 + y^2 = 5 \end{cases}$

2 $\begin{cases} x - 2y = 3 \\ x^2 - y^2 = 24 \end{cases}$

3 $\begin{cases} 3x - y = 2 \\ x^2 + y^2 = 20 \end{cases}$

4 $\begin{cases} x + y = 3 \\ 3x^2 - y^2 = \frac{9}{2} \end{cases}$

5 $\begin{cases} 3x + 2y = 1 \\ 3x^2 - y^2 = -4 \end{cases}$

6 $\begin{cases} x + y = 6 \\ x^2 + y^2 = 20 \end{cases}$

7 $\begin{cases} 5x - 3y = 10 \\ x^2 - y^2 = 6 \end{cases}$

8 $\begin{cases} 2x + y = 10 \\ xy = 12 \end{cases}$

9 $\begin{cases} 2x + 3y = 7 \\ x^2 + y^2 + 4y + 4 = 0 \end{cases}$

10 $\begin{cases} x - y + 4 = 0 \\ x^2 + 3y^2 = 12 \end{cases}$

11 $\begin{cases} 5x - y = 21 \\ y = x^2 - 5x + 4 \end{cases}$

12 $\begin{cases} x^2 - 25y^2 = 20 \\ 2x^2 + 25y^2 = 88 \end{cases}$

13 $\begin{cases} x - y^2 = 0 \\ x^2 + 2y^2 = 24 \end{cases}$

14 $\begin{cases} 3x^2 - 8y^2 = 40 \\ 5x^2 + y^2 = 81 \end{cases}$

15 $\begin{cases} 2x^2 - 3y^2 = 6 \\ 3x^2 + 2y^2 = 35 \end{cases}$

16 $\begin{cases} x^2 - y^2 = 7 \\ x^2 + y^2 = 25 \end{cases}$

17 $\begin{cases} x^2 + 9y^2 = 33 \\ x^2 + y^2 = 25 \end{cases}$

18 $\begin{cases} x^2 + 5y^2 = 70 \\ 3x^2 - 5y^2 = 30 \end{cases}$

19 $\begin{cases} 4x^2 - y^2 = 4 \\ 4x^2 + \frac{5}{3}y^2 = 36 \end{cases}$

20 $\begin{cases} x^2 - 2y^2 = 17 \\ 2x^2 + y^2 = 54 \end{cases}$

21 $\begin{cases} 2x^2 - 3y^2 = 20 \\ x^2 + 2y = 20 \end{cases}$

22 $\begin{cases} 4x^2 + 3y^2 = 43 \\ 3x^2 - y^2 = 3 \end{cases}$

23 $\begin{cases} x^2 - 2y^2 = 1 \\ x^2 + 4y^2 = 25 \end{cases}$

24 $\begin{cases} 2x^2 - 5y^2 + 8 = 0 \\ x^2 - 7y^2 + 4 = 0 \end{cases}$

25 $\begin{cases} x^2 + 4y = 8 \\ x^2 + y^2 = 5 \end{cases}$

26 $\begin{cases} 3x - 2y = 9 \\ 9x = y^2 \end{cases}$

27 $\begin{cases} x^2 + y^2 = 16 \\ x^2 - y^2 = -34 \end{cases}$

28 $\begin{cases} x^2 - 4y^2 = -15 \\ -x^2 + 3y^2 = 11 \end{cases}$

29 $\begin{cases} x^2 + y^2 = 25 \\ (x - 5)^2 + y^2 = 9 \end{cases}$

30 $\begin{cases} x^2 - y = 0 \\ x^2 + (y - 6)^2 = 36 \end{cases}$

REVIEW PROBLEM SET

In problems 1–4, identify the domain and range of each relation. Also, graph the relations.

1 $\{(5,-2),(-3,6),(5,4),(1,6)\}$

2 $\{(-5,0),(0,-5),(-5,-5),(0,0)\}$

3 $\{(x,y)\,|\,2y^2=x,\ x\in\{2,18\}\}$

4 $\{(x,y)\,\big|\,|y|=x,\ x\in\{0,1,2\}\}$

In problems 5–8, sketch the graph of each relation.

5 $y\le 5x$ **6** $x\ge -2y+\tfrac{1}{2}$

7 $2-3y>6x$ **8** $\tfrac{1}{3}y+\tfrac{1}{2}x<1$

In problems 9–16, indicate which relations are functions.

9 $\{(1,2),(2,3),(1,5),(6,7)\}$ **10** $\{(1,3),(3,1),(-2,3),(-4,1)\}$

11 $\{(-3,1),(-2,0),(-1,-1)\}$ **12** $\{(1,8),(2,5),(1,1),(2,6)\}$

13 $\{(x,y)\,|\,y=25-x^2\}$ **14** $\{(x,y)\,|\,y=7x-2\}$

15 $\{(x,y)\,|\,y=|x+3|\}$ **16** $\{(x,y)\,\big|\,|y|>|x|\}$

17 Figure 9.47 contains graphs of relations. Which graphs represent functions?

Figure 9.47

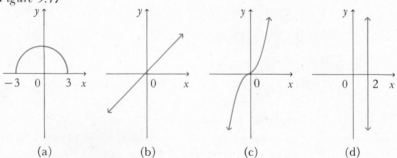

(a)   (b)   (c)   (d)

18 Figure 9.48 illustrates a function as a correspondence from x into y. What is the domain of f? What is the range of f? Find $f(3),f(1)$, and $f(5)$.

Figure 9.48

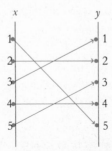

In problems 19–26, let $f(x) = 3x^2 + 2$. Find the values.

19 $f(-1)$ **20** $f(0)$

21 $f(1)$ **22** $f(\frac{2}{3})$

23 $f(a)$ **24** $f(b)$

25 $f(a + b)$ **26** $f(a + b) - f(a)$

In problems 27–30, let $f(x) = |x|$ and $g(x) = 2x - 3$. Find the value of x in each relation.

27 $f(x + 3) = 7$ **28** $f(x - 2) < 1$

29 $g(x - 5) > 2$ **30** $g(2x + 1) = 3$

In problems 31–36, express y as a function of $x[y = f(x)]$ in each case. Also, graph the function.

31 If y is directly proportional to x and if $y = 8$ when $x = 12$.

32 If y is directly proportional to x^2 and if $y = 18$ when $x = 3$.

33 If y is directly proportional to $\sqrt{x}$ and if $y = 16$ when $x = 16$.

34 If y is directly proportional to $\sqrt{x}$ and if $y = 9$ when $x = 16$.

35 If y is inversely proportional to x and if $y = 4$ when $x = 5$.

36 If y is inversely proportional to x and if $y = 12$ when $x = \frac{3}{4}$.

37 Hooke's law states that the extension of an elastic spring beyond its natural length is directly proportional to the force applied. If a weight of 8 pounds causes a spring to stretch from a length of 9 inches to a length of 9.5 inches, what weight will cause it to stretch to a length of 1 foot?

38 The power required to operate a fan is directly proportional to the speed of the fan. If 1 horsepower will drive a fan at a speed of 480 revolutions per minute, how fast will 8 horsepower drive it? What power will be required to give it a speed of 600 revolutions per minute?

In problems 39–43, let f be a linear function. Give conditions so that each equation is valid for all real numbers.

39 $3f(x) = f(3x)$ **40** $f(x + 7) = f(x) + f(7)$

41 $f(3x + 4) = 3f(x) + f(4)$ **42** $f(3) = 4$ and $f(5) = 6$

43 $7f(x) = f(7x + 1)$

44   Let f be a linear function. Is $f(3t + 2)$ linear? Prove your answer.

In problems 45–48, find the domain, range, and slope of each linear function and sketch each graph.

45 $f(x) = 2x - 2$ **46** $f(x) = \frac{3}{4}x + 7$

47 $f(x) = \frac{3}{2}x$ **48** $f(x) = -3(x + 1) + 4$

In problems 49–54, graph each quadratic function; determine the domain, the range, the extreme point, and the x and y intercepts.

49 $f(x) = 6x^2 - 5x - 4$ **50** $f(x) = 2x^2 - x - 6$

51 $f(x) = x^2 + 6x + 9$ **52** $f(x) = x^2 - 8x + 16$

53 $f(x) = -3 - 10x - 8x^2$ **54** $f(x) = 10 + 3x - x^2$

In problems 55–60, sketch the graph of each function, and find the domain and the range. Indicate whether the function is increasing or decreasing.

55 $f(x) = \frac{1}{2}x$ **56** $f(x) = 7^x$

57 $f(x) = 3(2^x)$ **58** $f(x) = -2(3^x)$

59 $f(x) = 4^x$ **60** $f(x) = (\frac{1}{3})^x$

In problems 61–64, sketch the graph of each function, and find the domain and range. Indicate whether the function is increasing or decreasing.

61 $g(x) = \log_{1/2}x$ **62** $g(x) = \log_2(x + 1)$

63 $f(x) = \log_3(x + 2)$ **64** $f(x) = \log_3|x|$

In problems 65–68, let $f(x) = 4^x$. Find the values.

65 $f(0)$ **66** $f(2)$

67 $f(-\frac{1}{2})$ **68** $f(\frac{5}{2})$

In problems 69–72, let $f(x) = \log_{16}x$. Find the values.

69 $f(32)$ **70** $f(64)$

71 $f(\sqrt[5]{2})$ **72** $f(\sqrt[3]{4})$

In problems 73–78, sketch the graph of each equation. Find the x intercepts and y intercepts and identify the graph.

73 $2x^2 + 2y^2 = 50$

74 $2x^2 + 3y^2 = 18$

75 $9x^2 + 16y^2 = 36$

76 $y^2 - x^2 = 4$

77 $y^2 = 4x + 13$

78 $x^2 = 2y + 1$

In problems 79–84, solve each system of equations algebraically; check your solutions by sketching the graph of each equation and approximating the points of intersection.

79 $\begin{cases} 3x - 4y = 25 \\ x^2 + y^2 = 25 \end{cases}$

80 $\begin{cases} 2x - y = 2 \\ x^2 + 2y^2 = 12 \end{cases}$

81 $\begin{cases} x + y^2 = 6 \\ x^2 + y^2 = 36 \end{cases}$

82 $\begin{cases} 3x^2 - 2y^2 = 35 \\ 7x^2 + 5y^2 = 43 \end{cases}$

83 $\begin{cases} x^2 + y^2 = 29 \\ x^2 - y^2 = 21 \end{cases}$

84 $\begin{cases} 3x^2 - 2y^2 = 180 \\ 2x^2 + 5y^2 = -108 \end{cases}$

CHAPTER 10

Sequences, Progressions, and the Binomial Theorem

In this chapter we consider special functions whose domains are sets of positive integers. Such functions are called sequences. Sequences play a key role in studying progressions and finite sums. The binomial theorem, which enables us to express any positive integral power of $x + y$ as a polynomial, is also covered here.

10.1 Sequences and Arithmetic Progressions

The notion of a sequence of numbers is extremely useful in the study of mathematics. A finite sequence of numbers is a list of such numbers. For instance, the list of numbers 1, 3, 5, 7, 9 is a finite sequence. The numbers 1, 3, 5, 7, and 9 are called, respectively, the *first, second, third, fourth,* and *fifth* terms of the sequence. A finite sequence is defined formally as a function whose domain is a finite set of integers. For instance, a function f defined by the rule $f(n) = n^2 - n$, $n = 1, 2, 3, 4, 5$, leads to the finite sequence 0, 2, 6, 12, 20. A function whose domain is the set of positive integers defines an infinite sequence. The function $f(n) = 2/n$, where n is a positive integer, is an example of an infinite sequence. Quite often, a sequence is defined by specifying its general term a_n. However, the notation $f(n)$ or $a(n)$ can sometimes be used in place of the subscript notation a_n. Thus, the sequence whose general term a_n is defined by $a_n = 2/n$ is written $a_1 = \frac{2}{1} = 2$, $a_2 = \frac{2}{2} = 1$, $a_3 = \frac{2}{3}$, $a_4 = \frac{2}{4} = \frac{1}{2}$, $a_5 = \frac{2}{5}$, and so forth.

EXAMPLES

Write the first five terms of the following sequences.

1 $a_n = (-1)^n$

SOLUTION.   The first five terms of the sequence $a_n = (-1)^n$ are $a_1 = -1$, $a_2 = 1$, $a_3 = -1$, $a_4 = 1$, and $a_5 = -1$.

2 $a_n = 3 - \dfrac{1}{n}$

SOLUTION. The first five terms of the sequence $a_n = 3 - 1/n$ are $a_1 = 2$, $a_2 = \frac{5}{2}$, $a_3 = \frac{8}{3}$, $a_4 = \frac{11}{4}$, and $a_5 = \frac{14}{5}$.

Arithmetic Progressions

A sequence in which the difference between successive terms remains constant is called an arithmetic progression or arithmetic sequence. For instance, consider the sequence whose general term is defined by $a_n = 1 + 2n$. The terms of this sequence are:

$$a_1 = 3, \ a_2 = 5, \ a_3 = 7, \ a_4 = 9, \ a_5 = 11, \text{ and } a_k = 1 + 2k, \ \ldots$$

Note that each term in the sequence, after the first term, is always 2 more than the preceding term; that is,

$$a_1 = 3$$
$$a_2 = 3 + 2$$
$$a_3 = (3 + 2) + 2$$
$$a_4 = [(3 + 2) + 2] + 2$$
$$\cdot$$
$$\cdot$$
$$\cdot$$
$$a_n = [(3 + 2) + 2] + 2 + \cdots + 2$$
$$\cdot$$
$$\cdot$$
$$\cdot$$

or, more briefly,

$$a_1 = 3$$
$$a_2 = 3 + 2$$
$$a_3 = 3 + 2 \cdot 2$$
$$a_4 = 3 + 3 \cdot 2$$
$$\cdot$$
$$\cdot$$
$$\cdot$$
$$a_n = 3 + (n - 1) \cdot 2$$
$$\cdot$$
$$\cdot$$
$$\cdot$$

This sequence, which can also be written as 3, 5, 7, 9, . . . , is an example of an *arithmetic progression*. 3 is called the *first term* of the progression, and 2 is called the *common difference.* Another example of an arithmetic progression is $\frac{1}{5}, \frac{2}{5}, \frac{3}{5}, \frac{4}{5}, \ldots$. Here, $\frac{1}{5}$ is the first term and $\frac{1}{5}$ is also the common difference.

In general, an *arithmetic progression* is a sequence of the following form:

$$a_1, a_1 + d, a_1 + 2d, a_1 + 3d, \ldots, a_1 + (n-1)d, \ldots$$

where a_1 is called the *first term* of the arithmetic progression, d is called the *common difference,* and $a_1 + (n-1)d$ is called the nth *term.* We usually denote the nth term by a_n. For example, $2, -1, -4, -7, \ldots$ is an arithmetic progression with first term 2 and common difference -3. The tenth term, a_{10}, of this sequence is

$$2 + (10 - 1)(-3) = -25$$

Now, let us investigate the sum of the first n terms of an arithmetic progression. For convenience, we will let S_n represent the sum of the first n terms. Then,

$$S_n = a_1 + (a_1 + d) + (a_1 + 2d) + (a_1 + 3d) + \cdots$$
$$+ [a_1 + (n-3)d] + [a_1 + (n-2)d]$$
$$+ [a_1 + (n-1)d]$$

First, note that the sum of the first term and the last term is the same as the sum of the second term and the next-to-last term, and so on. That is,

$$(a_1) + [a_1 + (n-1)d] = a_1 + a_n$$
$$(a_1 + d) + [a_1 + (n-2)d] = (a_1) + [a_1 + (n-1)d]$$
$$= a_1 + a_n$$
$$(a_1 + 2d) + [a_1 + (n-3)d] = (a_1) + [a_1 + (n-1)d]$$
$$= a_1 + a_n$$

and so on. Since there are $n/2$ such pairs of terms, we have $S_n = (n/2)(a_1 + a_n)$ or, equivalently,

$$S_n = \frac{n}{2}[2a_1 + (n-1)d]$$

EXAMPLES

3   Find the twentieth term and the sum of the first twenty terms of an arithmetic progression whose first term is 2 and whose common difference is 4.

SOLUTION. $a_1 = 2$, $d = 4$, and $n = 20$. Hence,

$$a_{20} = a_1 + (20 - 1)d = 2 + 19(4) = 78$$

Furthermore, by the formula for S_n,

$$S_{20} = \frac{n}{2}(a_1 + a_{20}) = \frac{20}{2}(2 + 78) = 800$$

4   The sum of the first ten terms of an arithmetic progression is 351 and the tenth term is 51. Find the first term and the common difference.

SOLUTION.   From the formula for S_n, we have

$$S_{10} = \frac{10}{2}(a_1 + a_{10})$$

That is, $351 = \frac{10}{2}(a_1 + 51)$ or, equivalently, $351 = 5 \cdot a_1 + 255$, so that $5a_1 = 96$ or $a_1 = 19.2$. Also, $a_{10} = a_1 + 9d$, so that $51 = 19.2 + 9d$, or, equivalently,

$$9d = 31.8 \qquad \text{so that} \qquad d = \frac{31.8}{9} = \frac{53}{15}$$

5   How many terms are there in the arithmetic progression for which $a_1 = 3$, $d = 5$, and $S_n = 255$?

SOLUTION.   From the formula for S_n, we have $255 = (n/2)(3 + a_n)$, so that $510 = n(3 + a_n)$. From the formula for a_n, we have

$$a_n = 3 + (n - 1)5 = 5n - 2$$

Hence, $510 = n(3 + 5n - 2)$; that is,

$$5n^2 + n - 510 = 0 \qquad \text{or} \qquad (5n + 51)(n - 10) = 0$$

so that $n = 10$ or $n = -\frac{51}{5}$. Hence, $n = 10$ is the number of terms, since n must be a positive integer.

PROBLEM SET 10.1

In problems 1–6, find the first five terms in each sequence function.

1 $f(n) = \dfrac{n(n+2)}{2}$ 　　　　**2** $f(n) = \dfrac{n+4}{n}$

3 $f(n) = \dfrac{n(n-3)}{2}$ 　　　　**4** $f(n) = \dfrac{3}{n(n+1)}$

5 $f(n) = (-1)^n + 3$ 　　　　**6** $f(n) = \dfrac{n^2 - 2}{2}$

In problems 7–14, determine which sequences are arithmetic progressions, and find the common difference d and S_{10} for each such progression.

7 2, 5, 8, 11, . . . 　　　　**8** 3, 5, 7, 9, . . .

9 7, 12, 17, 22, . . . 　　　　**10** $11a + 7b, 7a + 2b, 3a - 3b, \ldots$

11 67, 54, 41, 28, . . . 　　　　**12** $9a^2, 16a^2, 23a^2, 30a^2, \ldots$

13 5.7, 6.9, 8.1, 9.3, . . . 　　　　**14** 1.4, 4.5, 7.6, 10.7, . . .

15 Find the tenth and fifteenth terms of the arithmetic progression $-13, -6, 1, 8, \ldots$.

16 Find the twelfth and thirty-fifth terms of the arithmetic progression 19, 17, 15, 13, ...

17 Find the sixth and ninth terms of the arithmetic progression $a + 24b$, $4a + 20b, 7a + 16b, \ldots$.

18 Find the third and sixteenth terms of the arithmetic progression $7a^2 - 4b, 2a^2 + 7b, -3a^2 + 18b, \ldots$.

19 Find S_7 for the arithmetic progression $6, 3b + 1, 6b - 4, \ldots$.

20 Find S_{10} for the arithmetic progression $x + 2y, 3y, -x + 4y, \ldots$.

In problems 21–25, certain elements of an arithmetic progression are given. Find the indicated unknowns.

21 $a_1 = 6;\ d = 3;\ a_{10};\ S_{10}$

22 $a_1 = 38;\ d = -2;\ n = 25;\ S_n$

23 $a_1 = 17;\ S_{18} = 2{,}310;\ d;\ a_{18}$

24 $d = 3;\ S_{25} = 400;\ a_1;\ a_{25}$

25 $a_1 = 27;\ a_n = 48;\ S_n = 1{,}200;\ n;\ d$

10.2   Geometric Progressions

A sequence in which the quotient of successive terms remains constant is called a geometric progression. For instance, consider the sequence 3, 6, 12, 24, ... Here the pattern in the formation of this sequence can be seen if we rewrite the sequence in the following form:

$$3, \; 3(2), \; 3(2)^2, \; 3(2)^3, \; \ldots$$

The nth term a_n of this sequence is given by $a_n = 3(2)^{n-1}$. 3 is called the *first term* and 2 is called the *common ratio.* Note that each term in this geometric progression after the first term is found by multiplying the preceding term by 2. In general: A *geometric progression* is a sequence of the following form:

$$a, \; ar, \; ar^2, \; \ldots, \; ar^{n-1}, \; \ldots$$

where a is called the *first term,* r is called the *common ratio,* and the nth term a_n is given by $a_n = ar^{n-1}$. For example, $1, \frac{1}{2}, \frac{1}{4}, \frac{1}{8}, \ldots$ is a geometric progression in which $a = 1$, $r = \frac{1}{2}$, and $a_n = (\frac{1}{2})^{n-1}$.

Just as we did with arithmetic progressions, we shall derive a formula for S_n, the sum of the first n terms of a geometric progression. We start with the expression for S_n in expanded form,

$$S_n = a + ar + ar^2 + \cdots + ar^{n-1}$$

and then multiply both sides of that equation by r, thereby obtaining

$$rS_n = ar + ar^2 + ar^3 + \cdots + ar^n$$

Next, subtracting rS_n from S_n, we get $S_n - rS_n = a - ar^n$ so that $(1 - r)S_n = a - ar^n$, or

$$S_n = \frac{a - ar^n}{1 - r} = \frac{a(1 - r^n)}{1 - r} \qquad \text{for } r \neq 1$$

If $r = 1$, then

$$S_n = a + a + \overbrace{\cdots + a}^{n \text{ terms}} = na$$

EXAMPLES

1   Find the tenth term and the sum of the first ten terms of the geometric progression whose first term is $\frac{1}{2}$ and whose common ratio is 2.

SOLUTION. Here we have $a = \frac{1}{2}$ and $r = 2$. Using the formula for a_n, we have $a_{10} = ar^{10-1} = \frac{1}{2}(2^9) = 256$. Also, from the formula for S_n, we have

$$S_{10} = \frac{\frac{1}{2}(1 - 2^{10})}{1 - 2} = \frac{\frac{1}{2}(-1,023)}{-1} = 511.5$$

2 The sum of the first five terms of a geometric progression is $2\frac{7}{27}$, and the common ratio is $-\frac{1}{3}$. Find the terms of the geometric progression.

SOLUTION. From the formula for S_n, we have

$$2\frac{7}{27} = \frac{a[1 - (-\frac{1}{3})^5]}{1 - (-\frac{1}{3})}$$

so that

$$2\frac{7}{27} = \left(\frac{\frac{244}{243}}{\frac{4}{3}}\right) \cdot a$$

Hence, $a = 3$, and the terms of the geometric progression are

$$3, 3(-\tfrac{1}{3}), 3(-\tfrac{1}{3})^2, 3(-\tfrac{1}{3})^3, \ldots \qquad \text{or} \qquad 3, -1, \tfrac{1}{3}, -\tfrac{1}{9}, \ldots$$

PROBLEM SET 10.2

In problems 1–8, determine which sequences are geometric progressions and give the value of the common ratio and S_5 for each such progression.

1 $2, 6, 18, \ldots$

2 $1, \frac{1}{5}, \frac{1}{25}, \ldots$

3 $1, -2, 4, \ldots$

4 $\frac{4}{9}, \frac{1}{6}, \frac{1}{16}, \ldots$

5 $81, 54, 36, \ldots$

6 $147, -21, 3, \ldots$

7 $9, -6, 4, \ldots$

8 $64, -32, 16, \ldots$

In problems 9–14, find the indicated term of each geometric progression.

9 The tenth term of $-4, 2, -1, \frac{1}{2}, \ldots$

10 The eighth term of $\frac{1}{8}, \frac{1}{4}, \frac{1}{2}, \ldots$

11 The fifth term of $32, 16, 8, \ldots$

12 The eleventh term of $1, 1.03, (1.03)^2, \ldots$

13   The nth term of $1, 1 + a, (1 + a)^2, \ldots$

14   The twelfth term of $10^{-5}, 10^{-7}, 10^{-9}, \ldots$

15   Find the sixth and tenth terms and the sum of the first ten terms of the geometric progression $6, 12, 24, 48, \ldots$.

16   Find the sixth and eighth terms and the sum of the first eight terms of the geometric progression $2, 6, 18, \ldots$.

17   Find the fifth term of the geometric progression $3, 6, 12, \ldots$.

18   Find the eleventh term of the geometric progression $10, 10^2, 10^3, \ldots$ and the sum of the first eleven terms.

In problems 19–26, find the indicated element in each geometric progression with the given elements.

19 $a_1 = 2$; $n = 3$; $S_n = 26$; r

20 $r = 2$; $n = 5$; $a_n = -48$; a_1 and S_n

21 $a_1 = 3$; $a_n = 192$; $n = 7$; r

22 $a_6 = 3$; $a_9 = -81$; r and a_1

23 $a_5 = \frac{1}{8}$; $r = -\frac{1}{2}$; a_9 and S_8

24 $a_1 = 1$; $r = (1.03)^{-1}$; $a_9 = (1.03)^{-8}$; S_8

25 $a_1 = \frac{1}{16}$; $r = 2$; $a_n = 32$; n and S_n

26 $a_1 = 250$; $r = \frac{3}{5}$; $a_n = 32\frac{2}{5}$; n and S_n

27   Write the first three terms of a geometric progression in which the fourth term is 2 and the seventh term is 54.

28   Write the first four terms of a geometric progression in which the fifth term is $\frac{1}{7}$ and the seventh term is $\frac{4}{343}$.

10.3   Finite Sums

In considering sums of terms of an arithmetic or a geometric progression, it is convenient to employ a special notation for a *finite sum* of the terms, called the *summation notation* or the *sigma notation*. We shall use the symbol

$$\sum_{k=1}^{n} a_k$$

to represent the sum $a_1 + a_2 + a_3 + \cdots + a_n$, read "the sum of a_k from

$k = 1$ to n." The symbol Σ (the capital sigma of the Greek alphabet, corresponding to the letter S) denotes that a sum is to be taken. The symbols above and below the Σ notation indicate that k is an integer running from 1 to n inclusive; k is called the *index of summation*. There is no particular reason to use k for the index of summation—any letter will do, but i, j, k, and n are the most commonly used indices. For instance,

$$\sum_{k=1}^{n} 5^k = \sum_{i=1}^{n} 5^i = \sum_{j=1}^{n} 5^j = 5^1 + 5^2 + 5^3 + \cdots + 5^n$$

EXAMPLES

In examples 1–3, evaluate each finite sum.

1 $\displaystyle\sum_{k=1}^{3} (4k^2 - 3k)$

SOLUTION. $a_k = 4k^2 - 3k$. To find the indicated sum, we substitute the integers 1, 2, and 3 for k in succession and then add the resulting numbers. Thus,

$$\sum_{k=1}^{3} (4k^2 - 3k) = [4(1^2) - 3(1)] + [4(2^2) - 3(2)] + [4(3^2) - 3(3)]$$
$$= 1 + 10 + 27 = 38$$

2 $\displaystyle\sum_{i=3}^{6} i(i - 2)$

SOLUTION. $a_i = i(i - 2)$. Notice here that the index of summation begins with 3. To find the indicated sum, we substitute the integers 3, 4, 5, and 6 in succession and then add the resulting numbers. Thus,

$$\sum_{i=3}^{6} i(i - 2) = [3(3 - 2)] + [4(4 - 2)] + [5(5 - 2)] + [6(6 - 2)]$$
$$= 3 + 8 + 15 + 24 = 50$$

3 $\displaystyle\sum_{k=2}^{5} \frac{k - 1}{k + 1}$

SOLUTION. Here $a_k = (k - 1)/(k + 1)$, so that

$$\sum_{k=2}^{5} \frac{k - 1}{k + 1} = \left(\frac{2 - 1}{2 + 1}\right) + \left(\frac{3 - 1}{3 + 1}\right) + \left(\frac{4 - 1}{4 + 1}\right) + \left(\frac{5 - 1}{5 + 1}\right)$$
$$= \tfrac{1}{3} + \tfrac{2}{4} + \tfrac{3}{5} + \tfrac{4}{6} = \tfrac{21}{10}$$

4   Write the following finite sum in sigma notation:

$$1 + \tfrac{1}{2} + \tfrac{1}{4} + \tfrac{1}{8} + \tfrac{1}{16}$$

SOLUTION

$$1 + \tfrac{1}{2} + \tfrac{1}{4} + \tfrac{1}{8} + \tfrac{1}{16} = (\tfrac{1}{2})^0 + (\tfrac{1}{2})^1 + (\tfrac{1}{2})^2 + (\tfrac{1}{2})^3 + (\tfrac{1}{2})^4$$

$$= \sum_{k=0}^{4} (\tfrac{1}{2})^k$$

or, equivalently,

$$1 + \tfrac{1}{2} + \tfrac{1}{4} + \tfrac{1}{8} + \tfrac{1}{16} = \sum_{k=1}^{5} (\tfrac{1}{2})^{k-1} \qquad \text{(Why?)}$$

PROBLEM SET 10.3

In problems 1–14, find the numerical value of each finite sum.

1    $\displaystyle\sum_{k=1}^{5} k$

2    $\displaystyle\sum_{k=0}^{4} \frac{2^k}{k+1}$

3    $\displaystyle\sum_{i=1}^{10} 2i(i-1)$

4    $\displaystyle\sum_{k=0}^{4} 3^{2k}$

5    $\displaystyle\sum_{k=2}^{5} 2^{k-2}$

6    $\displaystyle\sum_{i=2}^{6} \frac{1}{i}$

7    $\displaystyle\sum_{k=1}^{3} (2k+1)$

8    $\displaystyle\sum_{k=1}^{5} (3k^2 - 5k + 1)$

9    $\displaystyle\sum_{i=1}^{4} \frac{i}{i+1}$

10    $\displaystyle\sum_{k=1}^{4} k^k$

11    $\displaystyle\sum_{k=1}^{100} 5$

12    $\displaystyle\sum_{i=3}^{7} (i+2)$

13    $\displaystyle\sum_{k=1}^{5} \frac{1}{k(k+1)}$

14    $\displaystyle\sum_{k=1}^{4} \frac{3}{k}$

In problems 15–18, express each finite sum in sigma notation.

15    $1 + 4 + 7 + 10 + 13$

16    $\tfrac{1}{2} + \tfrac{1}{4} + \tfrac{1}{8} + \tfrac{1}{16} + \tfrac{1}{32}$

17    $\tfrac{3}{5} + \tfrac{9}{25} + \tfrac{27}{125} + \tfrac{81}{625}$

18    $\tfrac{1}{6} + \tfrac{2}{11} + \tfrac{3}{16} + \tfrac{4}{21}$

In problems 19–23, determine whether each statement is true or false. Give the reason.

19 $\displaystyle\sum_{k=0}^{100} k^3 = \sum_{k=1}^{100} k^3$

20 $\displaystyle\sum_{k=1}^{100} 2 = 200$

21 $\displaystyle\sum_{k=0}^{100} (k+2) = \left(\sum_{k=0}^{100} k\right) + 2$

22 $\displaystyle\sum_{k=0}^{99} (k+1)^2 = \sum_{k=1}^{100} k^2$

23 $\displaystyle\sum_{k=0}^{100} k^2 = \left(\sum_{k=0}^{100} k\right)^2$

10.4 Binomial Theorem

In Section 2.4 we considered the special products $(a+b)^2$ and $(a+b)^3$. In this section we shall develop a formula for the expansion of $(a+b)^n$, where n is any positive integer. By actual multiplication, we can show that

$$(a+b)^1 = a+b$$
$$(a+b)^2 = a^2 + 2ab + b^2$$
$$(a+b)^3 = a^3 + 3a^2b + 3ab^2 + b^3$$
$$(a+b)^4 = a^4 + 4a^3b + 6a^2b^2 + 4ab^3 + b^4$$
$$(a+b)^5 = a^5 + 5a^4b + 10a^3b^2 + 10a^2b^3 + 5ab^4 + b^5$$

The above pattern holds for the expansion of $(a+b)^n$, where n is a positive integer. The following rules are used for that expansion:

1 There are $n+1$ terms. The "first term" is a^n and the "last term" is b^n.

2 The powers of a decrease by 1 and the powers of b increase by 1 for each term. In any case, the sum of the exponents of a and b is n for each term.

One method for displaying the coefficients in the expansion of $(a+b)^n$, for $n = 1, 2, 3, \ldots$, is the following array of numbers, known as *Pascal's triangle:*

$$(a + b)^0$$
$$(a + b)^1$$
$$(a + b)^2$$
$$(a + b)^3$$
$$(a + b)^4$$
$$(a + b)^5$$

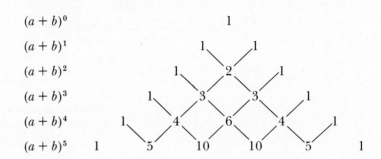

The coefficients in each line—except the first and last, which are always 1—can be found by adding the pairs of coefficients from the preceding line as indicated by the V's. For example,

indicates that 10 was obtained by adding 4 and 6. The pattern for determining the coefficients in the expansion of $(a + b)^n$ is easier to detect if the following notation for the product of all positive integers from 1 to n inclusive is used. The symbol $n!$ (read "n factorial" or "factorial n") is defined by $n! = 1 \cdot 2 \cdot 3 \cdots (n - 1)n$ or

$$n! = n(n - 1)(n - 2) \cdots 2 \cdot 1$$

Thus, $4! = 4 \cdot 3 \cdot 2 \cdot 1 = 24$ and $6! = 6 \cdot 5 \cdot 4 \cdot 3 \cdot 2 \cdot 1 = 720$.

We have defined $n!$ for positive integers n as

$$n! = n(n - 1)(n - 2) \cdots 4 \cdot 3 \cdot 2 \cdot 1$$

Therefore,

$$(n - 1)! = (n - 1)(n - 2)(n - 3) \cdots 4 \cdot 3 \cdot 2 \cdot 1$$

and, by multiplying both sides of this equation by n, we have the following recursive relationship:

$$n! = n[(n - 1)!]$$

Setting $n = 1$ in this relationship, we have

$$1! = 1[(1 - 1)!] \quad \text{or} \quad 1! = 1 \cdot 0!$$

Therefore, we shall define $0! = 1$.

Expressions involving factorial notation may be simplified as follows:

$$\frac{7!}{5!} = \frac{7 \cdot 6 \cdot 5 \cdot 4 \cdot 3 \cdot 2 \cdot 1}{5 \cdot 4 \cdot 3 \cdot 2 \cdot 1} = 7 \cdot 6 = 42$$

$$\frac{8!}{3! \cdot 5!} = \frac{8 \cdot 7 \cdot 6 \cdot 5 \cdot 4 \cdot 3 \cdot 2 \cdot 1}{(3 \cdot 2 \cdot 1)(5 \cdot 4 \cdot 3 \cdot 2 \cdot 1)} = 8 \cdot 7 = 56$$

In expanding $(a + b)^n$, the observations above will be emphasized, using factorial notation. A sum of terms with variable factors like a^n, $a^{n-1}b$, $a^{n-2}b^2$, $a^{n-3}b^3$, . . . , ab^{n-1}, b^n, with suitable coefficients, is obtained as follows:

1   The "first term" is a^n, and the coefficient is 1.

2   The "second term" contains $a^{n-1}b$, and the coefficient is $n/1! = n$.

3   The "third term" contains $a^{n-2}b^2$, and the coefficient is

$$\frac{n(n-1)}{2!} = \frac{n(n-1)}{2}$$

4   The "fourth term" contains $a^{n-3}b^3$, and the coefficient is

$$\frac{n(n-1)(n-2)}{3!} = \frac{n(n-1)(n-2)}{6}$$

These results can be generalized to obtain the *binomial theorem* (also known as the *binomial expansion*):

$$(a + b)^n = a^n + \frac{n}{1!}a^{n-1}b + \frac{n(n-1)}{2!}a^{n-2}b^2$$

$$+ \frac{n(n-1)(n-2)}{3!}a^{n-3}b^3 + \cdots$$

$$+ \frac{n(n-1)(n-2) \cdots (n-k+2)}{(k-1)!}a^{n-k+1}b^{k-1}$$

$$+ \cdots + b^n$$

where k is the number of the term. For example, by substituting x for a and $2y^2$ for b, we have

$$(x + 2y^2)^5 = [x + (2y^2)]^5$$

$$= x^5 + \frac{5}{1!}x^4(2y^2) + \frac{5 \cdot 4}{2!}x^3(2y^2)^2$$

$$+ \frac{5 \cdot 4 \cdot 3}{3!}x^2(2y^2)^3 + \frac{5 \cdot 4 \cdot 3 \cdot 2}{4!}x(2y^2)^4$$

$$+ \frac{5 \cdot 4 \cdot 3 \cdot 2 \cdot 1}{5!}(2y^2)^5$$

$$= x^5 + 10x^4y^2 + 40x^3y^4 + 80x^2y^6 + 80xy^8 + 32y^{10}$$

In order to write any particular term of the binomial expansion $(a + b)^n$, or to find the term where b has any particular exponent, observe that the kth term (denoted by u_k) in the binomial expansion is given by

$$u_k = \frac{n(n-1)(n-2)\,\cdots\,(n-k+2)}{(k-1)!}a^{n-k+1}b^{k-1}$$

For example, the sixth term u_6 of $(x^2 + 2y)^{12}$ is

$$\frac{12 \cdot 11 \cdot 10 \cdot 9 \cdot 8}{5 \cdot 4 \cdot 3 \cdot 2 \cdot 1}(x^2)^7(2y)^5 = 25{,}344x^{14}y^5$$

If we replace $(k - 1)$ by k in the expression for u_k, we obtain an expression for the $k + 1$ term, denoted by u_{k+1}, which contains the factor b^k. That is,

$$u_{k+1} = \frac{n(n-1)(n-2)\,\cdots\,(n-k+1)}{k!}a^{n-k}b^k$$

$$= \frac{n!}{k!(n-k)!}a^{n-k}b^k$$

The latter form of the coefficient results from multiplying the numerator and the denominator of the coefficient by $(n - k)!$.

Thus, the term involving y^4 in the expansion of $(x^2 + 2y)^{12}$ is given by

$$\frac{12 \cdot 11 \cdot 10 \cdot 9}{4!}(x^2)^8(2y)^4 = 7{,}920x^{16}y^4$$

EXAMPLES

1 Write in expanded form and simplify.

a) $\dfrac{7!}{5! \cdot 3!}$ b) $\dfrac{3! + 5!}{4! - 6!}$ c) $\dfrac{(n+1)!}{(n-1)!}$

SOLUTION

a) $\dfrac{7!}{5! \cdot 3!} = \dfrac{7 \cdot 6 \cdot 5!}{5! \cdot 3 \cdot 2 \cdot 1} = 7$

b) $\dfrac{3! + 5!}{4! - 6!} = \dfrac{3!(1 + 5 \cdot 4)}{3!(4 - 4 \cdot 5 \cdot 6)} = \dfrac{1 + 20}{4 - 120} = -\dfrac{21}{116}$

c) $\dfrac{(n+1)!}{(n-1)!} = \dfrac{(n+1)(n)[(n-1)!]}{(n-1)!} = (n+1)n = n^2 + n$

2 Expand $(x + y)^7$ by the binomial theorem.

SOLUTION. Letting $a = x$, $b = y$, and $n = 7$, we have

$$(x + y)^7 = x^7 + \frac{7}{1!}x^6y + \frac{7 \cdot 6}{2!}x^5y^2 + \frac{7 \cdot 6 \cdot 5}{3!}x^4y^3 + \frac{7 \cdot 6 \cdot 5 \cdot 4}{4!}x^3y^4$$

$$+ \frac{7 \cdot 6 \cdot 5 \cdot 4 \cdot 3}{5!}x^2y^5 + \frac{7 \cdot 6 \cdot 5 \cdot 4 \cdot 3 \cdot 2}{6!}xy^6 + y^7$$

$$= x^7 + 7x^6y + 21x^5y^2 + 35x^4y^3 + 35x^3y^4$$

$$+ 21x^2y^5 + 7xy^6 + y^7$$

3 Find the eighth term in the expansion of $(x - y)^{12}$.

SOLUTION. The kth term of $(a + b)^n$ is

$$\frac{n(n - 1)(n - 2) \cdots (n - k + 2)}{(k - 1)!}a^{n-k+1}b^{k-1}$$

Letting $a = x$, $b = -y$, $k = 8$, and $n = 12$, we have

$$n - k + 2 = 12 - 8 + 2 = 6$$

$$n - k + 1 = 12 - 8 + 1 = 5$$

and $k - 1 = 8 - 1 = 7$. Therefore, the eighth term is given by

$$\frac{12 \cdot 11 \cdot 10 \cdot 9 \cdot 8 \cdot 7 \cdot 6}{7!}x^5(-y)^7 = -792x^5y^7$$

4 Find and simplify the term involving x^7 in the expansion of $(2 - x)^{12}$.

SOLUTION. Using the formula for the $k + 1$ term, that is,

$$u_{k+1} = \frac{n!}{k!(n - k)!}a^{n-k}b^k$$

we have

$$\frac{12!}{7!(12 - 7)!}(2)^5(-x)^7 = \frac{12 \cdot 11 \cdot 10 \cdot 9 \cdot 8 \cdot 7!}{7! \cdot 5 \cdot 4 \cdot 3 \cdot 2 \cdot 1}(-32x^7)$$

$$= -25,344x^7$$

PROBLEM SET 10.4

In problems 1–10, write each expression in expanded form and simplify the results.

1 $\dfrac{4!}{6!}$

2 $\dfrac{10!}{5! \cdot 7!}$

3 $\dfrac{2!}{4! - 3!}$

4 $\dfrac{1}{4!} + \dfrac{1}{3!}$

5 $\dfrac{3! \cdot 8!}{4! \cdot 7!}$

6 $\dfrac{4! \cdot 6!}{8! - 5!}$

7 $\dfrac{0}{0!}$

8 $\dfrac{(n-2)!}{(n+1)!}$

9 $\dfrac{(n+1)!}{(n-3)!}$

10 $\dfrac{(n+k)!}{(n+k-2)!}$

In problems 11–18, expand each expression by using the binomial theorem and simplify each term.

11 $(x + 2)^5$

12 $(a - 2b)^4$

13 $(x^2 + 4y^2)^3$

14 $(1 - a^{-1})^5$

15 $(a^3 - a^{-1})^6$

16 $\left(1 - \dfrac{x}{y^2}\right)^5$

17 $\left(2 + \dfrac{x}{y}\right)^5$

18 $(x + y + z)^3$

In problems 19–22, use the binomial theorem to expand each expression and check the results using Pascal's triangle.

19 $(2z + x)^8$

20 $(x - 3)^8$

21 $(y^2 - 2x)^5$

22 $\left(\dfrac{1}{a} + \dfrac{x}{2}\right)^5$

In problems 23–26, find the first four terms of each expansion.

23 $(x^2 - 2a)^{10}$

24 $\left(2a - \dfrac{1}{b}\right)^6$

25 $\left(\sqrt{\dfrac{x}{2}} + 2y\right)^7$

26 $\left(\dfrac{1}{a} + \dfrac{x}{2}\right)^{11}$

In problems 27–34, find the first five terms in each expansion and simplify.

27 $(x + y)^{16}$

28 $(a^2 + b^2)^{12}$

29 $(a - 2b^2)^{11}$

30 $(a + 2y^2)^8$

31 $(x - 2y)^7$

32 $\left(1 - \dfrac{x}{y^2}\right)^8$

33 $(a^3 - a^2)^9$

34 $\left(x + \dfrac{1}{2y}\right)^{15}$

In problems 35–40, find the indicated term for each expression.

35 $\left(\dfrac{x^2}{2} + a\right)^{15}$, fourth term

36 $(y^2 - 2z)^{10}$, sixth term

37 $\left(2x^2 - \dfrac{a^2}{3}\right)^9$, seventh term

38 $(x + \sqrt{a})^{12}$, middle term

39 $\left(a + \dfrac{x^3}{3}\right)^9$,

term containing x^{12}

40 $\left(2\sqrt{y} - \dfrac{x}{2}\right)^{10}$,

term containing y^4

REVIEW PROBLEM SET

In problems 1–6, for each sequence, write the first four terms.

1 $f(n) = 3 + (-1)^{n+1}$

2 $f(n) = 2^n$

3 $f(n) = 5 - \dfrac{3}{n}$

4 $f(n) = \dfrac{2}{n + 1}$

5 $f(n) = \dfrac{n(4n + 1)}{5}$

6 $f(n) = (-1)^n 2^{n-1}$

In problems 7–12, for each arithmetic progression, find the indicated term and the indicated sum.

7 4, 9, 14, . . . ; ninth term and S_9

8 21, 19, 17, . . . ; tenth term and S_{10}

9 42, 39, 36, . . . ; eleventh term, and S_{11}

10 0.3, 1.2, 2.1, . . . ; fifteenth term and S_{15}

11 $\frac{1}{6}, \frac{1}{3}, \frac{1}{2}$, . . . ; twenty-fourth term and S_{24}

12 $\frac{1}{6}, \frac{1}{4}, \frac{1}{3}$, . . . ; thirtieth term and S_{30}

In problems 13–16, find the value of x so that each will be an arithmetic progression.

13 $2, 1 + 2x, 21 - 3x, \ldots$

14 $2x, \frac{1}{2}x + 3, 3x - 10, \ldots$

15 $3x, 2x + 1, x^2 - 4, \ldots$

16 $1, x + 1, 3x - 5, \ldots$

In problems 17–22, for each geometric progression, find the indicated term and the indicated sum.

17 $3, 12, 48, \ldots$; eighth term and S_8

18 $16, 8, 4, \ldots$; ninth term and S_9

19 $81, -27, 9, \ldots$; sixth term and S_6

20 $\sqrt{2}, 2, 2\sqrt{2}, \ldots$; tenth term and S_{10}

21 $3, -3\sqrt{2}, 6, \ldots$; eighteenth term and S_{18}

22 $2, -2\sqrt{2}, 4, \ldots$; twentieth term and S_{20}

In problems 23–26, determine the value of x so that each will be a geometric progression.

23 $x - 6, x + 6, 2x + 2, \ldots$

24 $\frac{1}{2}x, x + 2, 3x + 1, \ldots$

25 $x - 7, x + 5, 8x - 5, \ldots$

26 $x + 1, x + 2, x - 3, \ldots$

In problems 27–32, find the numerical value of each finite sum.

27 $\displaystyle\sum_{k=1}^{5} k(2k - 1)$
        28 $\displaystyle\sum_{k=5}^{10} (2k - 1)^2$

29 $\displaystyle\sum_{k=1}^{4} 2k^2(k - 3)$
        30 $\displaystyle\sum_{k=1}^{6} 3^{k+1}$

31 $\displaystyle\sum_{k=2}^{6} (k + 1)(k + 2)$
        32 $\displaystyle\sum_{k=4}^{7} \frac{1}{k(k - 3)}$

In problems 33–44, use the binomial theorem to expand each expression.

33 $(x + 2y)^4$

34 $(x - 3y)^4$

35 $(1 + x)^5$

36 $(2x + 1)^5$

37 $(1 - 2x)^6$

38 $(a - b)^6$

39 $(3x + y)^4$

40 $\left(x - \dfrac{1}{x}\right)^8$

41 $(3x + \sqrt{x})^5$

42 $\left(3y + \dfrac{1}{3\sqrt{y}}\right)^6$

43 $\left(2x + \dfrac{1}{y}\right)^3$

44 $\left(x^3 - \dfrac{1}{\sqrt{x}}\right)^9$

In problems 45–50, find the indicated term in each binomial expansion.

45 fifth term of $(x + y)^{10}$

46 sixth term of $(x - y)^{11}$

47 fifth term of $(2x + y)^{10}$

48 sixth term of $(x - 3y)^9$

49 fourth term of $(3x + y)^{11}$

50 third term of $(2x + y)^{20}$

APPENDIX A

Measurement

LENGTHS IN THE METRIC SYSTEM

meter (m)	millimeter (mm)	centimeter (cm)	decimeter	decameter	hectometer	kilometer (km)
1 m = $\Big\{$	1000	100	10	$0.1 = \dfrac{1}{10}$	$0.01 = \dfrac{1}{100}$	$0.001 = \dfrac{1}{1000}$

WEIGHTS IN THE METRIC SYSTEM

gram (g)	milligram (mg)	centigram (cg)	decigram	decagram	hectogram	kilogram (kg)
1 g = $\Big\{$	1000	100	10	$0.1 = \dfrac{1}{10}$	$0.01 = \dfrac{1}{100}$	$0.001 = \dfrac{1}{1000}$

LENGTHS IN THE ENGLISH SYSTEM

foot (ft)	inch (in)	yard (yd)	mile
1 ft = $\Big\{$	12	$\dfrac{1}{3}$	$\dfrac{1}{5,280}$

WEIGHTS IN THE ENGLISH SYSTEM

pound (lb)	ounce (oz)	ton
1 lb = $\Big\{$	16	$\dfrac{1}{2,000}$

CONVERSION FROM THE ENGLISH SYSTEM TO THE METRIC SYSTEM

1 inch = 2.54 centimeters,   1 meter = 39.37 inches,   1 pound = 0.4536 kilogram

APPENDIX B

Formulas from Geometry

B.1 Areas of Plane Figures

Consider the triangle and the rectangle represented in Figure B.1. The interior of the triangle is completely contained in the interior of the rectangle, while there are points in the interior of the rectangle not in the triangle. We say that the *area* covered by the rectangle is *larger than* that of the triangle.

Figure B.1

To find a way of measuring area, we must first agree on a basic *unit of measure.* For convenience, we shall select the square to be our basic unit of measurement. The number that represents the *area of a plane figure,* then, represents the *number of squares that are contained in the plane figure.* The length of the side of the square will depend on the relative size of the object being measured. We can use English units as well as metric units as standard units of measurement for areas.

Area of a Rectangle

The area of a rectangle is the number of basic units (squares) it contains. In Figure B.2 we have shown the number of squares that are contained in rectangle *ABCD*. By actually counting, we find that there are 4 rows of squares, each containing 5 squares, or a total of 20 squares. Repeating this process with many other rectangles, we find that the number of squares contained in a rectangle is always *equal to the product of the base of the rectangle and the side of the rectangle, called the height* (in the example above, $20 = 5 \times 4$). In general, the area of a rectangle is the product of its length and width, or $A = lw$.

Figure B.2

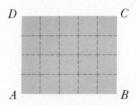

EXAMPLES

1   Find the area of a rectangle whose length is 5 centimeters and width is 7 centimeters.

SOLUTION. $A = lw = 5 \times 7 = 35$. Therefore, the area is 35 square centimeters.

2   How many square feet of carpeting will be required for a room that is 12 feet wide and 16 feet long?

SOLUTION.   The room is rectangular in shape, so that the area $A = lw$ or $A = 16 \times 12 = 192$. Therefore, the area is 192 square feet.

Area of a Square

The area of a square can be found by considering the fact that a square is also a rectangle. If we have a square of side s, then both its length and width are equal to s; hence, applying the formula for the area of a rectangle, we have

$$A = lw = ss = s^2$$

EXAMPLES

3   Find the area of a square whose side is 4.3 meters.

SOLUTION. $A = s^2 = (4.3)^2 = 18.49$. Therefore, the area is 18.49 square meters.

4   What is the cost of laying the cement for a square patio that is 18 feet wide at 40 cents per square foot?

SOLUTION.   Here $s = 18$ feet. We have $A = s^2$, so that $A = (18)^2 = 324$. Then the area is 324 square feet. The cost of the patio will be (40 cents) $\times$ (324 square feet) = \$129.60.

Area of a Parallelogram

The area of a parallelogram (Figure B.3) can be determined by the formula

$$A = bh$$

where b is the base and h is the height of the parallogram. The *height* of a parallelogram, then, is the *perpendicular line segment drawn between two parallel sides, one of which is the base.*

Figure B.3

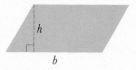

EXAMPLES

5   Find the area of a parallelogram whose base is 5 feet and whose height is 3 feet.

   SOLUTION.   The area of the parallelogram is expressed by $A = bh$ or $A = 5 \times 3 = 15$. Therefore, $A = 15$ square feet.

6   Find the height of the parallelogram whose area is 208 square inches and whose base is 26 inches.

   SOLUTION. $A = bh$. Now, $208 = 26 \times b$, so that $b = 8$. Therefore, the height is 8 inches.

Area of a Triangle

The area of a triangle is equal to one half the product of its base and height (Figure B.4). Symbolically, it is written $A = \frac{1}{2}bh$.

Figure B.4

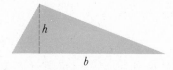

EXAMPLE

7   Find the area of a triangle whose base is 10 centimeters and whose height is 3.7 centimeters.

SOLUTION.   The area of the triangle is $A = \frac{1}{2}bh$, so that

$A = \frac{1}{2}(10) \times 3.7 = 18.5$ or
$A = 18.5$ square centimeters

The area of an equilateral triangle with side s is one fourth the product of the side squared and $\sqrt{3}$ (Figure B.5). Symbolically, $A = (s^2/4)\sqrt{3}$.

Figure B.5

EXAMPLE

8   Find the area of an equilateral triangle whose sides are 8 inches.

SOLUTION.   The area of an equilateral triangle is

$$A = \frac{s^2}{4}\sqrt{3} = \frac{(8)^2}{4}\sqrt{3} = 16\sqrt{3}$$

so that the area is $16\sqrt{3}$ square inches.

Area of a Trapezoid

The area of a trapezoid is equal to one half the product of its height and the sum of its bases (**Figure B.6**). Symbolically, $A = \frac{1}{2}h(b + b_1)$.

Figure B.6

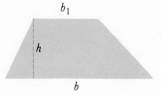

EXAMPLE

9   Find the area of a trapezoid whose bases are 12 meters and 8 meters and whose height is 5 meters (Figure B.7).

SOLUTION. $A = \frac{1}{2}h(b + b_1)$ or
$$A = \frac{1}{2}(5) \times (12 + 8) = 50$$

Therefore, the area is 50 square meters.

Figure B.7

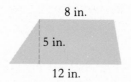

8 in.

5 in.

12 in.

Area of a Circle

Given a circle of radius r (Figure B.8), its area A is π times r^2, where $\pi = 3.1416$ . . .. In symbols, $A = \pi r^2$.

Figure B.8

r

EXAMPLE

10 Find the area of a circle whose radius is 10 centimeters.

SOLUTION. $A = \pi r^2$ or $A = \pi(10 \text{ inches})^2$ or
$$A = 3.1416 \times 100 = 314.16$$

Therefore, the area is 314.16 square centimeters.

PROBLEM SET B.1

1 Can you determine the base and the height of a rectangle if given its area?

2 How is the area of a parallelogram affected if its base and height are both doubled?

3 Given the base of a parallelogram and its area, can you determine its height?

4 Given the area of a triangle and the length of each side, can you find the lengths of its three heights?

5 If two triangles have the same area, must they have equal bases and equal heights?

6 Does a line segment joining the midpoints of the bases of a trapezoid divide it into two trapezoids of equal area?

7 Find the area of the following rectangles.
a) $h = 3$ meters, $b = 5$ meters
b) $h = 2\frac{1}{2}$ inches, $b = 3\frac{2}{3}$ inches
c) $h = 4.01$ centimeters, $b = 6.32$ centimeters

8 Find the areas of the following parallelograms.
a) $h = 4.32$ feet, $b = 5.7$ feet
b) $h = 6\frac{2}{3}$ inches, $b = 4\frac{1}{5}$ inches
c) $h = 2$ meters, $b = 3$ meters

9 Find the area of the following triangles.
a) $h = 6$ inches, $b = 10$ inches
b) $h = 7$ inches, $b = 4.2$ inches
c) $h = 16$ centimeters, $b = 11$ centimeters

10 Find the area of the following squares.
a) $s = 3.5$ inches
b) $s = 16$ feet
c) $s = 0.57$ meters

11 Find the area of the following trapezoids.
a) $h = 6$ feet, $b = 11$ feet, $b_1 = 4$ feet
b) $h = 7.5$ inches, $b = 4.3$ inches, $b_1 = 5.2$ inches
c) $h = 3\frac{1}{2}$ meters, $b = 4\frac{1}{3}$ meters, $b_1 = 5\frac{2}{3}$ meters

12 Find the area of the following circles.
a) $r = 5$ inches
b) $r = 7$ feet
c) $r = 3.1$ centimeters

13 Find the area of the regions bounded as shown in Figure B.9.

Figure B.9

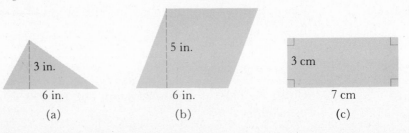

3 in.   5 in.   3 cm

6 in.   6 in.   7 cm

(a)   (b)   (c)

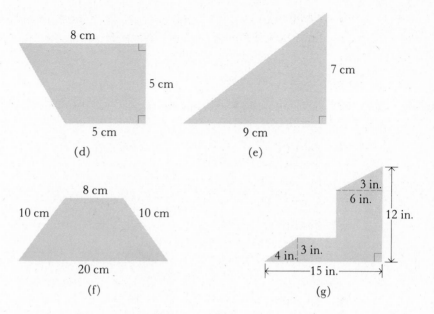

14   What is the cost of painting a rectangular wall 18 feet wide and 24 feet long at 80 cents per square foot?

15   A triangular plot of ground has a base of 120 meters and a height of 65 meters. What is the cost of the plot at $2.50 per square meter?

B.2   Perimeters of Plane Figures

The sum of the lengths of the sides on any plane figure is called the *perimeter*. The units of measurements of the perimeter are inches, feet, centimeters, and so on.

Perimeter of a Rectangle

The perimeter P of a rectangle is the sum of twice its length l and twice its width w (Figure B.10). Symbolically, $P = 2l + 2w$.

Figure B.10

EXAMPLES

1   Find the perimeter of a rectangle whose length is 8 centimeters and whose width is 5 centimeters.

SOLUTION. $P = 2l + 2w = 2(8) + 2(5) = 16 + 10 = 26$. Therefore, the perimeter is 26 centimeters.

2   A rectangle is four times as long as it is wide, and its perimeter is 60 feet. Find its width.

SOLUTION.   Let x feet be the width of the rectangle; then the length is $4x$ feet, so that $2x + 2(4x) = 60$ or $10x = 60$ or $x = 6$. Therefore, the width is 6 feet.

3   A rectangular field with length along a straight river is to be fenced off. If no fencing is needed along the river, what is the total cost of the fence if the length of the field is 3000 meters and the width is 1600 meters and the cost per foot is \$1.40.

SOLUTION.   The length l is given by $l = 3000 + 1600 + 1600 = 6200$, so that the length of the fence is 6200 meters. The total cost will be (6200 meters)(\$1.40) = \$8,680.

Perimeter of a Square

The perimeter P of a square is four times its side (Figure B.11). Symbolically, $P = 4s$.

Figure B.11

EXAMPLE

4   Find the perimeter of a square whose side is 8 centimeters.

SOLUTION. $P = 4s = 4(8) = 32$. Therefore, the perimeter is 32 centimeters.

Perimeter of a Parallelogram

The perimeter P of a parallelogram is the sum of the lengths of its sides (Figure B.12). Symbolically, $P = 2l + 2w$.

Figure B.12

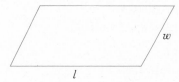

EXAMPLE

5 Find the perimeter of a parallelogram whose length is 10 centimeters and whose width is 6 centimeters.

SOLUTION. $P = 2l + 2w = 2(10) + 2(6) = 20 + 12 = 32$. Therefore, the perimeter is 32 centimeters.

Perimeter of a Triangle

The perimeter P of a triangle is the sum of the length of its sides (Figure B.13). Symbolically, $P = a + b + c$.

Figure B.13

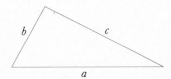

EXAMPLE

6 Find the perimeter of a triangle with sides 5 inches, 6 inches, and 8 inches.

SOLUTION. $P = a + b + c = 5 + 6 + 8 = 19$. Therefore, the perimeter is 19 inches.

Perimeter of a Trapezoid

The perimeter P of a trapezoid is $P = a + b + c + d$ (Figure B.14).

Figure B.14

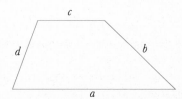

EXAMPLE

7   Find the perimeter of a trapezoid with sides 8 inches, 5 inches, 6 inches, and 5 inches.

SOLUTION. $P = a + b + c + d = 8 + 5 + 6 + 5 = 24$. Therefore, the perimeter is 24 inches.

Circumference of a Circle

The circumference C of a circle of radius r is 2π times r (Figure B.15). Symbolically, $C = 2\pi r$.

Figure B.15

EXAMPLE

8   Find the circumference of radius 5 centimeters.

SOLUTION. $C = 2\pi r = 2\pi(5) = 10\pi$. Therefore, the circumference is 10π or 31.4 centimeters.

PROBLEM SET B.2

1   Find the perimeter of a rectangle with the indicated length and width.
   a) $l = 8$ centimeters, $w = 2$ centimeters
   b) $l = 7$ inches, $w = 4$ inches
   c) $l = 3\frac{1}{5}$ feet, $w = \frac{3}{4}$ foot
   d) $l = 1\frac{1}{2}$ meters, $w = \frac{1}{2}$ meter

2   Find the perimeter of a square with the given side.
   a) $s = 10$ inches
   b) $s = 5$ centimeters
   c) $s = 3$ feet
   d) $s = 4$ meters
   e) $s = 2$ feet

3   Find the perimeter of the parallelogram $ABCD$ (Figure B.16) with the given dimensions.
   a) $l = 10$ inches, $w = 6$ inches
   b) $l = 8$ centimeters, $w = 4$ centimeters

c) $l = 3$ feet, $w = 2$ feet

d) $l = 2$ meters, $w = 1.5$ meters

Figure B.16

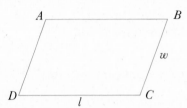

4 Find the perimeter of a triangle with the given sides.

a) 6 inches, 4 inches, and 8 inches

b) 5 meters, 4 meters, and 3 meters

c) 3 inches, 5 inches, and 4 inches

d) 8 inches, 8 inches, and 8 inches

e) 12 centimeters, 12 centimeters, and 8 centimeters

5 Find the perimeter of a trapezoid *ABCD* (Figure B.17) with the given dimensions.

a) $a = 10$ centimeters, $b = 4$ centimeters, $c = 6$ centimeters, and $d = 4$ centimeters

b) $a = \frac{1}{2}$ foot, $b = \frac{1}{4}$ foot, $c = \frac{1}{3}$ foot, and $d = \frac{1}{4}$ foot

c) $a = 2$ meters, $b = \frac{1}{2}$ meter, $c = 1$ meter, and $d = \frac{1}{2}$ meter

d) $a = 20$ inches, $b = 5$ inches, $c = 15$ inches, and $d = 5$ inches

Figure B.17

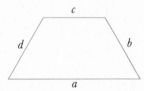

6 Find the circumference of a circle with the given radius.

a) $r = 4$ inches

b) $r = 5$ centimeters

c) $r = 3$ inches

d) $r = 2$ inches

e) $r = 7$ meters

7 The perimeter of a rectangular field whose length is $\frac{8}{5}$ of its width is 1,040 feet. Find its length and its width.

8 The length of a rectangle is 10 centimeters less than twice its width, and its perimeter is 106 centimeters. Find the length and the width.

B.3   Volumes and Surface Area

In this section we present some of the standard formulas for volume. The unit we shall use for measuring volume is the *cube*, whose edge is a unit length. For example, if the edge of the cube is 1 inch, the *unit of volume* is 1 cubic inch.

Volume and Surface Area of Prisms

Figure B.18

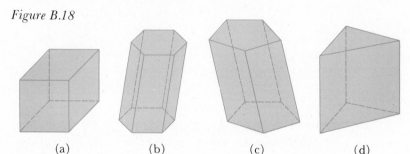

(a)   (b)   (c)   (d)

The volume of the prism (Figure B.18) is the product of the area of the base and the altitude. In symbols, $V = Ah$, where A is the area of the base and h is the altitude. The lateral surface area L.S. of the prism is the sum of the areas of its lateral faces, that is, L.S. $= Ph$, whereas the total surface area S is the sum of its lateral surface area and the area of its two bases, that is, $S = 2A + Ph$, where A is the area of the base, P is the perimeter of the base, and h is the altitude.

EXAMPLE

1   Find the lateral surface area, the total surface area, and the volume of a right triangular prism (the prism whose lateral faces are rectangles) if its altitude is 6 centimeters and its base is an equilateral triangle of side 8 centimeters (Figure B.19).

SOLUTION.   Since the perimeter $P = 8 + 8 + 8 = 24$ and the lateral area L.S. $= Ph$,

$$\text{L.S.} = 24 \times 6 = 144$$

The total surface area S is

$$S = 2A + Ph = 2(16\sqrt{3}) + 144 = 32\sqrt{3} + 144$$

The volume V is

$$V = Ah = (16\sqrt{3})(6) = 96\sqrt{3}$$

Therefore, the lateral surface area is 144 square centimeters, the total surface area is $32\sqrt{3} + 144$ square centimeters, and the volume is $96\sqrt{3}$ cubic centimeters.

Figure B.19

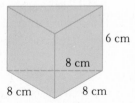

A prism whose faces and bases are rectangles is called a *rectangular solid* or a *rectangular parallelepiped* (box-shaped). Thus, if we denote the length of the rectangular solid by l, its width by w, and its height by h, its volume V is given by $V = lwh$ (Figure B.20). The total surface area S of such a solid is the sum of the area of the faces and bases; that is, $S = 2(lw + lh + wh)$.

Figure B.20

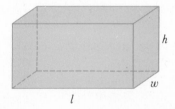

EXAMPLES

2   Find the volume and the total surface area of a rectangular box whose dimensions are 7, 5, and 3 units.

SOLUTION.   The volume is given by the formula

$$V = lwh = 7 \times 5 \times 3 = 105$$

The total surface area is given by the formula

$$S = 2(lw + lh + wh) = 2(7 \times 5 + 7 \times 3 + 5 \times 3) = 142$$

Therefore, the volume is 105 cubic units and the total surface area is 142 square units.

3   Find the volume of water required to fill a rectangular swimming pool of length 50 feet, width 30 feet, and depth 6 feet.

SOLUTION.   The volume of the water necessary to fill the swimming pool is $V = lwh = 50 \times 30 \times 6 = 9,000$. Therefore, the volume is 9,000 cubic feet.

4   A storage vault has a rectangular floor 72 feet by 48 feet. The walls are vertical and 15 feet high.
   a)   Find the total area of walls, floor, and ceiling.
   b)   Find the storage space of the room.

SOLUTION

a)   The total surface S is found by the formula

$$S = 2(lw + lh + wh)$$
$$= 2(72 \times 48 + 72 \times 15 + 48 \times 15)$$
$$= 10,512$$

Hence, the total area of walls, floor, and ceiling is 10,512 square feet.

b)   The storage space V is found by the formula

$$V = lwh$$
$$= 72 \times 48 \times 15$$
$$= 51,840$$

Therefore, the storage space is 51,840 cubic feet.

Volume and Surface Area of a Right Circular Cylinder

There are other solid figures besides prisms which have parallel congruent bases. An example of such a solid is a *right circular cylinder* whose bases are congruent circles that are perpendicular to its sides (Figure B.21).

Figure B.21

The formulas for the volume V, lateral surface area L.S. and the total surface area S of a right circular cylinder are given by $V = \pi r^2 h$, L.S. $= 2\pi rh$, and $S = 2\pi r^2 + 2\pi rh$, where r is the radius of the base and h is the altitude.

EXAMPLES

5   Find the lateral surface area, total surface area, and volume of a right circular cylinder of radius 3 inches and altitude 6 inches.

SOLUTION

$$\text{L.S.} = 2\pi rh = 2\pi(3)(6) = 36\pi$$

$$S = 2\pi r^2 + 2\pi rh$$
$$= 2\pi(3^2) + 36\pi = 18\pi + 36\pi$$
$$= 54\pi$$

$$V = \pi r^2 h = \pi(3^2)(6) = 54\pi$$

Therefore, the lateral surface area is 36π square inches, the total surface area is 54π square inches, and the volume is 54π cubic inches.

6   How many gallons of gasoline will a cylindrical tank hold that is 6 feet in diameter and 25 feet long if 1 cubic foot is equivalent to 7.5 gallons? Use $\pi = 3.14$.

SOLUTION. $V = \pi r^2 h = \pi(3^2)(25) = 706.50$, so that the volume is 706.50 cubic feet. Since 1 cubic foot is equivalent to 7.5 gallons, the number of gallons $= 706.50/7.5 = 94.2$ gallons.

Volume and Surface Area of the Pyramid

A *pyramid* is a three-dimensional figure whose base is a polygon and whose lateral faces are triangles (Figure B.22). If the base of the pyramid is a regular polygon and the pyramid has equal lateral edges, it is called a *regular pyramid* (Figure B.22a and b).

Figure B.22

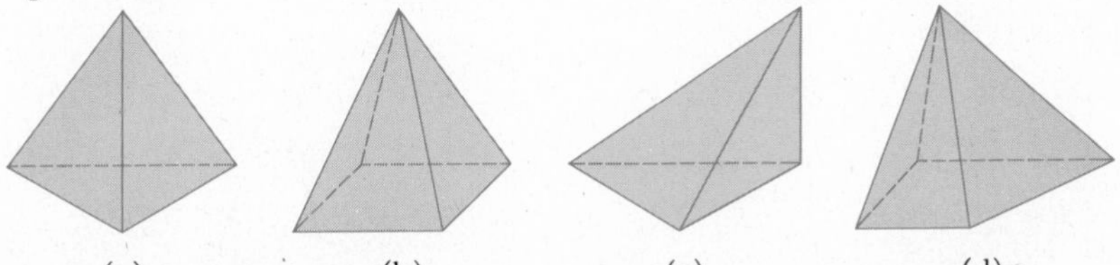

(a)   (b)   (c)   (d)

The volume V of a regular pyramid is expressed by the formula $V = \frac{1}{3}Ah$, where A is the area of the base and h is the altitude. Its lateral surface area L.S. is expressed by the formula $\text{L.S.} = \frac{1}{2}Pl$, where P is the perimeter of the base and l is the slant height (the height of a lateral face) (see Figure B.23).

Figure B.23

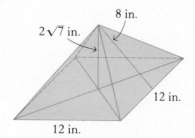

8 in.

$2\sqrt{7}$ in.

12 in.

12 in.

EXAMPLE

7 Find the lateral surface area and the volume of a regular square pyramid with a base edge of 12 inches, a height of $2\sqrt{7}$ inches, and a slant height of 8 inches (Figure B.23).

SOLUTION. $P = 4 \times 12 = 48$. The lateral surface area is $\frac{1}{2}Pl = \frac{1}{2} \times 48 \times 8 = 192$. The volume of the pyramid is $V = \frac{1}{3}Ah = \frac{1}{3} \times 144 \times 2\sqrt{7} = 96\sqrt{7}$. Therefore, the lateral surface area is 192 square inches and the volume is $96\sqrt{7}$ cubic inches.

Volume and Surface Area of a Right Circular Cone

The *right circular cone* resembles the pyramid except that the base of the cone is a circle (Figure B.24). The volume of a right circular cone is given by the formula $V = \frac{1}{3}\pi r^2 h$, where r is the radius and h is the height. The lateral surface area is given by the formula $\text{L.S.} = \pi r l$, where l is the slant height.

Figure B.24

h l

r

EXAMPLES

8   Find the lateral surface area and the volume of a right circular cone of radius 6 centimeters, altitude 8 centimeters, and slant height 10 centimeters (Figure B.25).

Figure B.25

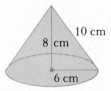

8 cm

10 cm

6 cm

SOLUTION

$$\text{L.S.} = \pi r l = \pi(6)(10) = 60\pi$$

$$V = \tfrac{1}{3}\pi r^2 h = \tfrac{1}{3}(36\pi)(8) = 96\pi$$

Therefore, the lateral surface area is 60π square centimeters and the volume is 96π cubic centimeters.

9   Find the volume of a conical funnel whose base radius is 4 inches and altitude is 8 inches.

SOLUTION. $V = \tfrac{1}{3}\pi r^2 h = \tfrac{1}{3}\pi(4^2)(8) = 128\pi/3$ cubic inches. Therefore, the volume is $128\pi/3$ cubic inches.

Volume and Surface Area of a Sphere

A *sphere* is the set of all points in space at a given distance from a given point (Figure B.26). Let r be the radius of the sphere; then the surface area S is expressed by the formula $S = 4\pi r^2$ and the volume V is expressed by the formula $V = \tfrac{4}{3}\pi r^3$.

Figure B.26

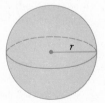

r

EXAMPLES

10   Find the surface area and the volume of a sphere of radius 3 centimeters.

SOLUTION

$$S = 4\pi r^2 = 4\pi(3^2) = 36\pi$$
$$V = \tfrac{4}{3}\pi r^3 = \tfrac{4}{3}(3^3) = 36\pi$$

Therefore, the surface area is 36π square centimeters, and the volume is 36π cubic centimeters.

11   How many cubic feet of gas is stored in a spherical storage tank of radius 10 feet?

SOLUTION. $V = \tfrac{4}{3}\pi r^3 = \tfrac{4}{3}\pi(10^3) = \tfrac{4}{3}\pi(1,000) = 4,000\pi/3$. Therefore, the volume is $4,000\pi/3$ cubic feet.

PROBLEM SET B.3

1   Find the volume and the total surface area of the rectangular prisms with the following dimensions.
a) $l = 3$ inches, $w = 4$ inches, and $h = 5$ inches
b) $l = 2$ meters, $w = 1$ meter, and $h = 3$ meters
c) $l = 3.5$ feet, $w = 2.5$ feet, and $h = 4$ feet
d) $l = 2$ centimeters, $w = 3$ centimeters, and $h = 6$ centimeters
e) $l = 2\sqrt{2}$ feet, $w = 3$ feet, and $h = 4$ feet

2   A right prism of altitude 10 inches has a base which is an equilateral triangle of side 6 inches. Find the volume and the total surface area of the prism.

3   Find the lateral surface area, total surface area, and the volume of the following cylinders.
a) $r = 2$ inches, $h = 4$ inches
b) $r = 3$ meters, $h = 4$ meters
c) $r = 4$ inches, $h = 6$ inches
d) $r = 1.5$ inches, $h = 6.5$ inches
e) $r = 2.5$ centimeters, $h = 5.3$ centimeters

4   Find the indicated missing dimension of each of the following right circular cylinders.
a) $r = 5$ centimeters, L.S. $= 16$ square centimeters, $h = ?$
b) $V = 320\pi$ cubic inches, $h = 5$ inches, $r = ?$
c) $S = 32\pi$ square inches, $h = 2$ inches, $r = ?$
d) $S = 32\pi$ square centimeters, $r = 4$ centimeters, $V = ?$

5   Find the total surface area and the volume of the following regular pyramids.
a)   Area of the square base $= 16$ square inches, the slant height $= 2\sqrt{10}$ inches, and the altitude $= 6$ inches.

b) The base is an equilateral triangle whose perimeter is 18 inches, the slant height is 10 inches, and the altitude $= \sqrt{97}$ inches.
c) A square base with an edge of 2 meters, height of 3 meters, and slant height of $\sqrt{10}$ meters.
d) A base that is an equilateral triangle with a side of 12 centimeters, slant height $= 2\sqrt{19}$ centimeters, and altitude $= 8$ centimeters.

6 Find the missing part for each of the following right circular cones.
a) $r = 3$ inches, $h = 4$ inches, $l = 5$ inches, L.S. $=$?, $S =$?, and $V =$?
b) $r = 5$ centimeters, $h = 12$ centimeters, $l = 13$ centimeters, L.S. $=$?, $S =$?, and $V =$?
c) $r = 2$ inches, $h = 4\sqrt{2}$ inches, L.S. $= 12\pi$ square inches, $S =$?, and $V =$?
d) $h = 3$ meters, $V = \pi$ cubic meters, $l = \sqrt{10}$ meters, L.S. $=$?, and $S =$?

7 Find the surface area and volume of a sphere with the given radius.
a) 5 inches
b) 2 inches
c) 6 centimeters
d) 4 inches
e) $3\sqrt{2}$ centimeters
f) $2\sqrt{3}$ inches

8 Find the volume of metal in a cylindrical pipe (open at both ends) of length 7 feet, inner radius 4 inches, and outer radius $4\frac{1}{4}$ inches.

9 How much will 1,000 cylindrical steel rods $\frac{5}{8}$ inch in diameter and 15 feet long weigh if 1 cubic foot of steel weighs 490 pounds?

10 If the radius of a given sphere is tripled, what is the relationship between the new volume and the original volume?

11 If each dimension of a rectangular prism is doubled:
a) How is its volume changed?
b) How is its lateral surface area changed?

APPENDIX C

Table of Common Logarithms

TABLE I **COMMON LOGARITHMS**

n	0	1	2	3	4	5	6	7	8	9
10	0000	0043	0086	0128	0170	0212	0253	0294	0334	0374
11	0414	0453	0492	0531	0569	0607	0645	0682	0719	0755
12	0792	0828	0864	0899	0934	0969	1004	1038	1072	1106
13	1139	1173	1206	1239	1271	1303	1335	1367	1399	1430
14	1461	1492	1523	1553	1584	1614	1644	1673	1703	1732
15	1761	1790	1818	1847	1875	1903	1931	1959	1987	2014
16	2041	2068	2095	2122	2148	2175	2201	2227	2253	2279
17	2304	2330	2355	2380	2405	2430	2455	2480	2504	2529
18	2553	2577	2601	2625	2648	2672	2695	2718	2742	2765
19	2788	2810	2833	2856	2878	2900	2923	2945	2967	2989
20	3010	3032	3054	3075	3096	3118	3139	3160	3181	3201
21	3222	3243	3263	3284	3304	3324	3345	3365	3385	3404
22	3424	3444	3464	3483	3502	3522	3541	3560	3579	3598
23	3617	3636	3655	3674	3692	3711	3729	3747	3766	3784
24	3802	3820	3838	3856	3874	3892	3909	3927	3945	3962
25	3979	3997	4014	4031	4048	4065	4082	4099	4116	4133
26	4150	4166	4183	4200	4216	4232	4249	4265	4281	4298
27	4314	4330	4346	4362	4378	4393	4409	4425	4440	4456
28	4472	4487	4502	4518	4533	4548	4564	4579	4594	4609
29	4624	4639	4654	4669	4683	4698	4713	4728	4742	4757
30	4771	4786	4800	4814	4829	4843	4857	4871	4886	4900
31	4914	4928	4942	4955	4969	4983	4997	5011	5024	5038
32	5051	5065	5079	5092	5105	5119	5132	5145	5159	5172
33	5185	5198	5211	5224	5237	5250	5263	5276	5289	5302
34	5315	5328	5340	5353	5366	5378	5391	5403	5416	5428
35	5441	5453	5465	5478	5490	5502	5514	5527	5539	5551
36	5563	5575	5587	5599	5611	5623	5635	5647	5658	5670
37	5682	5694	5705	5717	5729	5740	5752	5763	5775	5786
38	5798	5809	5821	5832	5843	5855	5866	5877	5888	5899
39	5911	5922	5933	5944	5955	5966	5977	5988	5999	6010
40	6021	6031	6042	6053	6064	6075	6085	6096	6107	6117
41	6128	6138	6149	6160	6170	6180	6191	6201	6212	6222
42	6232	6243	6253	6263	6274	6284	6294	6304	6314	6325
43	6335	6345	6355	6365	6375	6385	6395	6405	6415	6425
44	6435	6444	6454	6464	6474	6484	6493	6503	6513	6522
45	6532	6543	6551	6561	6571	6580	5690	6599	6609	6618
46	6628	6637	6646	6656	6665	6675	6684	6693	6702	6712
47	6721	6730	6739	6749	6758	6767	6776	6785	6794	6803
48	6812	6821	6830	6839	6848	6857	6866	6875	6884	6893
49	6902	6911	6920	6928	6937	6946	6955	6964	6972	6981
50	6990	6998	7007	7016	7024	7033	7042	7050	7059	7067
51	7076	7084	7093	7101	7110	7118	7126	7135	7143	7152
52	7160	7168	7177	7185	7193	7202	7210	7218	7226	7235
53	7243	7251	7259	7267	7275	7284	7292	7300	7308	7316
54	7324	7332	7340	7348	7356	7364	7372	7380	7388	7396

n	0	1	2	3	4	5	6	7	8	9
55	7404	7412	7419	7427	7435	7443	7451	7459	7466	7474
56	7482	7490	7497	7505	7513	7520	7528	7536	7543	7551
57	7559	7566	7574	7582	7589	7597	7604	7612	7619	7627
58	7634	7642	7649	7657	7664	7672	7679	7686	7694	7701
59	7709	7716	7723	7731	7738	7745	7752	7760	7767	7774
60	7782	7789	7796	7803	7810	7818	7825	7832	7839	7846
61	7853	7860	7868	7875	7882	7889	7896	7903	7910	7917
62	7924	7931	7938	7945	7952	7959	7966	7973	7980	7987
63	7993	8000	8007	8014	8021	8028	8035	8041	8048	8055
64	8062	8069	8075	8082	8089	8096	8102	8109	8116	8122
65	8129	8136	8142	8149	8156	8162	8169	8176	8182	8189
66	8195	8202	8209	8215	8222	8228	8235	8241	8248	8254
67	8261	8267	8274	8280	8287	8293	8299	8306	8312	8319
68	8325	8331	8338	8344	8351	8357	8363	8370	8376	8382
69	8388	8395	8401	8407	8414	8420	8426	8432	8439	8445
70	8451	8457	8463	8470	8476	8482	8488	8494	8500	8506
71	8513	8519	8525	8531	8537	8543	8549	8555	8561	8567
72	8573	8579	8585	8591	8597	8603	8609	8615	8621	8627
73	8633	8639	8645	8651	8657	8663	8669	8675	8681	8686
74	8692	8698	8704	8710	8716	8722	8727	8733	8739	8745
75	8751	8756	8762	8768	8774	8779	8785	8791	8797	8802
76	8808	8814	8820	8825	8831	8837	8842	8848	8854	8859
77	8865	8871	8876	8882	8887	8893	8899	8904	8910	8915
78	8921	8927	8932	8938	8943	8949	8954	8960	8965	8971
79	8976	8982	8987	8993	8998	9004	9009	9015	9020	9025
80	9031	9036	9042	9047	9053	9058	9063	9069	9074	9079
81	9085	9090	9096	9101	9106	9112	9117	9122	9128	9133
82	9138	9143	9149	9154	9159	9165	9170	9175	9180	9186
83	9191	9196	9201	9206	9212	9217	9222	9227	9232	9238
84	9243	9248	9253	9258	9263	9269	9274	9279	9284	9289
85	9294	9299	9304	9309	9315	9320	9325	9330	9335	9340
86	9345	9350	9355	9360	9365	9370	9375	9380	9385	9390
87	9395	9400	9405	9410	9415	9420	9425	9430	9435	9440
88	9445	9450	9455	9460	9465	9469	9474	9479	9484	9489
89	9494	9499	9504	9509	9513	9518	9523	9528	9533	9538
90	9542	9547	9552	9557	9562	9566	9571	9576	9581	9586
91	9590	9595	9600	9605	9609	9614	9619	9624	9628	9633
92	9638	9643	9647	9652	9657	9661	9666	9671	9675	9680
93	9685	9689	9694	9699	9703	9708	9713	9717	9722	9727
94	9731	9736	9741	9745	9750	9754	9759	9763	9768	9773
95	9777	9782	9786	9791	9795	9800	9805	9809	9814	9818
96	9823	9827	9832	9836	9841	9845	9850	9854	9859	9863
97	9868	9872	9877	9881	9886	9890	9894	9899	9903	9908
98	9912	9917	9921	9926	9930	9934	9939	9943	9948	9952
99	9956	9961	9965	9969	9974	9978	9983	9987	9991	9996

Answers to Selected Problems

Chapter 1

PROBLEM SET 1.1, Page 7

1. $5 \in A$ $5 \notin B$ $5 \in C$ $10 \notin C$
    $10 \in A$ $10 \in B$ $3 \notin C$ $12 \notin A$
    $2 \notin A$ $8 \in B$ $6 \in C$ $4 \notin B$

3. $M = \{a,c,e,h,m,s,t,u\}$ **5.** {March, May}

7. $\{3,4,5,6,7,8,9,10,11,12\}$ **9.** $\{x \mid x$ is a person who lives in Detroit$\}$

11. $\{x \mid x$ is a person listed in the Washington, D.C. telephone directory$\}$

13. finite   **15.** finite   **17.** false   **19.** true   **21.** false

23. true   **25.** $\{2,3\}$; proper, $\varnothing$, $\{2\}$, $\{3\}$

27. $\{a,b,c\}$; proper, $\varnothing$, $\{a\}$, $\{b\}$, $\{c\}$, $\{a,b\}$, $\{a,c\}$, $\{b,c\}$

29. $\{a,b,c,d\}$; proper, $\varnothing$, $\{a\}$, $\{b\}$, $\{c\}$, $\{d\}$, $\{a,b\}$, $\{a,c\}$, $\{a,d\}$,
    $\{b,c\}$, $\{b,d\}$, $\{c,d\}$, $\{a,b,c\}$, $\{a,b,d\}$, $\{a,c,d\}$, $\{b,c,d\}$

31. disjoint   **33.** one choice is $\{1,3,5,7, \ldots,29\}$

35. one choice is $\{x \mid x$ is an athlete$\}$ **37.** $\{1,2,3,4,5,7\}$

39. $\{2\}$ **41.** $\{2,3,5,6,7,8\}$ **43.** $\varnothing$ **45.** $\{8\}$

47. $\{2,3,5,7,8,9\}$ **49.** $\{6,3,5,8\}$ **51.** $\{6,3,5,8\}$

53. $\{5,7,8,9\}$ **55.** a) $\{6,7,8\}$ b) $\{6,7,8\}$ c) yes

PROBLEM SET 1.2, Page 16

1.

3.

5.

7. $\{0,1,2,3,4, \ldots\}$ **9.** $\{0\}$ **11.** 0.6

13. 1.5   **15.** 0.8   **17.** -1.25

19. $-2.\overline{3}$ **21.** $\dfrac{27}{100}$ **23.** $\dfrac{66}{25}$

25. $-\dfrac{1}{8}$ **27.** $-\dfrac{527}{10,000}$ **29.** $\dfrac{329}{1,000,000}$

31. $\dfrac{5}{9}$ **33.** $\dfrac{46}{99}$ **35.** $\dfrac{1}{2}$

37. $-\dfrac{7,355}{999}$

PROBLEM SET 1.3, Page 24

1. a) no, no   b) associative property   c) both

3. closure property for addition

5. commutative property for multiplication

7. associative property for addition   **9.** distributive property

11. distributive property and commutative property for multiplication

13. multiplicative inverse   **15.** Property 11c

17. Property 12   **19.** cancellation property for multiplication

21. Property 10   **23.** 26   **25.** 45   **27.** $\frac{58}{3}$

29. 208   **31.** $\frac{1}{4}$ **33.** 3

35. $a = b$ and $c = c$ so that $ac = bc$ by the multiplication property of equality

37. $(a + b)(c + d) = a(c + d) + b(c + d)$ (distributive property)
$= ac + ad + bc + bd$ (distributive property)

PROBLEM SET 1.4, Page 32

1. 8	**3.** 7	**5.** 12	**7.** $-\frac{2}{5}$	**9.** 20
11. 6	**13.** 4	**15.** -14	**17.** -12	**19.** 0
21. -91	**23.** -14	**25.** -39	**27.** 37	**29.** 20
31. -15	**33.** -16	**35.** 12	**37.** 0.059	**39.** $\frac{3}{2}$
41. 57	**43.** 6	**45.** $\frac{1}{3}$	**47.** -27	**49.** -7
51. 30	**53.** 210	**55.** 2	**57.** -2	**59.** 3
61. -13	**63.** 15	**65.** 1,500	**67.** 0	

REVIEW PROBLEM SET, Page 33

1. N **3.** F, Q **5.** F **7.** F

9. $\varnothing$ **11.** false   **13.** false   **15.** false

17. false   **19.** true   **21.** one choice is {all women}

23. one choice is {all states}   **25.** $\{c\}$ **27.** $\{a,c,d,e,f,g\}$

29. $\{a,c\}$ **31.** $\{a,c\}$ **33.** $\{a,b,c,d,g\}$

35.

$$\leftarrow \;\; \overset{\bullet}{\underset{-5}{|}} \;\; \overset{}{\underset{-4}{|}} \;\; \overset{\bullet}{\underset{-3}{|}} \;\; \overset{}{\underset{-2}{|}} \;\; \overset{\bullet}{\underset{-1}{|}} \;\; \overset{}{\underset{0}{|}} \;\; \overset{\bullet}{\underset{1}{|}} \;\; \overset{}{\underset{2}{|}} \;\; \overset{\bullet}{\underset{3}{|}} \;\; \overset{}{\underset{4}{|}} \;\; \overset{\bullet}{\underset{5}{|}} \;\; \rightarrow$$

37. 0.175   **39.** 0.68   **41.** $0.\overline{7}$ **43.** $0.3\overline{571428}$

45. 0.5   **47.** $\frac{4}{25}$ **49.** $\frac{1}{8}$ **51.** $\frac{87}{100}$

53. $\frac{31}{990}$ **55.** $\frac{3,019}{990}$ **57.** $\frac{629}{999}$

59. one choice is $\{0, \frac{1}{2}, 2, -4\}$ for $m \in \{0, 1, 2, 4\}$

61. $\{1\}$ **63.** $0.\overline{4}$ **65.** closure property for addition

67. commutative property for multiplication

69. associative property for multiplication

71. identity property for addition   **73.** Property 10

75. cancellation property for addition   **77.** Property 11b

79. 13   **81.** 2   **83.** -5 **85.** 2

87. -14 **89.** 14   **91.** 30   **93.** -12

95. 634   **97.** -40 **99.** -784 **101.** 105

103. -9 **105.** 8   **107.** -3

Chapter 2

PROBLEM SET 2.1, Page 45

1. 144   **3.** -108 **5.** 42   **7.** x^4

9. $2x^2y - 3x^3$ **11.** $3^5 = 243$ **13.** $(-2)^5 = -32$

15. x^{12} **17.** x^{12} **19.** x^{2n}

21. $2^6 = 64$ **23.** $(-2)^6 = 64$ **25.** x^{35}

27. y^{22} **29.** x^{12} **31.** $16x^4$ **33.** x^5y^5

35. $x^7y^7z^7$ **37.** $-8x^3$ **39.** $3^n x^n y^n$ **41.** $\dfrac{9}{16}$

43. $-\dfrac{8}{27}$ **45.** $\dfrac{x^4}{y^4}$ **47.** $-\dfrac{a^5}{b^5}$ **49.** $\dfrac{x^6}{y^6}$

51. $3^3 = 27$ **53.** $4^3 = 64$ **55.** x^5 **57.** y^5

59. x^{n-10} **61.** $81x^8y^{12}$ **63.** x^3y^4 **65.** $\dfrac{x^6}{y^3}$

67. $16a^{12}b^8$ **69.** $\dfrac{-16y^{10}z^9}{x}$ **71.** 1   **73.** -1

PROBLEM SET 2.2, Page 50

1. binomial; 1st degree; $3, -2$ **3.** monomial; 2nd degree; 4

5. trinomial; 2nd degree; $1, -5, 6$ **7.** binomial; 7th degree; $2, -13$

9. trinomial; 4th degree; $-1, -1, 13$

11. 7   **13.** 19   **15.** -1 **17.** 2

19. 21   **21.** 3   **23.** 2   **25.** 5

27. 4   **29.** 18   **31.** 9   **33.** not a polynomial

35. polynomial

PROBLEM SET 2.3, Page 55

1. $12x^2$ **3.** $11x^3$ **5.** $12xy$ **7.** $4x^2$

9. $4xy^2$ **11.** $8x + 7$ **13.** $7x^2 + 5x + 5$ **15.** $4x^2 - 2x - 2$

17. $5x^4 + x^3 + 4x^2 + x$ **19.** $16x^3 + x^2y - xy^2 - 3y^3$

21. $4x$ **23.** $2x^2$ **25.** $3xy$ **27.** $5x^3$ **29.** $-2x^2y$

31. $9x - 3y + 2z$ **33.** $6x^3 - 11x - 6$ **35.** $5x^2y + 3xy - 10xy^2$

37. $4x^4 - x^3 + 4x^2 + 2x - 5$ **39.** $4x^4y^4 - 7x^3y^3 + 5x^2y^2 + 3xy + 3$

41. $5x^2 + 2x - 2$ **43.** $2x^3 + 2x^2 - x + 2$

45. $x^2 + x + 9$ **47.** $-2xy - 7xz + 9yz$

49. $x^3 + 6x^2 + x - 1$

PROBLEM SET 2.4, Page 66

1. $6x^6$ **3.** $-30x^7$ **5.** $-28x^5y^7$ **7.** $12x^3y^3z^4$

9. $-24x^3y^3z^4$ **11.** $x^2 + x$ **13.** $x^3 + 2x^2$

15. $6x^3 - 12x$ **17.** $-2x^3y^2 - 3xy^4 - 2xy^2$

19. $8x^5y^2 - 6x^4y^3 + 2x^3y^4 - 2x^2y$ **21.** $x^2 + 2xy + y^2 - x - y$

23. $2x^3 + 5x^2y + 4xy^2 + y^3$ **25.** $x^5 + 2x^4 + 10x^2 - 9x + 12$

27. $x^4 - x^3 - 10x^2 + 4x + 24$ **29.** $x^5 - x^4y - 2x^3y^2 + 2x^2y^3 + xy^4 - y^5$

31. $x^2 + 3x + 2$ **33.** $x^2 + 9x + 20$ **35.** $2x^2 + 7xy + 3y^2$

37. $6x^2 + x - 1$ **39.** $5x^2 - x - 4$ **41.** $24x^2 - 2xy - 15y^2$

43. $28x^2 - 9xy - 9y^2$ **45.** $2x^2 - 11x + 12$ **47.** $10x^2 - 27xy + 5y^2$

49. $50x^2 - 115x + 56$ **51.** $x^2 + 2x + 1$ **53.** $4x^2 + 4xy + y^2$

55. $x^2 - 2xy + y^2$ **57.** $9x^2 - 30x + 25$ **59.** $16x^2 + 40xy + 25y^2$

61. $x^2 - y^2$ **63.** $x^2 - 49$ **65.** $4x^2 - 81$

67. $64x^2 - y^2$ **69.** $25x^2 - 36y^2$ **71.** $x^3 + 3x^2 + 3x + 1$

73. $8x^3 + 12x^2y + 6xy^2 + y^3$ **75.** $x^3 - 3x^2y + 3xy^2 - y^3$

77. $27x^3 - 54x^2y + 36xy^2 - 8y^3$ **79.** $125x^3 + 150x^2y + 60xy^2 + 8y^3$

81. $x^3 + 1$ **83.** $27 + y^3$ **85.** $x^3 - y^3$ **87.** $27x^3 + 8y^3$

89. $64x^3 - 125y^3$ **91.** $x^4 - y^6$ **93.** $x^4 + 6x^2 + 9$

95. $x^2 + 2xy + y^2 - 1$ **97.** $x^3 - 2x^2y - 4xy^2 + 8y^3$

99. $x^9 - 6x^6 + 12x^3 - 8$

PROBLEM SET 2.5, Page 72

1. $x(x - 1)$ **3.** $3x(3x + 1)$ **5.** $x(4x + 7y)$

7. $ab(a - b)$ **9.** $6pq(p + 4q)$ **11.** $6ab(b + 5a)$

13. $12x^2y(x - 4y)$ **15.** $2ab(a^2 - 4ab - 3b^2)$

17. $xy^2(x^2 + xy + 2y^2)$ **19.** $9mn(m + 2n - 3)$

21. $(3x + 5y)(2a + b)$ **23.** $(5x + 9ay + 9by)(a + b)$

25. $(m - 1)(x - y)$ **27.** $[7x + 14(2a + 7b) + (2a + 7b)^2](2a + 7b)$

29. $(xy + 2)[y(xy + 2)^2 - 5x(xy + 2) + 7]$ **31.** $(a + b)(x + y)$

33. $(x^4 - 1)(x + 3)$ **35.** $(y - 1)(z + 2)$ **37.** $(b^2 - d)(a - c)$

39. $(2a - b)(x - y)$ **41.** $(x - a)(x + b)$ **43.** $(a + b)(x + y + 1)$

45. $(x^2 + y)(2x + y - 1)$

PROBLEM SET 2.6, Page 77

1. $(x + y)^2$ **3.** $(x - 4)^2$ **5.** $(2x + 1)^2$

7. $(3y - 1)^2$ **9.** $(2x + 3y)^2$ **11.** $(3x - 7y)^2$

13. $(4y + 3)^2$ **15.** $(xy - 2z)^2$ **17.** $(x^2 + 7)^2$

19. $(ax^3 - 3)^2$ **21.** $(x - 2)(x + 2)$ **23.** $(1 - 3y)(1 + 3y)$

25. $(6 - 5a)(6 + 5a)$ **27.** $(4x - 5y)(4x + 5y)$ **29.** $(ab - c)(ab + c)$

31. $(1 - 10abc)(1 + 10abc)$ **33.** $(2x^2 - 1)(2x^2 + 1)$

35. $(x - 3y)(x + 3y)(x^2 + 9y^2)$ **37.** $(x - y)(x + y)(x^2 + y^2)(x^4 + y^4)$

39. $(x + y - a + b)(x + y + a - b)$ **41.** $(x + 1)^3$

43. $(x - y)^3$ **45.** $(x + 2)^3$ **47.** $(x - 3)^3$

49. $(x + 1)(x^2 - x + 1)$ **51.** $(4 - y)(16 + 4y + y^2)$

53. $(4x + y)(16x^2 - 4xy + y^2)$ **55.** $(2x - 3y)(4x^2 + 6xy + 9y^2)$

57. $(x - 2yz)(x^2 + 2xyz + 4y^2z^2)$ **59.** $(x^2 + 2)(x^4 - 2x^2 + 4)$

61. $(x - 1)(x^2 + x + 1)(x^6 + x^3 + 1)$ **63.** $(x + 1)(x^2 + 5x + 7)$

65. $(y + 3)(y^2 + 3)$ **67.** $2x(2x - y)(2x + y)$

69. $xy(x - 2)^2$ **71.** $x^2(2 - x)(4 + 2x + x^2)$

73. $(x + y - z - 3)(x + y + z + 3)$ **75.** $(x^2 - xy + y^2)(x^2 + xy + y^2)$

PROBLEM SET 2.7, Page 84

1. $(x + 1)(x + 3)$ **3.** $(x - 1)(x - 2)$ **5.** $(x + 7)(x + 8)$

7. $(x - 5)(x + 3)$ **9.** $(x - 7)(x - 9)$ **11.** $(x + 5y)(x + 6y)$

13. $(x - 9)(x + 2)$ **15.** $(x + 12y)(x - 10y)$ **17.** $(6 + x)(2 - x)$

19. $(9 + x)(4 - x)$ **21.** $(2x + 1)(x + 3)$ **23.** $(3x - 1)(x + 2)$

25. $(5x - 1)(x - 2)$ **27.** $(3x + y)(x + 2y)$ **29.** $(3x + 2)(2x + 3)$

31. $(3x - 2y)(2x + 3y)$ **33.** $(4x - 1)(3x + 5)$ **35.** $(8x - 5)(7x - 6)$

37. $(4 + x)(3 - 2x)$ **39.** $(5y - 4x)(y + 2x)$ **41.** $5x(x - 4)(x - 7)$

43. $2st(8t - s)^2$ **45.** $y^2(x + 3)(x + 7)$ **47.** $4m^2(n + 7)(n - 1)$

49. $wy(x - 2)(x - 7)$

PROBLEM SET 2.8, Page 89

1. $3x^3$ **3.** $-\dfrac{3x^2}{y^2}$ **5.** $\dfrac{2z^2}{x^2}$ **7.** $3y^2 - 2x$

9. $2xy^2 - 8y^2 + 2$ **11.** $2xy^2 + y - \dfrac{4}{xy^2} + \dfrac{3}{x^2y^3}$

13. $x - 2$ **15.** $3x - 1$ **17.** $2x^3 + x^2 - 6x + 8; \ R = -9$

19. $x^2 + xy - 4y^2$ **21.** $-3x^2 - 4xy - 12y^2; \ R = -42y^3$

23. $x^2 + xy + y^2$ **25.** $x + 2$ **27.** $x + 2$

29. $x^2 - 5x + 8 = (x - 3)(x - 2) + 2$

31. $x^3 + 3x^2 - 2x - 5 = (x^2 + x - 4)(x + 2) + 3$

33. $x^3 - x^2y - xy^2 + 2y^3 = (x^2 - 2xy + y^2)(x + y) + y^3$

35. $x^5 - y^5 = (x^4 + x^3y + x^2y^2 + xy^3 + y^4)(x - y)$

REVIEW PROBLEM SET, Page 91

1. x^{11} **3.** x^{12} **5.** y^{24} **7.** x^3y^6

9. $-27x^9y^3$ **11.** $\dfrac{16x^4}{y^{12}}$ **13.** x^3 **15.** $3y^{10}$

17. 2nd degree; $4, -39, 100$ **19.** 3rd degree; $81, -2, 25$

21. 2nd degree; $36, 4, 14$ **23.** 4th degree; $1, -3$

25. 89   **27.** 171   **29.** 18

31. $3x^2 - 2x + 3$ **33.** $x^2 + 4x + 6$

35. $4x^2 + 6x - 4$

37. $6x^5 - 12x^3$

39. $6x^3y^4 - 10x^2y^5 + 2xy^4$

41. $x^3 - 3x^2y + 3xy^2 - y^3$

43. $x^2 + 10x + 21$

45. $2x^2 + 7x - 15$

47. $4x^2 - 17xy + 15y^2$

49. $x^2 - 64$

51. $9x^2 - 12x + 4$

53. $x^2 + 12xy + 36y^2$

55. $x^3 + 125$

57. $1 + 3y + 3y^2 + y^3$

59. $7xy(x - 3y^2)$

61. $13x^2y^2(2x + 3x^3y^2 - 4y)$

63. $2(y - 2x)(y + z)$

65. $(x + 8)^2$

67. $(y - 11)(y + 11)$

69. $(x + 4)(x^2 - 4x + 16)$

71. $(y - 1)^3$

73. $(2x - 3)(2x + 3)(4x^2 + 9)$

75. $(3 + z)(x - y)$

77. $(x + y - a - b)(x + y + a + b)$

79. $(x^2 - x + 1)(x^2 + x + 1)$

81. $(x + 3y)(x - y)$

83. $(x - 9)(x + 4)$

85. $(3x + 2)(x + 5)$

87. $(2x + 3)(x - 2)$

89. $(5x - 4y)(4x - 3y)$

91. $6x^2$

93. $8x^2y^2 - 4xy$

95. $5x - 1$

97. $x - 2$

99. $x^{10} - x^5 + 1$

Chapter 3

PROBLEM SET 3.1, Page 100

1. 0

3. 0

5. 3

7. $-3, 3$

9. $-2, 1, 3$

11. not equivalent

13. equivalent

15. equivalent

17. not equivalent

19. equivalent

21. equivalent

23. 30

25. $15x^6y^2$

27. $15(x - y)$

29. $22x(x - 1)(x + 5)$

31. $x^3 - 9x$

33. $8(x + 2)$

35. $\dfrac{3}{7}$

37. $\dfrac{2}{7}$

39. $\dfrac{x}{y^2}$

41. $\dfrac{1}{5xy}$

43. $\dfrac{y^2c}{3x}$

45. $\dfrac{x + 1}{x - 1}$

47. $\dfrac{x - 1}{x + 2}$

49. $\dfrac{2x + 3}{3x}$

51. $\dfrac{x + 4}{x + 7}$

53. $\dfrac{x - 6}{3x + 4}$

55. $\dfrac{1}{x - y}$

57. $\dfrac{-x - y}{y + 2x}$

59. $\dfrac{x + 4}{x - 2}$

61. $\dfrac{2 - x}{2(x + 2)}$

63. $\dfrac{6(x^2 + 2x + 4)}{5(x + 2)(x^2 + 4)}$

65. $\dfrac{z + w}{z + y}$

67. $\dfrac{a}{b} = \dfrac{(-1)a}{(-1)b} = \dfrac{-a}{-b}$

PROBLEM SET 3.2, Page 107

1. $\dfrac{14}{51}$

3. $-\dfrac{7y}{20x}$

5. $\dfrac{y}{x^2}$

7. $\dfrac{5}{3x}$

9. $\dfrac{3}{5(x - 8)}$

11. $7a + 2$

13. $(x + 12)(x - 4)$

15. $\dfrac{(x + 1)(3x + 1)}{x - 9}$

17. $\dfrac{x + 5}{x - 5}$

19. $\dfrac{(a - b)(a + 2b)}{(a + b)(a - 2b)}$

21. $\dfrac{a^2 - b^2}{5}$

23. $-\dfrac{5}{6}$

25. $\dfrac{20}{3}$ **27.** $4x$ **29.** $\dfrac{21}{22}$

31. $\dfrac{x(x+1)}{2x-1}$ **33.** $\dfrac{(a+2)^2}{(a-2)^2}$ **35.** $\dfrac{3x-9}{(x+3)(x-y)}$

37. $\dfrac{-x(x^2+xy+y^2)}{y^2}$ **39.** $\dfrac{4x(x-1)}{(3x-5)(x+2)}$ **41.** $\dfrac{x+8}{x(x+1)(3x+2)}$

43. $\dfrac{15x^3y^4}{2b^6}$ **45.** $\dfrac{2(x+2)}{3(x-2)(x+1)}$ **47.** $\dfrac{1-a}{x(a-3)(1+x)}$

49. $\dfrac{7}{3x-6}$ **51.** $\dfrac{x+1}{x-1}$

PROBLEM SET 3.3, Page 117

1. $\dfrac{8}{11}$ **3.** $\dfrac{24}{x}$ **5.** $\dfrac{4}{13}$

7. $\dfrac{2x}{7}$ **9.** $\dfrac{3x}{3x+1}$ **11.** $\dfrac{4}{9}$

13. $\dfrac{54}{3-x}$ **15.** $-\dfrac{2b}{7a}$ **17.** $9-2x$

19. 0 **21.** $\dfrac{59}{56}$ **23.** $\dfrac{13}{15}$

25. $\dfrac{47x}{30}$ **27.** $\dfrac{y}{24}$ **29.** $\dfrac{3x+2}{x^2}$

31. $\dfrac{11x+13}{24}$ **33.** $\dfrac{5a^3-4b^2}{20a^2b^2}$ **35.** $\dfrac{15y-57}{(y-3)(y-5)}$

37. $\dfrac{2x+36}{x^2-9}$ **39.** $\dfrac{8x-46}{(x-7)(x-5)}$ **41.** $\dfrac{105}{180}, \dfrac{18}{180}, \dfrac{8}{180}$

43. $\dfrac{3x^2y}{84xyz}, \dfrac{35y}{84xyz}, \dfrac{98z^2}{84xyz}$ **45.** $\dfrac{5x}{x^2+3x}, \dfrac{4}{x^2+3x}$

47. $\dfrac{5(x^2-4)}{12x^2(x^2-4)}, \dfrac{14x^2}{12x^2(x^2-4)}, \dfrac{x^2(x+2)}{12x^2(x^2-4)}$

49. $\dfrac{8(x-2)(x+1)}{x(x-2)^2(x+1)}, \dfrac{4x(x+1)}{x(x-2)^2(x+1)}, \dfrac{x^2(x-2)}{x(x-2)^2(x+1)}$

51. $\dfrac{(2x-3)(x+5)}{x^2-25}, \dfrac{(x+1)(x-5)}{x^2-25}, \dfrac{x-1}{x^2-25}$

53. $\dfrac{2x-5}{x^2-25}$ **55.** $\dfrac{3-4x}{(x-3)(x+3)(x-2)}$ **57.** $\dfrac{2x^2+4}{x(x-6)(x+1)}$

59. $\dfrac{2x^2+14x-6}{(x+2)(x+7)(x+8)}$ **61.** $\dfrac{6}{(x+2)(x+5)(4-x)}$ **63.** $\dfrac{x^2+4x}{4-x^2}$

65. $\dfrac{9-x-17x^2}{3(x^2-9)}$ **67.** $\dfrac{3x^2-5x+13}{(x-3)^2(x+2)}$ **69.** $\dfrac{2}{x^2+1}$

PROBLEM SET 3.4, Page 124

1. $\dfrac{5}{6}$ **3.** $\dfrac{113y}{32}$ **5.** $\dfrac{8}{7}$ **7.** $\dfrac{1}{x-1}$ **9.** $\dfrac{1}{x+2}$

11. x **13.** $\dfrac{x^2-y^2}{x^2+y^2}$ **15.** 1 **17.** $-\dfrac{y}{2}$ **19.** $3-x$

21. $\dfrac{1+x^2}{x}$ **23.** $-\dfrac{2x}{x^2+1}$ **25.** $\dfrac{y}{y-2x}$

REVIEW PROBLEM SET, Page 125

1. $\dfrac{1}{5}$ **3.** $-\dfrac{2x}{3z}$ **5.** $\dfrac{8y}{7bx}$ **7.** $\dfrac{x+y}{y-x}$ **9.** $\dfrac{x-4}{x+6}$

11. $\dfrac{x+3}{x+1}$ **13.** $\dfrac{x-3}{x+2}$ **15.** $\dfrac{x+z}{3(x+y)}$ **17.** xy^2 **19.** $a(a-b)$

21. $\dfrac{a^3x^4}{z^3}$ **23.** $b-1$ **25.** $\dfrac{3-y}{x(x+1)}$ **27.** $\dfrac{a-b}{2a}$

29. $(x+y)(a+b)$ **31.** 1 **33.** $\dfrac{b^2(a-2b)}{a^2(a-b)}$ **35.** -1

37. $\dfrac{2x}{x+y}$ **39.** $\dfrac{c-b}{x+y}$ **41.** $\dfrac{x-2}{x-7}$ **43.** $\dfrac{x+1}{x-3}$

45. $\dfrac{(a+b)(x-y)}{(a-b)(x+y)}$ **47.** $\dfrac{1}{x-y}$ **49.** $\dfrac{(b-4)(3b-1)}{(5b-1)(b+6)}$ **51.** $\dfrac{2}{x}$

53. $\dfrac{5a-3b}{8}$ **55.** $\dfrac{x+30}{20}$ **57.** $\dfrac{5}{12x}$ **59.** $\dfrac{5z-6x+7y}{xyz}$

61. $\dfrac{5x^2+3x-2}{x^3}$ **63.** $\dfrac{2x+1}{(x+3)(x-2)}$ **65.** $\dfrac{15x^2-22xy-12y^2}{(7x+2y)(3x+4y)}$

67. $\dfrac{x+y}{xy}$ **69.** $\dfrac{5x-9}{(x-1)(x-3)(x-4)}$ **71.** $\dfrac{3x^2+20x-72}{(x-2)(x+3)(x-4)}$

73. $\dfrac{4x^2-22x-87}{(x+2)(x-7)}$ **75.** $-\dfrac{1}{3}$ **77.** $\dfrac{1}{a^2x}$ **79.** $-\dfrac{2x}{x+y}$ **81.** $\dfrac{10}{3}$

83. y **85.** $\dfrac{1}{x-1}$ **87.** $\dfrac{1}{y^2+2y-1}$ **89.** $\dfrac{6}{x}$

Chapter 4

PROBLEM SET 4.1, Page 138

1. 2 **3.** 1 **5.** 2 **7.** 2 **9.** 2 and 3

11. $\{5\}$ **13.** $\{2\}$ **15.** $\{-1\}$ **17.** $\{0\}$ **19.** $\{5\}$ **21.** $\{1\}$

23. $\{3\}$ **25.** $\{3\}$ **27.** $\{9\}$ **29.** $\{1\}$ **31.** $\{3\}$ **33.** $\{-\frac{2}{3}\}$

35. $\{6\}$ **37.** $\{5\}$ **39.** $\{4\}$ **41.** $\{2\}$ **43.** $\{15\}$ **45.** $\varnothing$

47. $\{-3\}$ **49.** $\{-\frac{4}{5}\}$

PROBLEM SET 4.2, Page 143

1. $\left\{\dfrac{b}{a}\right\}$ **3.** $\{b-a\}$ **5.** $\left\{\dfrac{c-b}{a}\right\}$ **7.** $\left\{\dfrac{bc+d}{a}\right\}$

9. $\left\{\dfrac{9b}{17}\right\}$ **11.** $\{1\}$ **13.** $\left\{\dfrac{2ab}{a+b}\right\}$ **15.** $\{a+b\}$

17. $\left\{\dfrac{2b+5}{7}\right\}$ **19.** $\left\{\dfrac{a^2+3a-1}{a-3}\right\}$ **21.** $l=\dfrac{V}{wh}$ **23.** $g=\dfrac{2s}{t^2}$

25. $R = \dfrac{E - Ir}{I}$ **27.** $r = \dfrac{C}{2\pi}$ **29.** $h = \dfrac{V}{\pi r^2}$ **31.** $w = \dfrac{P - 2l}{2}$

33. $m = \dfrac{L - a + d}{d}$ **35.** $t = \dfrac{I}{Pr}$ **37.** $l = \dfrac{V}{w^2}, \; l = 9$ inches

39. $n = \dfrac{A - P}{Pr}$ **41.** $R_1 = \dfrac{RR_2}{R_2 - R}$

PROBLEM SET 4.3, Page 150

1. 27   **3.** 32 and 34   **5.** 36   **7.** 4 and 14
9. 11, 14   **11.** 45 nickels, 15 dimes, 11 quarters
13. 43 nickels, 27 dimes   **15.** 30 quarters, 50 dimes
17. $32\frac{8}{11}$ minutes   **19.** 99   **21.** length $= 17$ inches, width $= 10$ inches
23. length $= 60$ meters, width $= 15$ meters   **25.** 10 feet
27. general partner received \$4,800, limited partner received \$4,000
29. \$2,000   **31.** \$16,250 and \$8,750   **33.** $73\frac{3}{5}$ pounds
35. $\frac{4}{5}$ quart   **37.** 12 hours

PROBLEM SET 4.4, Page 160

1. multiplication property   **3.** transitive property
5. multiplication property   **7.** addition property
9. multiplication property   **11.**

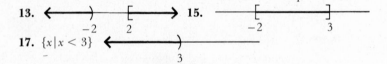

13. ◄────)──[────► $-2 \quad 2$ **15.** ──[────]── $-2 \quad 3$

17. $\{x \mid x < 3\}$ ◄────) 3

19. $\{x \mid x \le 4\}$ ◄────] 4

21. $\{x \mid x < \frac{7}{5}\}$ ◄────) $\frac{7}{5}$

23. $\{x \mid x \le -\frac{2}{3}\}$ ◄────] $-\frac{2}{3}$

25. $\{x \mid x \ge 2\}$ ──[────► 2

27. $\{x \mid x \ge \frac{9}{4}\}$ ──[────► $\frac{9}{4}$

29. $\{x \mid x < -1\}$ ◄────) -1

31. $\{x \mid x < 3\}$ ◄────) 3

33. $\{x \mid x \le 1\}$ ◄────] 1

35. $\{x|x \le \frac{13}{6}\}$

37. $\{x|x \le 8\}$

39. $\{x|x > -3\}$

41. $\{x|x \ge \frac{9}{2}\}$

43. $\{x|x < \frac{22}{3}\}$

45. $\{x|x \ge 12\}$

47. $\{x|x \ge \frac{91}{6}\}$

51. no; for example, $(-2)^2 < (-5)^2$, but $-2 > -5$

53. 9 or more   **55.** 7 or less

PROBLEM SET 4.5, Page 169

1. 12   **3.** 4   **5.** 5   **7.** 9   **9.** 13

11. $\{-5,5\}$ **13.** $\{-2,2\}$ **15.** $\{-1,5\}$ **17.** $\{-11,1\}$

19. $\{-2,5\}$ **21.** $\{-3,7\}$ **23.** $\{-3,6\}$

25. Equation is $|x-3| = 7$, solution set is $\{-4,10\}$

27. Equation is $|x+1| = 1$, solution set is $\{-2,0\}$

29. $\{x|-5 < x < 5\}$

31. $\{x|x < -2 \text{ or } x > 2\}$

33. $\{x|-4 \le x \le 4\}$

35. $\{x|x \in R\}$

37. $\{x|x < -7 \text{ or } x > 9\}$

39. $\{x|-9 < x < 5\}$

41. $\{x|-\frac{7}{2} \le x \le \frac{3}{2}\}$

43. $\{x|x \le -\frac{5}{2} \text{ or } x \ge 3\}$

45. $\{x \mid -1 < x < 4\}$

47. $\{x \mid x \le \frac{5}{7} \text{ or } x \ge 1\}$

49. $\{x \mid x < -\frac{9}{5} \text{ or } x > \frac{3}{5}\}$

REVIEW PROBLEM SET, Page 170

1. $\{6\}$ **3.** $\{-4\}$ **5.** $\{4\}$ **7.** $\{-\frac{2}{7}\}$ **9.** $\{\frac{20}{9}\}$

11. $\{6\}$ **13.** $\{1\}$ **15.** $\{3\}$ **17.** $\{25\}$ **19.** $\varnothing$ **21.** $\left\{\dfrac{a-c}{b}\right\}$

23. $\left\{\dfrac{b}{a-c}\right\}$ **25.** $\left\{\dfrac{7x^2 + 4c}{4}\right\}$ **27.** $\left\{\dfrac{2b}{x}\right\}$ **29.** $\left\{\dfrac{a^2 - ax + 3a}{a-x}\right\}$

31. $m = Gr^3$ **33.** $h = \dfrac{v^2 - gR}{g}$ **35.** $h = \dfrac{s - 2\pi r^2}{2\pi r}$

37. $a = \dfrac{s(x-1)}{x^n - 1}$ **39.** $R = \dfrac{rs}{r+s}$

41. $\{x \mid x > 2\}$

43. $\{x \mid x \ge -3\}$

45. $\{x \mid x < -4\}$

47. $\{x \mid x \le \frac{1}{18}\}$

49. $\{x \mid x > 1\}$

51. 10 **53.** 5 **55.** 7 **57.** $\{-7, 7\}$ **59.** $\{-7, 3\}$

61. $\{-8, 14\}$ **63.** $\{-\frac{16}{3}, \frac{8}{3}\}$ **65.** $\{-1, 4\}$

67. $\{x \mid -7 < x < 7\}$

69. $\{x \mid x \ge 9 \text{ or } x \le -9\}$

71. $\{x \mid -2 < x < 8\}$

73. $\{x \mid x < -4 \text{ or } x > \frac{1}{2}\}$

75. $\{x \mid x \le -2 \text{ or } x \ge \frac{18}{5}\}$

77. 119 and 124 **79.** 98

Chapter 5

PROBLEM SET 5.1, Page 182

1. $\dfrac{1}{25}$　　**3.** 16　　**5.** $\dfrac{16}{9}$　　**7.** $\dfrac{1}{x^{100}}$　　**9.** 7

11. x^5　　**13.** 1　　**15.** 1　　**17.** 1　　**19.** 14

21. $\dfrac{2y}{x}$　　**23.** 49　　**25.** $\dfrac{1}{81}$　　**27.** $\left(\dfrac{5}{2}\right)^5$

29. $\dfrac{1}{x^5}$　　**31.** $\dfrac{1}{2^6}$　　**33.** $\dfrac{1}{4^6}$　　**35.** $\dfrac{1}{y^{12}}$

37. $\dfrac{1}{81x^4}$　　**39.** $\dfrac{25}{y^8}$　　**41.** $\dfrac{x^3}{3^6}$　　**43.** $\dfrac{9}{25}$

45. $-\dfrac{x^6}{8}$　　**47.** $\dfrac{y^4}{x^4}$　　**49.** 64　　**51.** $\dfrac{7}{2}$

53. $\dfrac{1}{y^3}$　　**55.** $\dfrac{y^3}{x^{15}}$　　**57.** $\dfrac{x^2}{y^{10}}$　　**59.** $\dfrac{a^2c^3}{b^7}$

61. $\dfrac{5a^3}{8c}$　　**63.** $\dfrac{1}{x^4y^4}$　　**65.** $\dfrac{x^{28}z^4}{y^{16}}$　　**67.** 3.782×10^3

69. 3.8173×10^4　　**71.** 1.371×10^8　　**73.** 2.71312×10^{-4}

75. 3.19×10^{-2}　　**77.** 210　　**79.** 0.0000314

81. 11,300　　**83.** 0.00000541　　**85.** 31,270

PROBLEM SET 5.2, Page 190

1. 8　　**3.** $\dfrac{1}{16}$　　**5.** -32　　**7.** $\dfrac{1}{27}$

9. $\dfrac{1}{27}$　　**11.** $\dfrac{1}{32}$　　**13.** 0.008　　**15.** 2

17. 2　　**19.** $\dfrac{1}{2}$　　**21.** $x^{13/10}$　　**23.** 5

25. $\dfrac{1}{x^2}$　　**27.** $y^{2/3}$　　**29.** $16x^{12}$　　**31.** 2^7x^3

33. $\dfrac{x}{5}$　　**35.** $\dfrac{1}{16x^4}$　　**37.** $5^{17/21}$　　**39.** $x^{1/2}$

41. $x^{11/4}$　　**43.** $5^{5/3} \cdot 3^{1/2}$　　**45.** $\dfrac{2}{125}$　　**47.** $-\dfrac{x^3}{20y}$

49. $\dfrac{x^3y^4}{3}$　　**51.** $x^{(r^2+s)/rs}y^{(r+1)/s}$　　**53.** $x^{1/2}y^{1/4}z$

PROBLEM SET 5.3, Page 197

1. 11　　**3.** -5　　**5.** 3　　**7.** -2

9. -0.5　　**11.** x^2　　**13.** $-x^2$　　**15.** 4

17. $\sqrt[6]{3^5}$　　**19.** $\sqrt[4]{26^3}$　　**21.** $\sqrt[3]{85^2}$　　**23.** $\frac{1}{8}$

25. $\sqrt[11]{x^5}$　　**27.** $\sqrt[22]{x^3}$　　**29.** 8　　**31.** 3

33. x^2　　**35.** $-x\sqrt[3]{9x^2}$　　**37.** $2x\sqrt[3]{x}$　　**39.** $-3x^3\sqrt[3]{9x}$

41. $\dfrac{7}{3}$ **43.** 5   **45.** -2 **47.** $\sqrt[3]{3}$

49. $2x$ **51.** $-x$ **53.** 3   **55.** 16

57. x^3 **59.** x^2 **61.** $\sqrt[8]{32}$ **63.** x

65. x^2 **67.** $\sqrt[3]{4}$ **69.** $x^2\sqrt[8]{x^3y}$ **71.** x^5

73. $xy\sqrt[4]{x}$ **75.** $\sqrt[6]{9x^2y^4}$ **77.** $\sqrt[4]{49x^2}$ **79.** $\sqrt[10]{9x^8y^6}$

81. $\sqrt[12]{x^{10}y^{11}}$ **83.** $y^2\sqrt[15]{x^{13}y}$

PROBLEM SET 5.4, Page 205

1. $8\sqrt{7}$ **3.** $5\sqrt{2}$ **5.** $5\sqrt{2}$ **7.** $9\sqrt{2}$

9. $33\sqrt{5}$ **11.** $\sqrt{5}$ **13.** $2\sqrt{3x}$ **15.** $(6x+3)\sqrt{x}$

17. $16\sqrt[3]{3}$.   **19.** $(-15-x)\sqrt[3]{2xy}$ **21.** $2\sqrt{6}$

23. $x+3\sqrt{x}$ **25.** $1-\sqrt{3}$ **27.** $2x+15\sqrt{x}+7$

29. $33+20\sqrt{2}$ **31.** $4x-4\sqrt{xy}+y$ **33.** 11

35. 42   **37.** $9x-121$ **39.** $x-3$ **41.** $\dfrac{2\sqrt{3}}{3}$

43. $\dfrac{8\sqrt{11x}}{77x}$ **45.** $\dfrac{5\sqrt[3]{49}}{7}$ **47.** $\dfrac{5(\sqrt{5}+1)}{2}$

49. $2-\sqrt{2}$ **51.** $\sqrt{7}-\sqrt{21}+\sqrt{5}-\sqrt{15}$

53. $\dfrac{3\sqrt{xy}+2y}{9x-4y}$ **55.** $\dfrac{x^2-x\sqrt{y}}{x^2-y}$ **57.** $\dfrac{37\sqrt{10}+129}{227}$

59. $\dfrac{2\sqrt{35}+\sqrt{7}-3\sqrt{5}+8}{38}$

PROBLEM SET 5.5, Page 209

1. $\{\frac{11}{2}\}$ **3.** $\varnothing$ **5.** $\{39\}$ **7.** $\{5\}$

9. $\{\frac{5}{2}\}$ **11.** $\{1\}$ **13.** $\{20\}$ **15.** $\varnothing$

17. $\{1\}$ **19.** $\{30\}$ **21.** $\{4\}$ **23.** $\{3\}$

25. $\{29\}$ **27.** $\{8\}$ **29.** $\varnothing$

PROBLEM SET 5.6, Page 215

1. i **3.** -1 **5.** i **7.** 1

9. i **11.** $-i$ **13.** $15i$ **15.** $5+9i$

17. $-2x\sqrt{7}i$ **19.** $-1+9i$ **21.** $-5+8i$ **23.** $12+0i$ **25.** $1+i$

27. $0-16i$ **29.** $1+5i$ **31.** $11-3i$ **33.** $27-5i$ **35.** 41

37. $5-5i$ **39.** $x=3,\ y=-4$ **41.** $x=-3,\ y=3$

43. $2-4i$ **45.** $3i$ **47.** $\frac{3}{4}-\frac{1}{4}i$ **49.** $\frac{2}{5}-\frac{11}{5}i$

51. $-\frac{1}{2}-\frac{1}{2}i$ **53.** $\frac{11}{13}+\frac{16}{13}i$ **55.** $\frac{19}{29}-\frac{4}{29}i$ **57.** $\frac{1}{5}-\frac{2}{5}i$

REVIEW PROBLEM SET, Page 216

1. $\dfrac{1}{9}$ **3.** $\dfrac{1}{1,000}$ **5.** 64   **7.** 64   **9.** $\dfrac{y^2}{x^2}$

11. $\dfrac{1}{a^3 b^2}$ **13.** 1   **15.** 1   **17.** $\dfrac{1}{81}$ **19.** x^{21}

21. $-\dfrac{1}{2^{20}}$ **23.** x^{17} **25.** $\dfrac{1}{125 x^3}$ **27.** $\dfrac{b^4}{a^2}$

29. $\dfrac{81}{16}$ **31.** $\dfrac{x^2}{9}$ **33.** 25   **35.** $x^5 y^5$

37. $\dfrac{x^6}{27 y^9}$ **39.** 4.68×10^7 **41.** 1.2×10^{-6} **43.** 5,600

45. 0.000192   **47.** -8 **49.** $\dfrac{1}{8}$ **51.** -8

53. $\dfrac{1}{625}$ **55.** 4   **57.** $\dfrac{1}{128}$ **59.** 7

61. $x^{1/4}$ **63.** $\dfrac{1}{8}$ **65.** $x^{-1/2}$ **67.** $-2 x^2$

69. $4 x^2$ **71.** $\dfrac{x^{18}}{9}$ **73.** $\dfrac{x^6}{4}$ **75.** $\dfrac{x^{75}}{y^{12}}$

77. $-\dfrac{8 z^9}{x^3 y^6}$ **79.** $15^{-3/4}$ **81.** $x^{1/21}$ **83.** $x^8 y^{16/3}$

85. $2^{1/2}$ **87.** 13^{-1} **89.** $x^{1/3}$ **91.** $x^{11/6}$

93. $x^{4/5}$ **95.** $\sqrt[5]{100}$ **97.** $\sqrt{5^{-3}}$ **99.** $\sqrt[9]{121}$

101. $\sqrt[3]{y^{-2}}$ **103.** $\sqrt[4]{b^{-3}}$ **105.** $\dfrac{\sqrt{5 x y^2}}{5x}$ **107.** 5

109. $x \sqrt[20]{x}$ **111.** $5 \sqrt{5}$ **113.** $2 x^2$ **115.** $2 x y^2$

117. $\dfrac{1}{25}$ **119.** $\dfrac{1}{2}$ **121.** -4 **123.** $-x^2$

125. 2   **127.** x^3 **129.** $\sqrt{x}$ **131.** $2 x y^2 \sqrt[4]{2 x y^2}$

133. $3 x y z^2 \sqrt[3]{2 x y^2 z}$ **135.** $\dfrac{5 \sqrt{5 x y}}{2 x y}$ **137.** $\dfrac{3 x \sqrt[4]{5 x^3}}{2 y^2}$ **139.** $\sqrt{x}$

141. $\sqrt[36]{x^6 y^2 z}$ **143.** $\sqrt[4]{289 x^2}$ **145.** $\sqrt[10]{9 x^4 y^2}$ **147.** $8 \sqrt{2}$

149. $4 \sqrt{x}$ **151.** $6 \sqrt{2}$ **153.** $3 \sqrt{3}$ **155.** $10 \sqrt{2 x}$

157. $5 \sqrt{7}$ **159.** $\sqrt{6} + \sqrt{15}$ **161.** $3 \sqrt{2} - 2 \sqrt{6}$ **163.** $3 \sqrt{2}$

165. $2 x - \sqrt{x y} - y$ **167.** $12 + 4 \sqrt{5}$ **169.** 6   **171.** $3 x - 2$

173. $x y z - 2 x \sqrt{y z} + x$ **175.** $\dfrac{4 \sqrt{3}}{3}$ **177.** $\dfrac{5 \sqrt{2}}{8}$ **179.** $\dfrac{2 \sqrt{3 x y}}{3 y}$

181. $\dfrac{5 \sqrt{11} - 10}{7}$ **183.** $5 + \sqrt{15}$ **185.** $13 - 2 \sqrt{42}$

187. $\dfrac{x^2 + x \sqrt{y} - 2 y}{x^2 - 4 y}$ **189.** $\{25\}$ **191.** $\{17\}$ **193.** $\{-1\}$

195. $\varnothing$ **197.** $\{22\}$ **199.** $\left\{\dfrac{73}{5}\right\}$ **201.** $\varnothing$ **203.** i **205.** $-i$

207. i **209.** $-1 + 16 i$ **211.** $-4 + 4 i$ **213.** $-27 + 20 i$

215. $3 \sqrt{6} + 7 \sqrt{5} i$ **217.** $12 + 16 i$ **219.** $-82 - 49 i$

221. $\frac{14}{17} + \frac{5}{17} i$ **223.** $\frac{15}{13} + \frac{23}{13} i$

Chapter 6

PROBLEM SET 6.1, Page 226

1. $\{1,2\}$ **3.** $\{-\frac{1}{2},\frac{1}{2}\}$ **5.** $\{-\frac{2}{3},\frac{2}{3}\}$ **7.** $\{-\frac{2}{5},\frac{2}{5}\}$

9. $\{-\frac{1}{6},\frac{1}{6}\}$ **11.** $\{-\sqrt{3},\sqrt{3}\}$ **13.** $\{0,-2\sqrt{2}\}$ **15.** $\{-\frac{7}{3},-\frac{2}{3}\}$

17. $\{2,4\}$ **19.** $\{-3,1\}$ **21.** $\{-4,5\}$ **23.** $\{-7,3\}$

25. $\{0,7\}$ **27.** $\{-2,2\}$ **29.** $\{-1,\frac{5}{3}\}$ **31.** $\{-\frac{1}{2},\frac{2}{5}\}$

33. $\{-1,\frac{3}{2}\}$ **35.** $\{-\frac{4}{3},\frac{2}{3}\}$ **37.** $\{\frac{5}{4},4\}$ **39.** $\{-\frac{5}{2},\frac{4}{3}\}$

41. $\{-\frac{2}{5},\frac{7}{2}\}$ **43.** $\{-6,-\frac{1}{4}\}$ **45.** $\{-\frac{3}{2},11\}$ **47.** $\{-3a,5a\}$

49. $\{\frac{3}{7},4\}$ **51.** $\{-1,7\}$

53. $x^2 - 11x + 30 = 0$ **55.** $9x^2 + 3x - 2 = 0$

57. $x^2 + 3x + 2 = 0$ **59.** $x^2 + 4 = 0$

PROBLEM SET 6.2, Page 231

1. $\{-8,8\}$ **3.** $\{-9i,9i\}$ **5.** $\{1-\sqrt{5},\, 1+\sqrt{5}\}$

7. $\{6-5i,\, 6+5i\}$ **9.** $\{\frac{1}{2},\frac{7}{6}\}$ **11.** $\{-\frac{1}{4},\frac{7}{4}\}$ **13.** $\{-3,3\}$

15. $\{-5,5\}$ **17.** 4 **19.** $\frac{9}{4}$ **21.** 1

23. 100 **25.** 3 **27.** $\{-3,1\}$ **29.** $\{5,7\}$

31. $\{3-\sqrt{2}i,\, 3+\sqrt{2}i\}$ **33.** $\left\{\dfrac{4-\sqrt{69}i}{5},\, \dfrac{4+\sqrt{69}i}{5}\right\}$ **35.** $\left\{\dfrac{1}{2},\dfrac{3}{2}\right\}$

37. $\left\{\dfrac{7-\sqrt{205}}{6},\, \dfrac{7+\sqrt{205}}{6}\right\}$ **39.** $\left\{\dfrac{-7-\sqrt{31}i}{8},\, \dfrac{-7+\sqrt{31}i}{8}\right\}$

41. $\left\{\dfrac{-6-\sqrt{26}}{2},\, \dfrac{-6+\sqrt{26}}{2}\right\}$ **43.** $\left\{\dfrac{-5-2\sqrt{10}}{5},\, \dfrac{-5+2\sqrt{10}}{5}\right\}$

45. $\left\{\dfrac{2}{3},2\right\}$ **47.** $\left\{\dfrac{1-\sqrt{2}}{3},\, \dfrac{1+\sqrt{2}}{3}\right\}$

49. $\left\{\dfrac{-b-\sqrt{b^2-12ab}}{2a},\, \dfrac{-b+\sqrt{b^2-12ab}}{2a}\right\}$

PROBLEM SET 6.3, Page 237

1. $\{1,4\}$ **3.** $\{2,4\}$ **5.** $\left\{\dfrac{-3-\sqrt{31}i}{2},\, \dfrac{-3+\sqrt{31}i}{2}\right\}$

7. $\{-5,7\}$ **9.** $\left\{\dfrac{5-\sqrt{17}}{4},\, \dfrac{5+\sqrt{17}}{4}\right\}$ **11.** $\left\{-\dfrac{1}{2},\dfrac{5}{3}\right\}$

13. $\left\{\dfrac{1-\sqrt{34}i}{5},\, \dfrac{1+\sqrt{34}i}{5}\right\}$ **15.** $\left\{\dfrac{1-\sqrt{31}i}{4},\, \dfrac{1+\sqrt{31}i}{4}\right\}$

17. $\left\{\dfrac{4-\sqrt{5}i}{7},\, \dfrac{4+\sqrt{5}i}{7}\right\}$ **19.** $\left\{\dfrac{-1-\sqrt{119}i}{15},\, \dfrac{-1+\sqrt{119}i}{15}\right\}$

21. $\left\{\dfrac{1-2\sqrt{5}}{2},\, \dfrac{1+2\sqrt{5}}{2}\right\}$ **23.** $\left\{\dfrac{5-\sqrt{21}}{2},\, \dfrac{5+\sqrt{21}}{2}\right\}$

25. $\{3-\sqrt{13}, 3+\sqrt{13}\}$ **27.** $\left\{\dfrac{-R-\sqrt{R^2-\dfrac{4L}{C}}}{2L}, \dfrac{-R+\sqrt{R^2-\dfrac{4L}{C}}}{2L}\right\}$

29. $-\frac{5}{6}, -1$ **31.** $100, 0$ **33.** $7, 13$ **35.** $\frac{2}{3}, -\frac{8}{3}$

37. $-2, 3$ **39.** $2, \frac{1}{3}$ **41.** $\frac{12}{7}, -2$ **43.** $-\frac{3}{5}, \frac{2}{5}$

45. Real and unequal   **47.** Real and equal   **49.** Complex

51. Real and unequal   **53.** Complex   **55.** Real and unequal

57. Real and unequal   **59.** Complex

PROBLEM SET 6.4, Page 241

1. $\{-3, -2, 2, 3\}$ **3.** $\{-2, 2, -i, i\}$ **5.** $\left\{-\dfrac{1}{2}, \dfrac{1}{4}\right\}$

7. $\left\{-\dfrac{\sqrt{5}}{5}, -\dfrac{1}{2}, \dfrac{1}{2}, \dfrac{\sqrt{5}}{5}\right\}$ **9.** $\{1, 4\}$ **11.** $\{-27, 64\}$

13. $\{-1, 0, 1\}$ **15.** $\{-3, -1, 1\}$ **17.** $\left\{-2, -1, \dfrac{1}{3}, \dfrac{2}{3}\right\}$

19. $\left\{-4, \dfrac{-1-\sqrt{13}}{2}, \dfrac{-1+\sqrt{13}}{2}, 3\right\}$ **21.** $\left\{-\dfrac{1}{4}\right\}$ **23.** $\{-3\}$

25. $\{-19, 61\}$ **27.** $\left\{-\dfrac{7}{2}, 3\right\}$ **29.** $\{4\}$ **31.** $\{0, 2\}$ **33.** $\{4\}$

35. $\{0, 3\}$ **37.** $\{27\}$ **39.** $\{7\}$ **41.** $\{5\}$ **43.** $\{1\}$

PROBLEM SET 6.5, Page 248

1. $\{x \mid -2 < x < 1\}$

3. $\{x \mid -1 \le x \le \frac{1}{2}\}$

5. $\{x \mid x \le -1 \text{ or } x \ge 2\}$

7. $\{x \mid x \le -2 \text{ or } x \ge 6\}$

9. $\varnothing$ **11.** $\{x \mid -2 \le x \le \frac{1}{3}\}$

13. $\{x \mid x \le -\frac{3}{5} \text{ or } x \ge 2\}$

15. $\{x \mid x < -3, \text{ or } x > 1\}$

17. $\{x \mid -2 \le x \le 3\}$

19. $\{x \mid 1 < x < 4\}$

21. $\{x \mid x \le -1 \text{ or } x > 2\}$

23. $\{x \mid -2 < x \le 4\}$

25. $\{x \mid x < \frac{1}{5} \text{ or } x > 2\}$

27. $\{x \mid x \le 1 \text{ or } x > 3\}$

29. $\{x \mid x < 0 \text{ or } x > \frac{1}{2}\}$

PROBLEM SET 6.6, Page 252

1. $-6, -5$ or $5, 6$ **3.** 5 and 9   **5.** $\frac{1}{3}$ or 3

7. $\dfrac{3 + \sqrt{34}}{5}, \dfrac{3 - \sqrt{34}}{5}$ **9.** 6 feet by 10 feet

11. 7 inches by 14 inches   **13.** 30 feet

15. 30 feet by 44 feet   **17.** 180 acres   **19.** \$3

21. 5 seconds; 25 seconds   **23.** 6 kilometers per hour

REVIEW PROBLEM SET, Page 254

1. $\{-7, 12\}$ **3.** $\{-4, 3\}$ **5.** $\{0, 2\}$

7. $\left\{\dfrac{1}{2}\right\}$ **9.** $\{-4, 2\}$ **11.** $\left\{-\dfrac{4}{9}, 1\right\}$

13. $\{-a - b, -a + b\}$ **15.** $\left\{-\dfrac{2}{5}, \dfrac{3}{4}\right\}$ **17.** $x^2 - 4x - 5 = 0$

19. $10x^2 - x - 3 = 0$ **21.** $\{-12, 12\}$ **23.** $\{-4i, 4i\}$ **25.** $\left\{\dfrac{3}{5}\right\}$

27. $\left\{-5, \dfrac{17}{3}\right\}$ **29.** $\{3 - \sqrt{2}, 3 + \sqrt{2}\}$ **31.** $\{4 - \sqrt{7}, 4 + \sqrt{7}\}$

33. $\left\{\dfrac{1-i}{3}, \dfrac{1+i}{3}\right\}$ **35.** $\left\{\dfrac{3-2i}{2}, \dfrac{3+2i}{2}\right\}$ **37.** $\left\{-\dfrac{3}{2}, \dfrac{1}{2}\right\}$

39. $\{-3c, 2c\}$ **41.** $\{-5 - \sqrt{38}, -5 + \sqrt{38}\}$

43. $\left\{\dfrac{-1 - \sqrt{11}i}{2}, \dfrac{-1 + \sqrt{11}i}{2}\right\}$ **45.** $\left\{\dfrac{1 - \sqrt{17}}{8}, \dfrac{1 + \sqrt{17}}{8}\right\}$

47. $\left\{\dfrac{-5 - \sqrt{161}}{4}, \dfrac{-5 + \sqrt{161}}{4}\right\}$ **49.** $\left\{-2, -\dfrac{1}{3}\right\}$

51. $\left\{\dfrac{5 - \sqrt{7}i}{4}, \dfrac{5 + \sqrt{7}i}{4}\right\}$ **53.** $\left\{\dfrac{-3 - \sqrt{65}}{4}, \dfrac{-3 + \sqrt{65}}{4}\right\}$

55. $\left\{\dfrac{k - 1 - \sqrt{k^2 - 6k + 1}}{2}, \dfrac{k - 1 + \sqrt{k^2 - 6k + 1}}{2}\right\}$ **57.** $\{4\}$ **59.** $\{0\}$

61. $4,4$ **63.** $-1,1$ **65.** $\dfrac{2}{7}, -\dfrac{5}{7}$ **67.** real and unequal

69. real and unequal   **71.** complex   **73.** complex

75. real and equal   **77.** $\{-2,-1,1,2\}$

79. $\{-1, -\frac{1}{4}, \frac{1}{4}, 1\}$

81. $\left\{-\dfrac{\sqrt{3}}{3}, \dfrac{\sqrt{3}}{3}, -2i, 2i\right\}$ **83.** $\{-27, 1\}$ **85.** $\{-3, -2, 1, 2\}$

87. $\{-1, 7\}$ **89.** $\{x \mid -5 < x < 3\}$

91. $\{x \mid x \le -2 \text{ or } x \ge 5\}$

93. $\{x \mid -\frac{1}{3} < x < \frac{5}{2}\}$

95. $\{x \mid x \le \frac{1}{3} \text{ or } x > \frac{7}{5}\}$

97. $6, 7, 8$ **99.** 10 feet   **101.** 12   **103.** 46

Chapter 7

PROBLEM SET 7.1, Page 267

15. Q_{I} **17.** Q_{II} **19.** Q_{III} **21.** none

23. none   **25.** 10   **27.** $\sqrt{17}$ **29.** 5

31. 5   **33.** $\dfrac{\sqrt{41}}{2}$ **35.** 5   **37.** $\sqrt{26}$

39. 13   **41.** a) $|\overline{P_1 P_2}| = 4\sqrt{2},\ |\overline{P_2 P_3}| = 2\sqrt{2},\ |\overline{P_1 P_3}| = 6\sqrt{2}$ b) yes

43. a) $(-1, 7)$ b) $(-\frac{7}{2}, \frac{7}{2})$ **45.** $7\sqrt{2}, \sqrt{73}, \sqrt{17}$

47. $7, 2\sqrt{17}, \sqrt{89}$

PROBLEM SET 7.2, Page 279

1. linear   **3.** nonlinear   **5.** linear   **7.** $(-1,-2),(0,1)$

9. $(1,2),(5,10)$ **11.** $(1,4),(2,1)$

13.

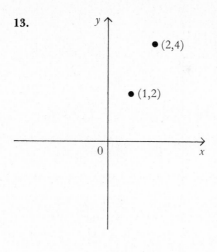

15.

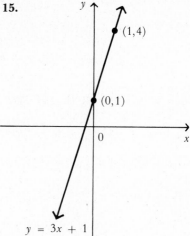

17.

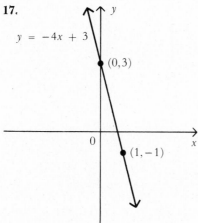

19.

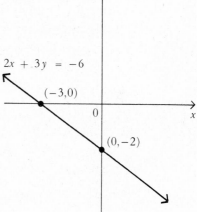

21.

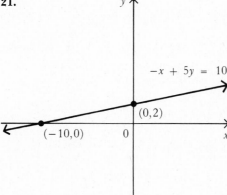

23. x intercept $= \frac{5}{3}$, y intercept $= 5$ **25.** x intercept $= \frac{9}{2}$, y intercept $= -3$

27. x intercept $= \frac{2}{7}$, y intercept $= 2$ **29.** x intercept $= 0$, y intercept $= 0$

31. x intercept $= 2$, y intercept $= -1$ **33.** x intercept $= \frac{3}{2}$, y intercept $= 3$

35. -1 **37.** $\frac{2}{3}$ **39.** $-\frac{1}{2}$ **41.** noncollinear

43. collinear   **45.** perpendicular   **47.** parallel

PROBLEM SET 7.3, Page 286

1. $y - 2 = -3(x + 1)$ **3.** $y - 1 = 5(x - 3)$

5. $y - 1 = -\frac{13}{7}x$ **7.** $y = -5$ **9.** $y = \frac{1}{2}x + \frac{7}{2}$

11. $y = -\frac{3}{2}x + 1$ **13.** $y = -3x + 5$ **15.** $x = -3$ **17.** $y = 4$

19. $y = \frac{2}{3}x - \frac{1}{3}$, $x = \frac{1}{2}$, $y = -\frac{1}{3}$ **21.** $y = -2x + 5$, $x = \frac{5}{2}$, $y = 5$

23. $y = 4x + 5$, $x = -\frac{5}{4}$, $y = 5$ **25.** $y = 2x + 0$, $x = 0$, $y = 0$

27. $y + 1 = -2(x - 3)$ **29.** $y - 1 = -\frac{1}{2}(x + \frac{1}{2})$

31. $y = -x + 1$ **33.** $y - 5 = -\frac{1}{3}(x - 1)$ **35.** $y - 2 = 2(x + 3)$

37. $y - 1 = \frac{3}{2}(x - 1)$ **39.** $\frac{x}{5} + \frac{y}{6} = 1$ **41.** $\frac{x}{-3} + \frac{y}{-11} = 1$

43. $y = \frac{1}{2}x + 3$ **45.** $y = -\frac{2}{3}x - \frac{2}{3}$ **47.** $y + 2 = 6(x - 1)$ **49.** $y = 6$

PROBLEM SET 7.4, Page 294

1. independent

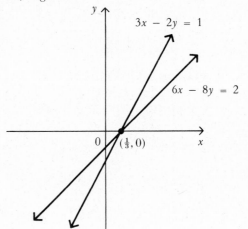

3. inconsistent

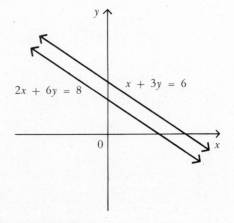

5. independent

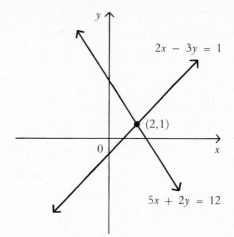

$2x - 3y = 1$

$(2,1)$

$5x + 2y = 12$

7. $\{(4,3)\}$ **9.** $\{(1,-1)\}$ **11.** $\{(1,5)\}$ **13.** $\{(\frac{7}{3},-\frac{4}{3})\}$

15. $\{(5,7)\}$ **17.** $\{(1,1)\}$ **19.** $\{(3,2)\}$ **21.** $\{(-5,3)\}$

23. $\{(2,-1)\}$ **25.** $\{(-\frac{3}{2},\frac{27}{2})\}$ **27.** $\{(\frac{2}{5},\frac{21}{5})\}$ **29.** $\{(\frac{22}{25},\frac{21}{25})\}$

31. $\{(7,9)\}$ **33.** $\{(10,24)\}$ **35.** $\left\{\left(\dfrac{3a+b}{10},\dfrac{a-3b}{10}\right)\right\}$

37. $\{(\frac{1}{13},\frac{1}{17})\}$ **39.** $\{(\frac{1}{4},-\frac{1}{5})\}$ **41.** $\{(-1,\frac{1}{3})\}$

PROBLEM SET 7.5, Page 298

1. $\{(2,3,-1)\}$ **3.** $\{(4,3,2)\}$ **5.** $\{(2,-1,3)\}$ **7.** $\{(1,1,1)\}$

9. $\{(3,-1,4)\}$ **11.** $\{(2,2,1)\}$ **13.** $\{(5,2,1)\}$ **15.** $\{(\frac{94}{35},\frac{1}{7},\frac{1}{5})\}$

17. $\left\{\left(2a+b-c,-a-\dfrac{b}{3}+\dfrac{2c}{3},\dfrac{-2b}{3}+\dfrac{c}{3}\right)\right\}$ **19.** $\{(3,4,6)\}$

PROBLEM SET 7.6, Page 303

1. 26   **3.** 28   **5.** $-\frac{13}{2}$ **7.** 0

9. 1   **11.** 116   **13.** 1   **15.** 18

17. -438 **19.** 63   **21.** -1 **23.** 2 or 8

25. $\frac{23}{2}$ **27.** $\{x \mid x < -2 \text{ or } x > 2\}$

PROBLEM SET 7.7, Page 310

1. $\{(\frac{1}{3},\frac{2}{3})\}$ **3.** $\{(0,0)\}$ **5.** $\{(\frac{85}{4},\frac{35}{4})\}$ **7.** $\{(-11,7)\}$

9. $\{(5,-6)\}$ **11.** $\{(1,2,3)\}$ **13.** $\{(\frac{19}{4},-\frac{3}{4},-\frac{29}{4})\}$

15. $\{(-5,-\frac{14}{3},-\frac{16}{3})\}$ **17.** $\{(2,1,-1)\}$ **19.** $\{(-7,8,1)\}$ **21.** $\{(1,2,3)\}$

PROBLEM SET 7.8, Page 314

1. 7 and 5   **3.** 72   **5.** 1,200 bushels at \$2.25; 1,800 bushels at \$2.50

7. \$1,000 at 5%; \$3,000 at 6%   **9.** Joan is 10, Doreen is 7

11. 3 hours and 5 hours   **13.** length = 160 rods, width = 120 rods

15. chain, $25; watch, $75; ring, $125

17. $4,000 at 6%; $10,000 at 7%; $6,000 at 8%

19. 10 inches, 15 inches, 20 inches

REVIEW PROBLEM SET, Page 316

1. Q_I **3.** Q_{II} **5.** none   **7.** Q_{IV}

9. 13   **11.** 5   **13.** $\sqrt{53}, \sqrt{53}, \sqrt{106}$

15.   **17.**

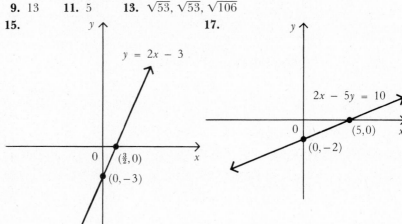

19. $x = 2, y = 4$ **21.** $x = -4, y = 3$ **23.** 1

25. $-\frac{1}{2}$ **27.** $y - 1 = 3(x - 1)$ **29.** $y = -2$

31. $y = 2x + 4$ **33.** $y + 1 = -\frac{3}{2}(x + 3)$ **35.** $\{(2, -2)\}$

37. $\varnothing$ **39.** $\{(2, -1)\}$ **41.** $\{(-1, 3, 2)\}$ **43.** $\{(\frac{1}{3}, \frac{1}{2})\}$

45. 11   **47.** 26   **51.** 60 at 13 cents, 100 at 18 cents

53. 20 nickels, 5 dimes, 10 quarters

Chapter 8

PROBLEM SET 8.1, Page 324

1. $\log_5 125 = 3$ **3.** $\log_{10} 100,000 = 5$ **5.** $\log_4 \frac{1}{16} = -2$

7. $\log_6 \frac{1}{36} = -2$ **9.** $\log_9 3 = \frac{1}{2}$ **11.** $\log_{100} 0.001 = -\frac{3}{2}$

13. $\log_7 1 = 0$ **15.** $\log_x a = 3$ **17.** $9^2 = 81$ **19.** $27^{2/3} = 9$

21. $10^{-3} = 0.001$ **23.** $(\frac{1}{3})^{-2} = 9$ **25.** $10^{0.64} = 4.35$

27. $(\sqrt{16})^{1/2} = 2$ **29.** $x^0 = 1$ **31.** 25   **33.** 32

35. 3   **37.** $\frac{1}{8}$ **39.** 243   **41.** $\{\frac{5}{2}\}$

43. $\{8\}$ **45.** $\{\frac{4}{3}\}$ **47.** $\{\frac{9}{4}\}$ **49.** $\{2\}$

51. 6   **53.** $\frac{1}{2}$ **55.** $-\frac{1}{2}$ **57.** 2

59. -3 **61.** $\frac{5}{2}$ **63.** -4 **65.** 6

67. 0   **69.** -3 **71.** $\{2\}$ **73.** $\{4\}$

75. $\{9\}$ **77.** $\{5\}$

PROBLEM SET 8.2, Page 328

1. $\log_5 \frac{9}{8}$ **3.** $\log_7 \frac{1}{6}$ **5.** $\log_5 16$ **7.** $\log_3 \frac{2}{25}$

9. $\log_a \frac{y}{3}$ **11.** $\log_2 5 + \log_2 7$ **13.** $4\log_4 3 + 2\log_4 5$

15. $\log_2 x - \log_2 y$ **17.** $5\log_4 2 - 3\log_4 5$ **19.** $\frac{1}{5}\log_3 x + \frac{1}{5}\log_3 y$

21. 4   **23.** 4   **25.** 3   **27.** 0.7781

29. 1.2552   **31.** 0.1761   **33.** 0.6990   **35.** 1.9084

37. -0.3010 **39.** $\{7\}$ **41.** $\{\frac{1}{9}\}$

PROBLEM SET 8.3, Page 336

1. 2.5011   **3.** 4.7275   **5.** 1.2330   **7.** 0.8331

9. 0.0719   **11.** -0.5017 **13.** -1.1469 **15.** -3.7496

17. -5.1549 **19.** 3.1889   **21.** 1.9007   **23.** 0.7253

25. -0.2422 **27.** -1.2723 **29.** 2.59   **31.** 19.7

33. 553   **35.** 3,560   **37.** 0.543   **39.** 0.00793

41. 0.0235   **43.** 0.000325   **45.** 0.696   **47.** 0.00712

49. 1.397   **51.** 34.88   **53.** 0.01682   **55.** 0.00003448

PROBLEM SET 8.4, Page 341

1. 1.0877   **3.** -2.8084 **5.** 1.6433   **7.** -1.9964

9. -0.6270 **11.** -0.7611 **13.** 188   **15.** 311

17. 0.00230   **19.** 0.00626   **21.** 1.21   **23.** 15.2

25. 4.03   **27.** 0.0183   **29.** 43.2   **31.** 0.228

33. 113   **35.** 0.247   **37.** 2.21   **39.** 4.63

41. 0.598   **43.** 12.1   **45.** 1.13   **47.** 0.358

49. 0.958   **51.** $\{2.81\}$ **53.** $\{-1.85\}$ **55.** $\{0.961\}$

57. $\{2.13\}$ **59.** $\{0.165\}$

PROBLEM SET 8.5, Page 346

1. 1.46 centimeters   **3.** 102.5 cubic inches   **5.** 19.5 cubic inches

7. a) $1,961   b) $1,967   c) $1,971   d) $1,977

9. $10,700   **11.** 11.9 years   **13.** 5.7%   **15.** $3,089   **17.** $40,985

REVIEW PROBLEM SET, Page 347

1. $\log_2 32 = 5$ **3.** $\log_8 32 = \frac{5}{3}$ **5.** $\log_{27} 9 = \frac{2}{3}$

7. $10^2 = 100$ **9.** $17^0 = 1$ **11.** $10^{-1} = \frac{1}{10}$

13. 9   **15.** $\frac{1}{8}$ **17.** $\frac{1}{100}$ **19.** $\{-\frac{1}{3}\}$

21. $\{\frac{5}{3}\}$ **23.** $\{\frac{1}{2}\}$ **25.** 2   **27.** 0

29. $-\frac{3}{2}$ **31.** $\frac{1}{2}$ **33.** -4 **35.** $\frac{7}{4}$

37. $\{2\}$ **39.** $\{\frac{17}{7}\}$ **41.** $\{2\}$ **43.** $\log_2 \frac{2}{9}$

45. $\log_5 \frac{1}{6}$ **47.** $\log_a \frac{x^5}{y^3}$ **49.** $6\log_2 3 + 7\log_2 4$

51. $\log_2 x + 5 \log_2 y$

53. $2 \log_a x - 4 \log_a y$

55. -3

57. 13.2

59. -1.64

61. -0.06

63. 2.56

65. $\left\{ \dfrac{1 + \sqrt{5}}{2} \right\}$

67. 2.9263

69. 1.8876

71. 0.5024

73. -0.3251

75. -1.5901

77. -2.2321

79. 5.9425

81. 1.0539

83. 3.7891

85. 5.58

87. 89.8

89. $1,083$

91. 0.0283

93. 0.0952

95. 0.000392

97. 0.06474

99. 3.7758

101. 1.6470

103. $1,480$

105. 15.8

107. $41,100$

109. 0.702

111. 1.67

113. 3.51

115. 6.61

117. $\{1.47\}$

119. $\{0.88\}$

121. $\$11,570$

Chapter 9

PROBLEM SET 9.1, Page 357

1. domain $= \{1,2,5\}$
range $= \{-3,1,2,4\}$

3. domain $= \{0\}$
range $= \{-7,3,6\}$

5. domain $= \{-3,4\}$
range $= \{-1,-2\}$

7. domain $= \{7,9,16\}$
range $= \{-\sqrt{7}, -3, -4, \sqrt{7}, 3, 4\}$

9. domain $= \{0.2\}$
range $= \{-2,0,2\}$

11. relation $\{(0,0), (2,2), (4,4), (6,6), (8,8)\}$
domain $= \{0,2,4,6,8\}$
range $= \{0,2,4,6,8\}$

13.

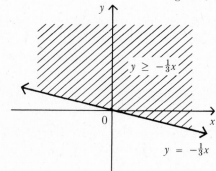

15.

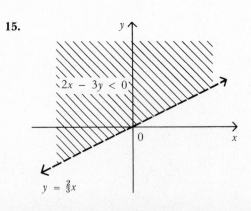

17.

19.

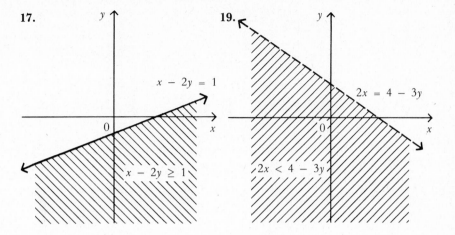

$x - 2y = 1$

$x - 2y \geq 1$

$2x = 4 - 3y$

$2x < 4 - 3y$

PROBLEM SET 9.2, Page 366

1. function   **3.** function   **5.** function

7. not a function   **9.** function

11. domain = set of all real numbers; $f(1) = 4, f(3) = 10, f(5) = 16$

13. domain = $\{1,3,5\}$ **15.** domain = $\{x \mid x \geq 1\}$

$\quad f(1) = -1, f(3) = 8, f(5) = 0 \qquad f(1) = 0, f(3) = \sqrt{2}, f(5) = 2$

17. domain = $\{x \mid x \neq 0\}$ **19.** domain = $\{x \mid x > -2\}$

$\quad f(1) = 1, f(3) = \dfrac{1}{3}, f(5) = \dfrac{1}{5} \qquad f(1) = \dfrac{1}{\sqrt{3}}, f(3) = \dfrac{1}{\sqrt{5}}, f(5) = \dfrac{1}{\sqrt{7}}$

21. domain = set of all real numbers

range = set of all real numbers

$f(0) = -7, f(-1) = -10, f(1) = -4$

$f(a) = 3a - 7, f(a + h) = 3a + 3h - 7$

$\dfrac{f(a + h) - f(a)}{h} = 3$

23. domain = set of all real numbers

range = $\{2\}$

$f(0) = 2, f(-1) = 2, f(1) = 2$

$f(a) = 2, f(a + h) = 2$

$\dfrac{f(a + h) - f(a)}{h} = 0$

25. domain = set of all real numbers

range = $\{y \mid y \geq 0\}$

$f(0) = 0, f(-1) = 2, f(1) = 2$

$f(a) = 2a^2, f(a + h) = 2a^2 + 4ah + 2h^2$

$\dfrac{f(a + h) - f(a)}{h} = 4a + 2h$

27. $P(x) = 4x$ **29.** $V(x) = x^3$

PROBLEM SET 9.3, Page 370

1. $y = f(x) = 2x,\ f(x+2) = 2x + 4,\ f(2) + f(3) = 10,\ \dfrac{f(x+h) - f(x)}{h} = 2$

3. $y = \dfrac{1}{2}x^3$ **5.** $y = 3x^3$ **7.** $\dfrac{14}{1{,}331}x^3$ **9.** $\pm\sqrt{\dfrac{54}{35}}$

11. $T = \dfrac{0.01x}{y}$ **13.** $y = \dfrac{192}{x^3}$ **15.** $3:2,\ 27:8$ **17.** $S = 6x^2,$
                                                              864 square inches

PROBLEM SET 9.4, Page 377

1. domain = set of all real numbers
range = set of all real numbers
slope = -3

3. domain = set of all real numbers
range = set of all real numbers
slope = $-\frac{3}{4}$

5. domain = set of all real numbers
range = $\{-1\}$
slope = 0

7. domain = set of all real numbers
range = set of all real numbers
slope = 2

9. domain = set of all real numbers
range = set of all real numbers
slope = 4

11. $f(x) = 2x + 1$ **13.** $f(x) = -\frac{4}{3}x + 4$ **15.** $f(x) = 3$
17. $f(x) = mx$ **19.** $f(x) = \frac{1}{5}x + \frac{14}{5}$

21. domain = set of all real numbers
range = $\{y \mid y \geq -3\}$
extreme point = $(0, -3)$
x intercepts = $\pm\dfrac{\sqrt{6}}{2}$
y intercept = -3

23. domain = set of all real numbers
range = $\{y \mid y \leq 0\}$
extreme point = $(-1, 0)$
x intercept = -1
y intercept = -1

25. domain = set of all real numbers
range = $\{y \mid y \geq -\frac{1}{4}\}$
extreme point = $(-\frac{5}{2}, -\frac{1}{4})$
x intercepts = $-3, -2$
y intercept = 6

27. domain = set of all real numbers
range = $\{y \mid y \geq -\frac{9}{8}\}$
extreme point = $(\frac{3}{4}, -\frac{9}{8})$
x intercepts = $0, \frac{3}{2}$
y intercept = 0

29. domain = set of all real numbers
range = $\{y \mid y \geq -1\}$
extreme point = $(-2, -1)$
x intercepts = $-3, -1$
y intercept = 3

31.

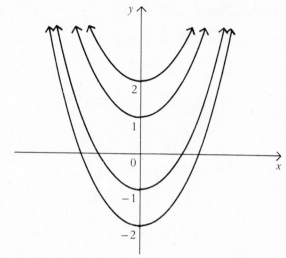

PROBLEM SET 9.5, Page 383

	Domain	Range	Increasing	Decreasing
1.	$\{x \mid x \in R\}$	$\{y \mid y > 0\}$	yes	no
3.	$\{x \mid x \in R\}$	$\{y \mid y > 0\}$	yes	no
5.	$\{x \mid x \in R\}$	$\{y \mid y < 0\}$	yes	no
7.	$\{x \mid x \in R\}$	$\{y \mid y > 0\}$	yes	no
9.	$\{x \mid x \in R\}$	$\{y \mid y > 0\}$	yes	no

11. 3   **13.** 4   **15.** Any nonzero real number

	Domain	Range	Increasing	Decreasing
19.	$\{x \mid x > 0\}$	$\{y \mid y \in R\}$	no	yes
21.	$\{x \mid x > 0\}$	$\{y \mid y \in R\}$	no	yes
23.	$\{x \mid x > 0\}$	$\{y \mid y \in R\}$	yes	no

25. 2   **27.** 4   **29.** 9

PROBLEM SET 9.6, Page 389

1. parabola: domain $= \{x \mid x \in R\}$, range $= \{y \mid y \geq 0, y \in R\}$, x intercept $= 0$, y intercept $= 0$; function

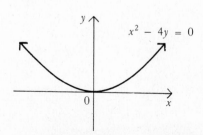

3. parabola: domain = $\{x \mid x \in R\}$, range = $\{y \mid y \leq 0, y \in R\}$, x intercept = 0, y intercept = 0; function

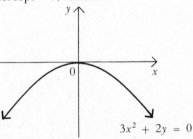

$3x^2 + 2y = 0$

5. circle: domain = $\{x \mid -2 \leq x \leq 2\}$, range = $\{y \mid -2 \leq y \leq 2\}$, x intercepts = -2 and 2, y intercepts = -2 and 2

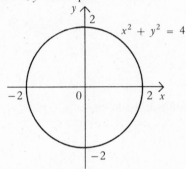

$x^2 + y^2 = 4$

7. circle: domain = $\{x \mid -4 \leq x \leq 4\}$, range = $\{y \mid -4 \leq y \leq 4\}$, x intercepts = -4 and 4, y intercepts = -4 and 4

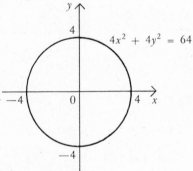

$4x^2 + 4y^2 = 64$

9. ellipse: domain = $\{x \mid -3 \leq x \leq 3\}$, range = $\{y \mid -2 \leq y \leq 2\}$, x intercepts = -3 and 3, y intercepts = -2 and 2

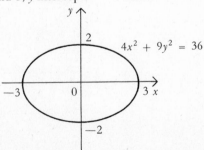

$4x^2 + 9y^2 = 36$

11. ellipse: domain = $\{x|-4 \leq x \leq 4\}$, range = $\{y|-5 \leq y \leq 5\}$,
x intercepts $= -4$ and 4, y intercepts $= -5$ and 5

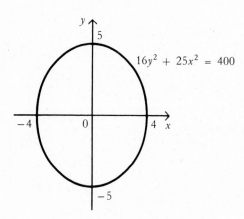

$$16y^2 + 25x^2 = 400$$

13. ellipse: domain = $\{x|-4 \leq x \leq 4\}$. range = $\{y|-2 \leq y \leq 2\}$,
x intercepts $= -4$ and 4, y intercepts $= -2$ and 2

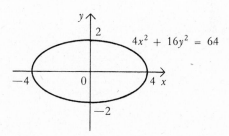

$$4x^2 + 16y^2 = 64$$

15. hyperbola: domain = $\{x|x \leq -\frac{1}{6} \text{ or } x \geq \frac{1}{6}\}$, range = $\{y|y \in R\}$,
x intercepts $= -\frac{1}{6}$ and $\frac{1}{6}$, y intercepts $=$ none

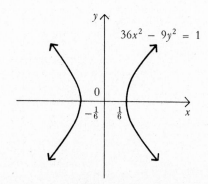

$$36x^2 - 9y^2 = 1$$

17. hyperbola: domain = $\{x \mid x \in R\}$, range = $\{y \mid y \leq -1 \text{ or } y \geq 1\}$, x intercepts = none, y intercepts = -1 and 1

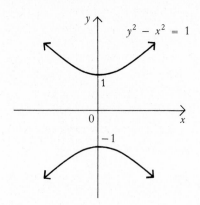

$$y^2 - x^2 = 1$$

19. hyperbola: domain = $\{x \mid x \leq -3 \text{ or } x \geq 3\}$, range = $\{y \mid y \in R\}$, x intercepts = -3 and 3, y intercepts = none

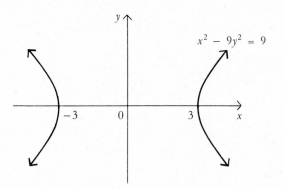

$$x^2 - 9y^2 = 9$$

PROBLEM SET 9.7, Page 393

1. $\{(-1,-2),(2,1)\}$ **3.** $\{(2,4),(-0.8,-4.4)\}$

5. $\{(-1+2i, 2-3i),(-1-2i, 2+3i),(-6, \sqrt{6}i),(-6,-\sqrt{6}i)\}$, no point of intersection

7. $\{(\frac{7}{2},\frac{5}{2}),(\frac{11}{4},\frac{5}{4})\}$

9. $\{(2+3i, 1-2i),(2-3i, 1+2i)\}$, no point of intersection

11. $\{(5,4)\}$ **13.** $\{(4,2),(4,-2)\}$ **15.** $\{(3,2),(-3,2),(3,-2),(-3,-2)\}$

17. $\{(-2\sqrt{6},1)(-2\sqrt{6},-1)(2\sqrt{6},1),(2\sqrt{6},-1)\}$

19. $\{(-2,2\sqrt{3}),(-2,-2\sqrt{3}),(2,2\sqrt{3}),(2,-2\sqrt{3})\}$

21. $\left\{(4,2),(-4,2),\left(\frac{4\sqrt{15}}{3},-\frac{10}{3}\right),\left(\frac{-4\sqrt{15}}{3},-\frac{10}{3}\right)\right\}$

23. $\{(3,2),(-3,2),(-3,-2),(3,-2)\}$

25. $\{(2i,3)(-2i,3),(-2,1),(2,1)\}$

27. $\{(3i,5),(3i,-5),(-3i,5),(-3i,-5)\}$, no point of intersection

29. $\left\{\left(\frac{41}{10},\frac{3\sqrt{91}}{10}\right),\left(\frac{41}{10},-\frac{3\sqrt{91}}{10}\right)\right\}$

REVIEW PROBLEM SET, Page 394

1. domain $= \{-3, 1, 5\}$, range $= \{-2, 4, 6\}$

3. domain $= \{2, 18\}$, range $= \{-3, -1, 1, 3\}$

5.

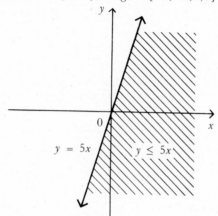

7.

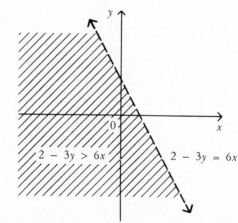

9. not a function   **11.** function   **13.** function

15. function   **17.** (a), (b), (c)   **19.** 5

21. 5   **23.** $3a^2 + 2$ **25.** $3a^2 + 6ab + 3b^2 + 2$

27. $-10, 4$ **29.** $x > \frac{15}{2}$ **31.** $f(x) = \frac{2}{3}x$

33. $f(x) = 4\sqrt{x}$ **35.** $f(x) = 20/x$ **37.** 192 pounds

39. $f(x) = mx$ **41.** $f(x) = mx$

43. $f(x) = mx + m/6$, where $m = 6b$

45. domain $= \{x \mid x \in R\}$, range $= \{y \mid y \in R\}$, slope $= 2$

47. domain $= \{x \mid x \in R\}$, range $= \{y \mid y \in R\}$, slope $= \frac{3}{2}$

49. domain $= \{x \mid x \in R\}$, range $= \{y \mid y \geq -\frac{121}{24}\}$,
     extreme point $= (\frac{5}{12}, -\frac{121}{24})$, x intercepts $= -\frac{1}{2}$ and $\frac{4}{3}$, y intercept $= -4$

51. domain $= \{x \mid x \in R\}$, range $= \{y \mid y \geq 0\}$, extreme point $= (-3, 0)$,
     x intercept $= -3$, y intercept $= 9$

53. domain $= \{x \mid x \in R\}$, range $= \{y \mid y \leq \frac{1}{8}\}$, extreme point $= (-\frac{5}{8}, \frac{1}{8})$,
     x intercepts $= -\frac{3}{4}$ and $-\frac{1}{2}$, y intercept $= -3$

	Domain	*Range*	*Increasing*	*Decreasing*
55.	$\{x \mid x \in R\}$	$\{y \mid y > 0\}$	no	yes
57.	$\{x \mid x \in R\}$	$\{y \mid y > 0\}$	yes	no
59.	$\{x \mid x \in R\}$	$\{y \mid y > 0\}$	yes	no
61.	$\{x \mid x > 0\}$	$\{y \mid y \in R\}$	no	yes
63.	$\{x \mid x > -2\}$	$\{y \mid y \in R\}$	yes	no

65. 1   **67.** $\frac{1}{2}$ **69.** $\frac{5}{4}$ **71.** $\frac{1}{20}$

73. circle: x intercepts $= -5$ and 5; y intercepts $= -5$ and 5

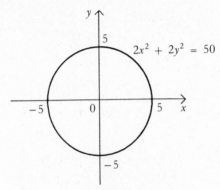

75. Ellipse: x intercepts $= -2, 2$; y intercepts $= -\frac{3}{2}, \frac{3}{2}$

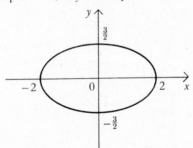

77. Parabola: x intercept $= -\frac{13}{4}$; y intercepts $= -\sqrt{13}, \sqrt{13}$

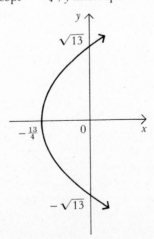

79. $\{(3,-4)\}$ **81.** $\{(6,0),(-5,\sqrt{11}),(-5,-\sqrt{11})\}$

83. $\{(5,2),(5,-2),(-5,2),(-5,-2)\}$

Chapter 10

PROBLEM SET 10.1, Page 403

1. $\frac{3}{2}, 4, \frac{15}{2}, 12, \frac{35}{2}$

3. $-1, -1, 0, 2, 5$

5. $2, 4, 2, 4, 2$

7. A.P., $d = 3$, $s_{10} = 155$

9. A.P., $d = 5$, $s_{10} = 295$

11. A.P., $d = -13$, $s_{10} = 85$

13. A.P., $d = 1.2$, $s_{10} = 111$

15. $a_{10} = 50$, $a_{15} = 85$

17. $a_6 = 16a + 4b$; $a_9 = 25a - 8b$

19. $63b - 63$

21. $a_{10} = 33$, $s_{10} = 195$

23. $d = 13\frac{5}{51}$; $a_{18} = 239\frac{2}{3}$

25. $n = 32$, $d = \frac{21}{31}$

PROBLEM SET 10.2, Page 405

1. G.P., $r = 3$, $s_5 = 242$ **3.** G.P., $r = -2$, $s_5 = 11$

5. G.P., $r = \frac{2}{3}$, $s_5 = 211$ **7.** G.P., $r = -\frac{2}{3}$, $s_5 = \frac{55}{9}$

9. $\frac{1}{128}$ **11.** 2   **13.** $(1 + a)^{n-1}$

15. $a_6 = 192$, $a_{10} = 3{,}072$, $s_{10} = 6{,}138$ **17.** 48

19. 3 or -4 **21.** 2 or -2 **23.** $a_9 = \frac{1}{128}$, $s_8 = \frac{85}{64}$

25. $n = 10$, $s_n = \frac{1{,}023}{16}$ **27.** $\frac{2}{27}, \frac{2}{9}, \frac{2}{3}$

PROBLEM SET 10.3, Page 408

1. 15   **3.** 660   **5.** 15   **7.** 15   **9.** $\frac{163}{60}$

11. 500   **13.** $\frac{5}{6}$ **15.** $\sum_{i=1}^{5} [3(i-1) + 1]$ or $\sum_{i=0}^{4} (3i + 1)$

17. $\sum_{k=1}^{4} \left(\frac{3}{5}\right)^k$ **19.** true   **21.** false   **23.** false

PROBLEM SET 10.4, Page 414

1. $\frac{1}{30}$ **3.** $\frac{1}{9}$ **5.** 2   **7.** 0   **9.** $(n+1)n(n-1)(n-2)$

11. $x^5 + 10x^4 + 40x^3 + 80x^2 + 80x + 32$

13. $x^6 + 12x^4y^2 + 48x^2y^4 + 64y^6$

15. $a^{18} - 6a^{14} + 15a^{10} - 20a^6 + 15a^2 - 6a^{-2} + a^{-6}$

17. $32 + 80\left(\frac{x}{y}\right) + 80\left(\frac{x}{y}\right)^2 + 40\left(\frac{x}{y}\right)^3 + 10\left(\frac{x}{y}\right)^4 + \left(\frac{x}{y}\right)^5$

19. $256z^8 + 1{,}024z^7x + 1{,}792z^6x^2 + 1{,}792z^5x^3 + 1{,}120z^4x^4 + 448z^3x^5$
$+ 112z^2x^6 + 16zx^7 + x^8$

21. $y^{10} - 10y^8x + 40y^6x^2 - 80y^4x^3 + 80y^2x^4 - 32x^5$

23. $x^{20} - 20x^{18}a + 180x^{16}a^2 - 960x^{14}a^3$

25. $\left(\dfrac{x}{2}\right)^{7/2} + 14\left(\dfrac{x}{2}\right)^{3}y + 84\left(\dfrac{x}{2}\right)^{5/2}y^2 + 280\left(\dfrac{x}{2}\right)^{2}y^3$

27. $x^{16} + 16x^{15}y + 120x^{14}y^2 + 560x^{13}y^3 + 1{,}820x^{12}\,y^4$

29. $a^{11} - 22a^{10}b^2 + 220a^9b^4 - 1{,}320a^8b^6 + 5{,}280a^7b^8$

31. $x^7 - 14x^6y + 84x^5y^2 - 280x^4y^3 + 560x^3y^4$

33. $a^{27} - 9a^{26} + 36a^{25} - 84a^{24} + 126a^{23}$

35. $\dfrac{455}{4{,}096}x^{24}a^3$ **37.** $\dfrac{672}{729}x^6a^{12}$ **39.** $\dfrac{14}{9}a^5x^{12}$

REVIEW PROBLEM SET, Page 415

1. 4, 2, 4, 2

3. $2, \frac{7}{2}, 4, \frac{17}{4}$

5. $1, \frac{18}{5}, \frac{39}{5}, \frac{68}{5}$

7. $a_9 = 44,\ s_9 = 216$

9. $a_{11} = 12,\ s_{11} = 297$

11. $a_{24} = 4,\ s_{24} = 50$

13. 3   **15.** 3 or -2

17. $a_8 = 49{,}152;\ s_8 = 65{,}535$

19. $a_6 = -\frac{1}{3}.\ s_6 = 60\frac{2}{3}$

21. $a_{18} = -768\sqrt{2},\ s_{18} = 1{,}533(1 - \sqrt{2})$

23. -2 or 24   **25.** $\frac{1}{7}$ or 10   **27.** 95   **29.** 20

31. 160   **33.** $x^4 + 8x^3y + 24x^2y^2 + 32xy^3 + 16y^4$

35. $1 + 5x + 10x^2 + 10x^3 + 5x^4 + x^5$

37. $1 - 12x + 60x^2 - 160x^3 + 240x^4 - 192x^5 + 64x^6$

39. $81x^4 + 108x^3y + 54x^2y^2 + 12xy^3 + y^4$

41. $243x^5 + 405x^{9/2} + 270x^4 + 90x^{7/2} + 15x^2 + x^{5/2}$

43. $8x^3 + \dfrac{12x^2}{y} + \dfrac{6x}{y^2} + \dfrac{1}{y^3}$

45. $210x^6y^4$ **47.** $13{,}440x^6y^4$ **49.** $1{,}082{,}565x^8y^3$

Appendix B

PROBLEM SET B.1, Page 425

1. no   **3.** yes   **5.** no

7. a) 15 square meters   c) 25.3432 square centimeters

8. a) 24.624 square feet   c) 6 square meters

9. a) 30 square inches   c) 88 square centimeters

10. a) 12.25 square inches   c) 0.3249 square meters

11. a) 45 square feet   c) $17\frac{1}{2}$ square meters

12. a) 25π square inches   c) 9.61π square centimeters

13. a) 9 square inches   c) 21 square centimeters
e) $31\frac{1}{2}$ square centimeters   g) 84 square inches

15. \$9,750

PROBLEM SET B.2, Page 430

1. a) 20 centimeters   c) 7.9 feet

2. a) 40 inches   c) 12 feet   e) 8 feet

3. a) 32 inches   c) 10 feet

4. a) 18 inches   c) 12 inches   e) 32 centimeters

5. a) 24 centimeters   c) 4 meters

6. a) 8π inches   c) 6π inches   e) 14π meters

7. length = 320 feet, width = 200 feet

PROBLEM SET B.3, Page 438

1. a) $V = 60$ cubic inches, $S = 94$ square inches
   c) $V = 35$ cubic feet, $S = 65.5$ square feet
   e) $V = 24\sqrt{2}$ cubic feet, $S = 28\sqrt{2} + 24$ square feet

3. a) L.S. $= 16\pi$ square inches, $S = 24\pi$ square inches,
   $V = 16\pi$ cubic inches   c) L.S. $= 48\pi$ square inches,
   $S = 80\pi$ square inches, $V = 96\pi$ cubic inches
   e) L.S. $= 26.5\pi$ square centimeters, $S = 39\pi$ square centimeters,
   $V = 33.125\pi$ cubic centimeters

4. a) $\dfrac{8}{5\pi}$ centimeters   c) $\sqrt{17} - 1$ inches

5. a) $S = 16(\sqrt{10} + 1)$ square inches, $V = 32$ cubic inches
   c) $S = 2(\sqrt{10} + 2)$ square meters, $V = 4$ cubic meters

6. a) L.S. $= 15\pi$ square inches, $S = 24\pi$ square inches,
   $V = 12\pi$ cubic inches

   c) $S = 16\pi$ square inches, $V = \dfrac{16\pi\sqrt{2}}{3}$ cubic inches

7. a) $S = 100\pi$ square inches, $V = \dfrac{500\pi}{3}$ cubic inches

   c) $S = 144\pi$ square centimeters, $V = 288\pi$ cubic centimeters
   e) $S = 72\pi$ square centimeters, $V = 72\sqrt{2}\pi$ cubic centimeters

9. 15,659.4 pounds   **11.** a) 8 times larger

Index

Abscissa, 260
Absolute inequality, 165
Absolute value, 26, 162
    definition of, 163
    equations involving, 163
Absolute value function, 365
Absolute value inequalities, 165
Addition
    associative property of, 19, 51, 57
    commutative property of, 18, 51, 57
    of complex numbers, 211
    of fractions, 109
    of polynomials, 51
    of radical expressions, 199
    of signed numbers, 26
Addition law of inequalities, 155
Additive inverse property, 22
Algebra, 39
    of polynomials, 39
Algebraic expressions, 39
Annuity
    definition of, 344
    present value of, 345
Applications
    of first-degree equations, 144
    geometric, 146
    investment, 147
    of logarithms, 342
    mixture, 148
    in physics, 149
    quadratic equations of, 249
    of systems of linear equations, 311
    of uniform motion, 149
Area
    of circles, 425
    of parallelograms, 423
    of rectangles, 421
    of squares, 422
    surface, 432
    of trapezoids, 424
    of triangles, 423
Arithmetic progression(s), 400
    common difference of, 401

    definition of, 400
    mth term of, 401
    sum of n terms of, 401
Associative property, 19
    of addition, 19, 51
    of multiplication, 20, 57
Axis of a coordinate system, 260

Base
    of exponential functions, 379
    of logarithms, 321
    of powers, 39, 186
Basic properties of real numbers, 17
Binomial(s)
    cube of, 73
    definition of, 48
    factorization of, 73
    products of, 62
    square of, 73
Binomial coefficients, 410
Binomial expansion, 411
Binomial theorem, 411

Cancellation property, 23
    for addition, 23
    for multiplication, 23
Cartesian coordinate system, 259
Center
    of circles, 385
    of ellipses, 385
Characteristic of a logarithm, 330
Circle, 385
    area of, 425
    equation of, 385
Closure property, 17
    for addition, 17
    for multiplication, 18
Coefficient, 48
Coincides with, 288
Collinear, 268, 280
Combined factoring methods, 76

Common factor, 69
  binomial, 71
Common logarithm, 330
Commutative property, 18
  of addition, 18, 51
  of mutliplication, 18, 57
Complete factorization, 69
Completing-the-square method, 227
Complex fraction, 120
Complex number(s), 210
  addition of, 221
  conjugate of, 213
  division of, 213
  equality of, 211
  form of, 210
  imaginary part of, 211
  multiplication of, 211
  real part of, 211
  subtraction of, 211
Compound interest, 343
Conditional equations, 132
Conditional inequalities, 157
Cone
  altitude of, 436
  circular, 436
  lateral surface of, 436
  volume of, 436
Conic section, 384
Conjugate, 213
Constant function, 371
Constant of variation, 367
Coordinate(s), 10, 260
Coordinate system, 259
Counting numbers, 9
Cramer's rule, 305
Cube root, 185
Cylinder
  altitude of, 434
  lateral surface area of, 435
  right circular, 395, 434
  volume of, 435

Degree
  of monomials, 48
  of polynomials, 48, 49
  of zero, 48
Denominator
  least common, 115
  rationalization of, 203
Dependent equations, 290

Determinant(s), 299
  in Cramer's rule, 305
  definition of, 300
  element of, 300
  expansion of, 301
  second-order, 300
  third-order, 301
Difference
  definition of, 28
  of squares, 73
  of two complex numbers, 211
Direct variation, 367
Discriminant, 236
Disjoint sets, 5
Distance
  between two points, 162, 263
  formula, 263
Distributive property, 20
Dividend, 88
Division
  of fractions, 104
  of polynomials, 85
  property of exponents, 41
  of radical expressions, 202
  of signed numbers, 26
Divisor, 88
Domain
  of absolute value functions, 365
  of constant functions, 371
  of exponential functions, 379
  of functions, 359
  of identity functions, 365
  of linear functions, 371
  of logarithmic functions, 381
  of polynomial functions, 371
  of quadratic functions, 375
  of relations, 354

Element
  of determinants, 300
  of sets, 1
Ellipse, 385
Empty set, 2
Enumeration, 1
Equality, 23
  addition property of, 23, 132
  of complex numbers, 211
  of fractions, 96
  multiplication property of, 23
  of sets, 4

Equation(s), 131
  absolute value in, 163
  applications of, 144
  conditional, 132
  dependent, 290
  equivalent, 132
  exponential, 322
  first-degree
    in one variable, 13
    in two variables, 268
    in three variables, 296
  fractional, 134
  graphing of, 269
  identities, 132
  inconsistent, 290
  independent, 290
  linear, 133, 268
  literal, 139
  logarithmic, 323, 327
  product of roots of, 235
  quadratic, 223
  quadratic in form, 238
  radical, 207, 240
  roots of, 225
  second-degree
    in one variable, 223
    in two variables, 384
  solution of, 131
  solution set of, 131
  sum of roots of, 235
  systems of linear, 288
  systems of quadratic, 389
Equivalent equations, 132
Equivalent expressions, 96
Evaluating polynomial, 49
Exponent(s), 39
  negative, 176
  positive integers as, 41
  positive rational, 185
  property of, 41, 187
  rational, 186
  zero, 176
Exponential equation, 322
Exponential function, 379
Expression, 39
Extraneous solution, 207
Extreme point, 374

Factor(s), 69
  common, 69

  of polynomials, 69
  prime, 69
Factorial notation, 410
Factoring, 69
  combined methods of, 76
  common, 69
  complete, 69
  of differences of cubes, 73
  of differences of squares, 73
  by grouping, 71
  of polynomials, 69
  of quadratic trinomials, 79
  of second-degree polynomials, 79
  solving equations by, 223
  by special products, 73
  of sums of cubes, 73
Finite sequence, 399
Finite set, 3
First-degree equations, 133
  applications of, 144
Formulas, 139
Formulas from geometry, 421
  area, 421
  circumference, 430
  perimeter, 427
  surface area, 432
  volume, 432
Fraction(s), 95
  addition of like, 109
  addition of unlike, 112
  additive inverse, 110
  complex, 120
  definition of, 95
  division of, 104
  equality of, 96
  equivalent, 96
  fundamental principle of, 98
  least common denominator of, 115
  lowest terms of, 99
  multiplication of, 103
  not defined, 96
  reduced to lowest terms, 99
  subtraction of like, 109
  subtraction of unlike, 112
  with zero denominator, 96
Fractional equations, 134
Function(s), 358
  decreasing, 382
  definition of, 359, 360
  domain of, 359
  exponential, 379

Function(s), (cont.)
  increasing, 382
  linear, 371
  logarithmic, 381
  notation for, 363
  polynomial, 371
  quadratic, 373
  range of, 359
  sequences, 399
Fundamental principle of fractions, 98

Geometric applications, 146
Geometric formulas, 421
Geometric progression(s)
  definition of, 404
  nth term of, 404
  ratio of, 404
  sum of n terms of, 404
Graph(s)
  of equations, 268
  exponential, 379
  of inequalities, 157, 243, 355
  line, 271
  of linear functions, 371
  logarithmic, 381
  of number lines, 10
  of ordered pairs, 260
  of quadratic functions, 373
Graphical solutions
  of inequalities, 156
  of systems of linear equations, 290
Grouping factoring, 71

Hyperbola, 384
  definition of, 387

Identities, 132
Identity function, 365
Identity property, 21
  for addition, 21
  for multiplication, 21
Imaginary part of complex numbers, 211
Inconsistent equations, 290
Independent equations, 290
Index
  of radicals, 192
  of summation, 407
Inequalities, 153
  absolute, 165

  addition law of, 155
  conditional, 157
  equivalent, 157
  first-degree, in one variable, 157
  graphs of, 156
  multiplication law of, 155
  properties of, 153
  quadratic, 242
  solution sets of, 158, 242
  solutions of, 158, 242
  symbolism for, 153
  and transitive law, 154
  and trichotomy law, 154
  unconditional, 157
Infinite set, 3
Integers
  negative, 11
  positive, 11
  set of, 11
Intercept of a graph, 273
Interest, compound, 343
Interpolation, 333
Intersection of sets, 6
Inverse
  additive, 22
  multiplicative, 22
Inverse variation, 367
Investment applications, 147
Irrational numbers, 16
  set of, 16

Joint variation, 369

Law(s)
  addition of inequalities, 155
  multiplication of inequalities, 155
  transitive, 154
  trichotomy, 154
  zero-factorial, 410
Least common denominator, 115
Length of a line segment, 162, 263
Less than, 153
Line(s)
  equations of, 268, 280
  graphs of, 268
  horizontal, 273
  number, 10
  parallel, 277
  perpendicular, 277

slope of, 274
vertical, 273
Line segment(s)
  length of, 162, 263
  midpoints of, 268
Linear equation(s)
  definition of, 133, 268
  graph of, 271
  point-slope form for, 281
  slope-intercept form for, 283
  standard form for, 271
  systems of, 288
Linear function(s)
  definition of, 371
  as direct variation, 367
  graph of, 371
  intercepts of graphs of, 273
  slope of graphs of, 274
Linear system(s)
  solution of, 288
    by addition–subtraction, 291
    by determinants, 305
    by substitution, 299
  solution set of, 288
Literal equations, 139
Logarithm(s)
  applications of, 342
  base of, 321
  characteristic of, 330
  common, 330
  computations using, 337
  mantissa of, 330
  properties of, 325
  table of common, 440
Logarithmic equations, 323, 327
Logarithmic function, 381
Lowest terms, 99

Mantissa, 330
Mixture applications, 148
Monomial
  definition of, 48
  degree of, 48
Multiplication
  associative property of, 19, 51, 57
  commutative property of, 18, 51, 57
  of complex numbers, 211
  of fractions, 103
  and law of inequalities, 155
  of polynomials, 57

of radical expressions, 201
of signed numbers, 29
using logarithms for, 337
Multiplication property of exponents, 41
Multiplication property of polynomials, 132
Multiplicative-inverse property, 22

nth root, 192
Natural numbers, 9
  set of, 9
Negative number(s)
  as exponents, 176
  on line graph, 11
Notation
  factorial, 410
  function, 363
  logarithmic, 321
  scientific, 181
  set-builder, 2
  sigma, 406
  summation, 406
Null set, 2
Number(s)
  absolute value of, 26, 162
  complex, 210
  counting, 9
  graph of, 10
  imaginary, 211
  integers, 11
  irrational, 16
  line, 10
  natural, 9
  negative, 11
  ordered pairs of, 260
  positive, 11
  rational, 12
  real, 16
  roots of, 185
  whole, 9
Numerical coefficients, 48

Operations of signed numbers, 25
Ordered pair(s)
  components of, 260
  graph of, 260
  meaning of, 259
  as solutions of equations, 268
Ordinate, 260
Origin, 260

Parabola, 374, 384
Parallel lines, 277
Parallelogram
  altitude of, 423
  area of, 423
  perimeter of, 429
Pascal's triangle, 409
Perpendicular lines, 277
Physics applications, 149
Points
  collinear, 268, 280
  distance between, 162, 263
  extreme, 374
Point-slope form, 281
Polynomial(s), 39
  addition of, 51
  completely factored form of, 69
  definition of, 39
  degree of, 48
  division of, 85
  factoring of, 55
  in more than one variable, 49
  multiplication of, 57
  in one variable, 47
  subtraction of, 51
  zero, 48
Positive number(s), 11
Power(s)
  definition of, 39
  of power properties, 41
  of product properties, 41
  of quotient properties, 41
Prime factor, 69
Principal square root, 185, 192
  qth root, 185
Prism
  altitude of, 432
  area of, 432
  bases of, 432
  lateral area of, 432
  lateral faces of, 432
  right rectangular, 432
  right triangular, 432
  volume of, 432
Product(s)
  of complex numbers, 211
  cube of binomials, 64
  of fractions, 103
  involving radicals, 201
  of polynomials, 57
  of real numbers, 18
  special, 62

  sum and difference of, 63
  with zero factors, 23
Progression(s)
  arithmetic, 400
  geometric, 404
Proper subset, 4
Properties
  associative, 51, 57
  basic, of real numbers, 17
  cancellation, 23
  closure, 17
  commutative, 51, 57
  distributive, 51
  division of exponents, 41
  of exponents, 41
  of identities, 21
  of integral exponents, 178
  inverse, 22
  of logarithms, 326
  multiplication of exponents, 41
  power of powers of exponents, 41
  power of products of exponents, 41
  power of quotients of exponents, 41
  of radicals, 193
  of rational exponents, 187
  reflexive, 23
  substitution, 23
  symmetric, 23
  transitive, 23
  zero, 23
Pyramid
  altitude of, 436
  area of, 436
  base of, 435
  lateral area of, 436
  regular, 435
  volume of, 436

Quadrant, 261
Quadratic equation(s)
  applications of, 249
  definition of, 223
  discriminant of, 236
  in form, 238
  formula, 223
  roots, kind of, 236
  solution of
    by completing the square, 227
    by extraction of roots, 228
    by factoring, 223

by formula, 232
sum and product of roots, 235
Quadratic function(s)
  definition of, 373
  graph of, 373
Quadratic inequalities, 242
Quadrilateral parallelogram, 423
  rectangle, 421
  square, 422
  trapezoid, 424
Quotient(s)
  definition of, 32
  of complex numbers, 213
  of fractions, 105
  of polynomials, 88
  of signed numbers, 32

Radical(s)
  addition of, 199
  definition of, 192
  division of, 202
  index of, 192
  multiplication of, 201
  operations of, 199
  properties of, 193
  similar, 199
  subtraction of, 199
Radical equations, 207, 240
Radical expressions, 199
Radicand, 192
Radius of a circle, 385
Range of a function, 359
Ratio, of geometric progressions, 404
Rational exponent, 186
Rational expression, 95
Rational numbers, 12
Rationalizing denominators, 203
Real axis, 16
Real line, 16
Real numbers, 16
Reciprocal, 22
Rectangle
  area of, 421
  perimeter of, 427
Rectangular solid, 433
Reducing fractions to lowest terms, 99
Reflexive property, 23
Relation(s)
  domain of, 354
  graph of, 354
  range of, 354

as set of ordered pairs, 354
Remainder, 88
Replacement set, 39
Right prism, 432
Rise of a line, 275
Root(s)
  cube, 185
  of equations, 225
  kinds of, for quadratic equations, 236
  nth, 192
  of numbers, 192
  square, 185
Run of a line, 275

Scientific notation, 181
Second-degree equations, 223
Second-degree systems, solution of, 390
  by addition–subtraction, 392
  by substitution, 391
Segment, length of, 162, 263
Sequence(s)
  definition of, 399
  functions of, 399
Set(s)
  disjoint, 5
  elements of, 1
  empty, 2
  enumeration of, 1
  equality of, 4
  finite, 3
  infinite, 3
  of integers, 11
  intersection of, 6
  of irrational numbers, 16
  of natural (counting) numbers, 9
  null, 2
  of rational numbers, 12
  of real numbers, 16
  solution of, 131
  union of, 6
  universal, 5
  of whole numbers, 9
Set-builder notation, 2
Similar terms, 52
Slope of a line, 274
Slope-intercept form, 283
Solid, rectangular, 433
Solution(s)
  of equations, 131
  extraneous, 207
  of inequalities, 158

Solution(s), *(cont.)*
  ordered pairs as, 288
  of systems, 288
Solution set
  of equations, 13
  factoring of, 73
  of inequalities, 158
  special products of, 62
  of systems, 288
Sphere
  surface area of, 437
  volume of, 437
Square of a binomial, 62, 63
Square root, principal, 185, 192
Subset, 4
  proper, 4
Substitution, solution by, 290, 296
Substitution property, 23
Subtraction
  of complex numbers, 211
  of fractions, 109, 112
  of polynomials, 51
  of signed numbers, 26
Sum
  of an arithmetic progression, 401
  of complex numbers, 211
  of a geometric progression, 404
  of polynomials, 51
  of two real numbers, 17
Summation, index of, 407
Summation notation, 406
Symmetric property, 23
System(s)
  of linear equations
    applications of, 311
    in two variables, 288
    in three variables, 296
  of second-degree equations, 389
  solution set of, 288

Table, common logarithms, 440
Term(s), 48
  addition of similar, 52
  of arithmetic progressions, 400
  of expressions, 48
  of geometric progressions, 404
  of sequences, 399
  similar, 52
Terminating decimals, 13
Transitive law of inequalities, 154

Transitive property, 23
Triangle
  area of, 423
  Pascal's, 409
  perimeter of, 429
Trichotomy law, 154
Trinomial(s)
  definition of, 48
  factoring of, 79

Uniform-motion applications, 149
Union of sets, 6
Unit
  of area, 421
  on number lines, 10
  of volume, 432
Universal set, 5
Universe, 5

Value, absolute, 26, 162
Variable(s), 39
Variation
  constant of, 367
  direct, 367
  as a function, 367
  inverse, 368
  joint, 369
Vertex of parabolas, 374
Volume, 432
  of cones, 436
  of cylinders, 434
  of parallelipipeds, 433
  of prisms, 432
  of pyramids, 435
  of rectangular solids, 433
  of spheres, 437
  unit of, 432

Whole numbers, 9
Word problems, 144, 249, 311, 342

Zero
  additive property of, 23
  division by, 96
  as exponent, 176
Zero factorial, 410